DELTA
EMPIRE

MAKING THE MODERN SOUTH

David Goldfield, Series Editor

DELTA EMPIRE

LEE WILSON AND THE TRANSFORMATION OF AGRICULTURE IN THE NEW SOUTH

Jeannie Whayne

02-01-12

To Rena Russell,

In appreciation of your

Dyess connection

Jeannie Whayne

LOUISIANA STATE UNIVERSITY PRESS
BATON ROUGE

Published with the assistance of the V. Ray Cardozier Fund

Published by Louisiana State University Press
Copyright © 2011 by Louisiana State University Press
Manufactured in the United States of America
First printing

Designer: Michelle A. Neustrom
Typeface: Whitman, text; Baurer Bodoni, display
Printer: McNaughton & Gunn, Inc.
Binder: Acme Bookbinding

Library of Congress Cataloging-in-Publication Data

Whayne, Jeannie M.
    Delta empire : Lee Wilson and the transformation of agriculture in the new South / Jeannie Whayne.
        p. cm. — (Making the modern South)
    Includes bibliographical references and index.
    ISBN 978-0-8071-3855-7 (cloth : alk. paper) — ISBN 978-0-8071-3856-4 (pdf) — ISBN 978-0-8071-3857-1 (epub)
— ISBN 978-0-8071-3858-8 (mobi)
    1. Wilson, Lee, 1865–1933. 2. Farmers—Arkansas—Mississippi County—Biography. 3. Plantation owners—
Arkansas—Mississippi County—Biography. 4. Businessmen—Arkansas—Mississippi County—Biography.
5. Agriculture—Arkansas—Mississippi County—History. 6. Plantations—Arkansas—Mississippi County—History.
7. Plantation life—Arkansas—Mississippi County—History. 8. Social change—Arkansas—Mississippi County—
History. 9. Mississippi County (Ark.)—History. 10. Mississippi County (Ark.)—Biography. I. Title.
    F417.M58W47 2011
    976.7'05092—dc22
    [B]

                                                                                    2011013606

Two previously published essays of mine proved useful in the writing of *Delta Empire*. "The Changing Face of Sharecropping and Tenant Farming in the Twentieth Century" (pages 41–46 in *Revolution of the Land: Southern Agriculture
in the Twentieth Century,* edited by Connie Lester [Starkville: Mississippi State University, 2002]) focused on the experiences of William "Snake" Toney, particularly on his ability to use his relationship with the Wilsons to his advantage,
although only up to a point. "Robert E. Lee Wilson and the Making of a Post–Civil War Plantation" (pages 95–117 in
Southern Elite and Social Change: Essays in Honor of Willard B. Gatewood [Fayetteville: University of Arkansas Press,
2002]) provided several background details, some of which called for revision in the later work, such as exactly how
much acreage Wilson inherited.

Contents

Illustrations follow page 140

Acknowledgments

This book had its origins in a research paper I wrote for Steve Hahn at the University of California, San Diego, in the early 1980s. In that paper I engaged the historiography on the evolution of plantation agriculture and the emergence of the tenancy system in the post–Civil War South, and I grappled with Lee Wilson's contradictory reputation regarding the treatment of black labor, a question that remains a major preoccupation in this book. When I began to think of dissertation topics, my mind naturally turned to the Wilson plantation, but a couple of insurmountable obstacles presented themselves, and then an encounter with a snake in the basement of a Mississippi County jail convinced me to look elsewhere for a dissertation topic. No company records existed, or so it seemed at the time, and a county official indicated that county records were unavailable to me. Oscar Fendler, a longtime attorney representing the Wilson family, gained me entry to the basement of the county jail so that I could examine the records discarded there. Thus began an adventure that Fendler, who died a few years ago, never tired of recalling, though he only heard the story from me—I think. The jailer held a flashlight, a guard stood by with a rifle as two black prisoners in jail jumpsuits picked up the books and held them in the light for me to examine. Finding nothing of interest, I noticed another stack of books across the room and started to move toward them. Years later it occurred to me that the entire charade—aside from the snake which slithered by at that moment and could not have been choreographed—was intended to discourage me. It worked. I chose another topic.

As it happens, by the time I got to that basement, I was already disinclined to write on the Wilson plantation. I had visited the company offices only to find that apparently no company records existed, a fact that would greatly circumscribe any scholarly undertaking. After returning to Arkansas in 1990 as an assistant professor at the University of Arkansas, I encountered Mike Wilson, the head of the company, at a book signing in Mississippi County (thanks to Mary Gay Shipley of "That Bookstore in Blytheville") for a book I co-edited with Willard Gatewood, *Arkansas Delta: A Land of Paradox.* By then Mike Wilson had decided that he wanted me to write a history of the Wilson plantation. I had some understandable misgivings. Any book I wrote, I explained to Mike, would be critical of certain aspects of the company's operation. He insisted he understood that and believed that it was important to cover all aspects of the company's history. He wanted the unvarnished truth, and I came to understand that he meant what he said. For example, I shared with him information about an encounter in 1932 between Mississippi County deputies working out of Wilson, Arkansas, and a black man, Virgil Branch. Lee Wilson had the deputies fired for their treatment of Branch, and I was interpreting that as one example of Wilson's benevolence. Mike corrected me. Branch worked for a man who represented an institution to which Wilson owed hundreds of thousands of dollars. Lee Wilson's response to the incident was entirely self-interested. I am enormously grateful to Mike for his commitment to the truth. Unfortunately, he died suddenly before I could put a finished copy in his hands. I dedicate this book to his memory.

Mike Wilson saw to it that a number of company ledgers were donated to the archives at the University of Arkansas, but this book would not have been possible had it not been for his later discovery of a treasure trove of correspondence files located in the company headquarters. Mike called me some time in the late 1990s to tell me that when workmen removed a malfunctioning air-conditioning unit to replace it, they discovered a false wall and a room full of boxes. Michael Dabrishus, then the director of the University of Arkansas's Special Collections Division, helped arrange the donation of the papers to his division. After Mike Wilson's death, his brother Steve stepped to the helm of the company alongside his sister, Midge, and Mike's son, Perry. They provided additional information and assistance for which I am particularly grateful. The company records provide the basis of my understanding of the Wilson operation, but another important piece fell into place when Dr. Eldon Fairley, now also sadly deceased, of the Mississippi County Historical Society rescued the county records from the snake-infested basement of the Mississippi County jail and made them available to me at the historical society's museum in Osceola.

I have a number of other people to thank for their assistance in bringing this book to fruition. Elliott West, Randall Woods, Bob McMath, Charles Robinson, Mike Pierce, and Calvin White, all colleagues of mine in the history department at the University of Arkansas, commented on papers I presented to the department's Phi Alpha Theta chapter or elsewhere. I thank my chair,

Lynda Coon, for her many kindnesses as I finished this project. Other friends at the university, especially those in the department and in the Faculty Senate, contributed in direct and subtle ways. I am particularly grateful to Chuck Culver for his assistance with early twenty-first-century statistics on agricultural production. I am deeply grateful to Margaret "Peggy" Guccione of the Geosciences Department for sharing her research on the sunk lands. I thank individuals like Ken Barnes and Ben Johnson who gave me opportunities to present papers at the Arkansas Association of College History Teachers and the Arkansas Historical Association. I am grateful to Fred Williams, Gil Fite, Claire Strom, David Danbom, Doug Helms, and others of the Agricultural History Society, where I gave presentations, and to Connie Lester and Jim Giesen, who invited me to speak at conferences at Mississippi State. Talks at Arkansas State University, thanks to invitations by Clyde Milner and Ruth Hawkins, gave me opportunities to present alongside Nan Woodruff and Grif Stockley, who provided cogent and thoughtful criticisms. And an invitation from David Stricklin and Bobby Roberts to speak at Central Arkansas Library System's Butler Center in Little Rock and another to address a class at the Clinton School helped me to hone my arguments. I am grateful to my very good friend Carol Perel for extending an invitation to speak at the Cotton Museum in Memphis, which led to an introduction to Lee Wilson Wesson. Wesson provided some important information about his part of the Wilson family, which was most helpful. I thank the individuals at Crystal Bridges Museum of American Art who gave me an opportunity to make a presentation in concert with Bob Ford of Theater Squared. Bob wrote a play on an incident that took place on the Wilson plantation, an incident related to me by William "Snake Toney" of St. Louis, another person to whom I owe a great debt. An opportunity to present at a Rural Studies Conference in Brighton, England, enabled me to secure helpful comments from a very different group of scholars possessing an international perspective. The comments of scholars like Jarod Roll and Richard Follett greatly enriched this project.

I want to acknowledge the contribution of Mary S. Hoffschwelle, who generously shared her research on Rosenwald schools with me, providing insight into Wilson's role in fostering a certain kind of education for the children of his black laborers. Will Guzmán's willingness to exchange research notes on the lynching of Henry Lowery was most appreciated. I am grateful to Mary Rolinson for providing leads on the United Negro Improvement Association (UNIA) divisions in Arkansas and to John Giggie for sharing his many insights about black religion and the importance of the WPA church records.

My special thanks go to Willard Gatewood, a longtime mentor and distinguished professor of history at the University of Arkansas who is now emeritus. Gatewood read two drafts of the manuscript and provided much sage advice. I am very grateful to Patrick Williams, editor of the *Arkansas Historical Quarterly*, who encouraged me to rethink some of the key arguments I make in the book. Gretchen Gearhart gave the manuscript a careful reading and saved me from some embarrassing oversights. I particularly appreciate the work of my eagle-eyed and astute copy editor, Jo Ann Kiser. I'm also especially grateful to David Goldfield, the editor of the series, and to Pete Daniel, who was the outside reader of the manuscript. Their patience with me as I whittled down a supersized manuscript to an acceptable length was monumental. Also monumental was the patience of my co-authors of *Arkansas: A Narrative History*. Tom DeBlack, Morris Arnold, and George Sabo have shown uncommon forbearance as I delayed rewriting my sections of that textbook for a new edition we have planned.

I am especially grateful to the folks at the University of Arkansas's Special Collections Division: Director Tom Dillard, Tim Nutt, Andrea Cantrell, and Geoffrey Stark. My friend Beth Juhl, the Web-services librarian, spent many hours poring over arcane questions that only a person with the patience of Job would entertain. I thank also my other friends at the library—Juana Young, Luti Salisbury, and Lora Lennertz. My thanks go as well to the staffs at the Arkansas History Commission, the Mississippi County Historical Society, and the Osceola and Blytheville Courthouses, especially Lib Shippen. I am grateful to my friends Norwood Greech and Tri Watkins in Lepanto. Norwood brought an artist's understanding to today's landscape, and Tri helped me to understand how plantation agriculture functions in the early twenty-first century. Bob and Jo Ann Pugh provided the same kind of insights about agriculture in southeast Arkansas. I am particularly grieved that Jo Ann did not live to see the publication of this book. She was a unique and wonderful person and is greatly missed. Graduate students Jennifer Koenig, Matt Kirkpatrick, Jamie Forrester, Krista Jones, and Chet Cornell did a good deal of research in the Wilson papers for me or contributed in other ways. Over the years, many graduate and undergraduate students who took courses with me asked perceptive questions, contributing in innumerable ways. One of my former students, Michael Lindsey, recently provided some crucial information that enabled me to better understand some of Lee Wilson's financial dealings.

My family, of course, deserves special mention. My brother, sister, nieces, nephews, and in-laws have served as inspirations to me. Finally, my husband,

John Cook, merits my greatest thanks. He read many drafts of this book and offered solid advice, for which I am very grateful. In the end, however, it is the sound of his guitar playing in the background that I most cherish, and I believe that some of that sweet sound made its way into this manuscript, greatly enriching it as he has enriched my life.

DELTA
EMPIRE

INTRODUCTION

Robert E. "Lee" Wilson founded a plantation empire in the lush and dangerous lower Mississippi River Valley in the late nineteenth century, parlaying a four-hundred-acre inheritance into a fifty-thousand-acre farming operation. Born in Mississippi County, Arkansas, in the last weeks of the Civil War to a frontier planter of few pretensions, Lee Wilson began life under inauspicious circumstances. The death of his father in nearby Memphis, Tennessee, in 1870 and of his mother eight years later in a yellow fever epidemic left him an orphan with nothing but his Arkansas land remaining to him. He crossed the river in 1880, when only fifteen years old, camped in the cypress swamp he inherited, and began harvesting timber. Fired by a determination to make his own way in the world, he convinced the county court to recognize his standing as an adult two years later and began borrowing money and buying up seemingly worthless swamplands. By 1889, he had moved into agricultural production, grown cotton for export, and employed over a hundred sharecroppers. On the eve of his death in 1933, he had 2,500 employees spread over fourteen plantations and ranked as the largest cotton producer in the South. His story tells us much about the South, particularly the evolution of the plantation system in the post–Civil War period, when planters renegotiated labor and race relations. The Wilson saga also makes clear that the unification of different classes of whites in the 1890s at the expense of African Americans was largely illusory and was frequently challenged. Conflicts over different visions of development, particularly in the environmentally vulnerable lower Mississippi River Valley, created further instability and persisted well into the twentieth century.[1]

Like Thomas Sutpen, in William Faulkner's *Absalom, Absalom*, Lee Wilson carved out a sizable plantation from an impenetrable swamp, seemingly willing it into existence. Sutpen, however, epitomized the region's inability to move beyond the social and economic legacies of slavery, and his personal excesses and failures doomed the dynasty he sought to create. Lee Wilson single-

1

mindedly but more successfully pursued his own version of a New South in the post–Civil War era, one founded on lumber and cotton. He eventually amassed his estate and organized it as a highly efficient business enterprise capable of sustaining a dynasty into the twenty-first century. The only other comparable entities in the Mississippi River Valley, the Percy family dynasty, the Delta Pine and Land Company, and the Chapman and Dewey Land and Lumber Company, failed to survive the transition to post-plantation agriculture after World War II. Though the reasons for the Wilson concerns' successful negotiation of the postwar period are complex and varied, in the final analysis, Lee Wilson created an entity capable of adjusting to new realities. Perhaps most significantly, his heirs, beginning with his grandson, proved exceptionally skilled at leading the operation through the remarkable transformation that occurred during the post–World War II era. In 2010, however, the heirs surrendered to the next phase of southern agriculture, a phase witnessing the rise of disinterested and faceless investors who purchase agricultural lands as a part of their investment portfolios and hire expert farmers to manage their estates. For 164 years, from the time that Josiah Wilson purchased a homestead in the Mississippi County swamps to 2010, the Wilsons have been involved in the evolution of southern agriculture. This latest phase of southern agriculture will develop without their participation.[2]

The first member of the Wilson clan to operate a plantation in Arkansas, Josiah, depended on slave labor; Lee Wilson, like others of his generation, turned to tenancy and sharecropping. He possessed a genius for manipulating new ideas and opportunities when they presented themselves. He accepted implicitly Henry Grady's New South message, for example, looking to the North for investment capital. But he borrowed more than money from the North; he also adapted certain ideas. He employed Progressive Era strategies for organizing his business, and impatient with taxes imposed on corporations, he dissolved his corporation and created a trust based on a Massachusetts model that allowed him to work outside government constraints. He recognized opportunity when government could serve his interests, however. He embraced New Deal programs that provided for fresh infusions of cash through the Reconstruction Finance Corporation and the Agricultural Adjustment Administration, and not a moment too soon. His company, like the country itself, was facing the worst economic crisis of its history.[3]

Wilson also borrowed something from the country's leading industrialists and, indeed, he was a kind of "rural Rockefeller," in that he pursued both horizontal and vertical integration of his enterprise. He bought up other farms

and plantations, both large and small, and absorbed them into his own operation. He functioned as Lee Wilson & Company, with various smaller entities
reporting to his head office in the town of Wilson or functioning as semi-
independent companies. His lumber operation included lumber camps, sawmills, box factories, and lumber distribution outlets in Memphis and St. Louis.
His plantation, like many other such larger operations, included mercantile
establishments, cotton gins, and cotton oil mills, but he also expanded into the
cotton factorage business so that he could better market his cotton to textile
industries. He built short-line railroads to bring his lumber and cotton to processing and distribution centers, but he also purchased a regional intrastate
railroad and eventually secured a seat on the board of directors of the St. Louis
and San Francisco "Frisco" Railroad. Finally, in order to adequately finance his
operations, he formed the Bank of Wilson, which positioned him to establish
important relationships with "corresponding" banks and financial institutions
in St. Louis, Chicago, and New York, and he arranged to issue shares of Lee
Wilson & Company stock in order to secure fresh infusions of cash.

African American labor played a crucial role in his success, and he liked to
believe that he appropriately rewarded them for helping him "to make what I
have." Embracing what C. Vann Woodward referred to as a conservative position on race relations, Wilson accepted the premise of black inferiority, but
he eschewed the violent racism that arose in the late nineteenth century and
instead provided black laborers with better-than-average housing, health care,
and schools. He created an environment in which black families could find
some peace and security, but only on his terms, and Wilson did not necessarily
set the tone for race relations in the broader community. When he constructed
a state-of-the-art black school to show his appreciation to the African Americans who labored for him, arsonists burned it to the ground in the early morning hours of the day it was to be dedicated, exposing the limits of Wilson's
power but, more important, demonstrating the depth of their own animosity to African American progress and their determination to maintain white
supremacy.[4]

Given the triumph of extreme racism in the region, the Wilson plantation
seemed an attractive alternative, an island of safety in a sea of unrestrained
violence. But the safety Wilson offered sometimes proved undependable, and
the semblance of security came with a steep price. Those who labored for
Wilson endured a high degree of regimentation and became ensnared in an
inescapable web of debt. His store managers maintained meticulous records
of every advance made to employees, and his office personnel negotiated chat-

tel mortgages with those who owned livestock and implements that required them to produce cotton in exchange for advances. However, despite the constraints his laborers worked under and because his plantation provided some safety against white night riders, he maintained a steady supply of black labor. He actually benefited from the extreme racism that prevailed in the Jim Crow South, which drove workers his way.[5]

The 1890s marked the nadir of race relations in the South. The intemperate political rhetoric of the era and the economic atmosphere inflamed landless white farmers, who competed with black labor for places on the expanding plantations of the Arkansas delta. They resorted to night-riding activities to expel African Americans from plantations and create a "white man's" country, something that did not work to the advantage of planters like Wilson whose organizations thrived on cheap and malleable black labor. In their efforts to drive away black labor, some night riders destroyed the property belonging to planters, prompting white authorities to launch campaigns to subdue them. In this context, class warfare emerged between different groups of whites and left African Americans under the dubious protection of plantation owners.[6]

The conflicts that prevailed among poor and rich whites involved more than the struggle concerning black labor, however. It also included an intense and violent contest over ownership of some of the richest undeveloped land remaining in America. Lee Wilson figured prominently in this battle, for he not only purchased cut-over lands at rock bottom prices; he also acquired farms at the courthouse steps when their owners failed to keep up with their tax payments. He became notorious for his aggressive acquisition of land and even more infamous for his support of drainage efforts that would turn his swamplands into productive agricultural acreage. He subscribed to a particularly capitalist vision of development, one not universally supported in the region. Many small farmers, already struggling with a tax burden that placed them too close to the margins for comfort, objected to the imposition of the further assessments necessary to fund the creation of drainage districts. Armed confrontations occurred over the issue, but Wilson ultimately prevailed, and tens of thousands of acres of swamps were "reclaimed" and made ready for use. When the Interior Department opened the "sunk lands" of Mississippi, Poinsett, and Craighead counties to homesteading, Wilson went back to court, claiming much of this valuable acreage by the arcane legal principle of riparian rights. Although Wilson and other planter litigants lost at the Supreme Court, a loophole allowed them to approach Congress for a solution in the late 1910s and early 1920s, and by that time, a very different Interior Department provided no succor for landless men or small farmers.

By the third decade of the twentieth century, Wilson stood as a force to be reckoned with, but he was about to face a series of challenges that put in jeopardy the enterprise he spent a lifetime building. The flood of 1927, the devastating drought of 1930–31, and the Great Depression tested his ability to navigate the obstacles that came his way. While his successful efforts to construct levees for flood control protected him against the greatest flood to ever strike the Mississippi River Valley, class conflict again erupted as small farmers dynamited his levees in order to save themselves from disaster. He ultimately prevailed in this struggle, too. The drought of 1930–31 presented a different kind of environmental disaster. His crops wilted in the fields, his creditors hounded him relentlessly, and he suffered a serious illness that briefly induced him to surrender management of his company to his son and heir. As the Great Depression deepened, Wilson found it necessary to tell his doctors "to go to hell" and to leave his young mistress at his vacation home in Hot Springs to reassume direct control over his company's affairs. His relationships with financiers and government officials, which he had cultivated for decades, coupled with his own shrewdness, enabled him to take advantage of new federal programs and save himself from bankruptcy.[7]

Although Lee Wilson embraced the Agricultural Adjustment Administration's plow-up program in 1933, he was not destined to see the benefits that program brought to his company. He died that fall, leaving the new management of the company to deal with a challenge from an unprecedented direction. Tenants and sharecroppers on one of the Wilson plantations filed suit just weeks after Wilson's death, heralding a new era of militancy on the part of plantation laborers. The emergence of federal programs contributed to this phenomenon and also funneled cash into the hands of planters, many of whom used the funds to begin a transition away from labor-intensive and toward capital-intensive agriculture, but the New Deal programs marked only the beginning of this dramatic transformation. World War II accelerated the trend as a migration of blacks, long in the making, intensified, and many rural white southerners joined them as they moved into war industries or the armed services. The ground was shifting under the South's plantation system, and Lee Wilson & Company, under the capable management of the founder's general farm manager, Jim Crain, successfully negotiated the period.

Unlike William Faulkner's Thomas Sutpen, Lee Wilson created a dynasty capable of long outliving him, but his son was not the man destined to lead the company through the Great Depression and the challenges of World War II. Wilson shared one other important similarity with Sutpen; he endured a series of personal crises, most of them involving his children. His son, R. E. L. "Roy"

Wilson Jr., resembles Sutpen's tormented son but, in fact, he more closely approximates yet another Faulknerian character, Quinton Compson. Young Compson attended Harvard and, brooding over the past, drowned himself in the Charles River; Roy Wilson attended Yale and, suffering debilitating bouts of malaria, drowned himself in a sea of alcohol. Roy voluntarily surrendered control over the company to Jim Crain, a rough-hewn, self-made man, who served as an aggressive and ambitious "place keeper" until the third genera- tion of Wilsons came of age. The transition from Crain to the legal heirs of the founder was not without controversy, and Wilson's hopes of keeping the estate entirely intact proved unsuccessful. However, R. E. Lee "Bob" Wilson III assumed control of Lee Wilson & Company in 1950. Characterizing himself as "not a (plantation) baron but a businessman," Bob Wilson led the company through the second phase of the transformation of post–Civil War plantation agriculture, a phase that began after New Deal programs and World War II: the neo-plantation described by historian Jack Kirby. His heirs would further that phase, creating a post-plantation model dedicated to a mix of three com- modity crops. A third phase—portfolio agriculture—is now in the making, but it will occur under the direction of new owners.[8]

By the time Bob Wilson turned the company over to his children, it was a smoothly running, efficient business enterprise of twenty-three thousand acres with a sufficient infrastructure to support the cultivation of at least three commercial commodities: cotton, soybeans, and rice. The sale of the com- pany to an investment firm in late 2010 has some significant, though as yet uncertain, implications. The transaction mirrors a trend occurring elsewhere across the country, a trend that separates the farming practice from those who own the land by more than mere geography. Investors who trust their portfolios to management firms have little knowledge of farming or interest in the well-being of agricultural communities. The possible consequences to the environment and the economic viability of local institutions are incalcu- lable. One thing is certain. The demographic revolution that accompanied the rise of the neo-plantation and the recent emergence of portfolio agriculture obscure the struggles of a vanished population. The vast flat fields of crops conceal the problematic landscape that Josiah encountered in the antebel- lum period and that Lee Wilson wrestled with in the late nineteenth century. The dramatic reclamation efforts of Wilson and his cohorts, who cleared the forests and drained the land, began in an era when few voices raised con- cerns about the environmental consequences of development along the lower Mississippi River Valley. The managers of today's portfolio plantations will have to wrangle with the implications of the use of a variety of chemicals they consider a necessary component of their operations. The Wilson family's antebellum ancestor, Josiah Wilson, who first carved out a homestead in the Arkansas swamps and whose story begins this book, would find their twenty-first-century world unrecognizable.[9]

1

THE SHAPING OF
THE LAND

Josiah Wilson, an unconventional but enterprising West Tennessee planter and businessman, early recognized great potential in the northeastern Arkansas swamps. He looked beyond their reputation for lawlessness and their appearance of limited agricultural possibilities. More cautious men saw only danger and unpromising circumstances in those wild and earthquake-damaged lands, and by planting their expectations in more solid ground elsewhere, they missed an opportunity that only shrewd and calculating eyes could see. Part of that vanguard of southern men moving ever westward in the antebellum period, laying down plantations founded on the labor of enslaved blacks, Wilson had taken a long road to the Mississippi River with his father, a road that began in Alabama four decades earlier. Over the course of the early nineteenth century, William L. Wilson moved his family from Alabama to Kentucky—where Josiah was born in 1805—and on to western Tennessee, where the patriarch died in 1837. In the mid-1840s Josiah began disposing of property in and around the small Mississippi River town of Randolph with a single purpose in mind: to stake a claim in the eastern Arkansas swamps. He could see the Mississippi County shore from the Randolph cemetery on the Chickasaw bluffs of Tennessee, where he had buried his parents, and he must have known that great hardship and potential disaster awaited him there. With the town of Randolph dying, however, he had to make a decision, and he made one that would bring both tragedy and riches to his own small family and even further tragedy and greater fortune to the generations of Wilsons that followed him.[1]

Inspired in part by the need to secure the interests and loyalties of the farmers and businessmen living in Tennessee and Kentucky, President Thomas Jefferson had authorized the purchase of the Louisiana Territory in 1803. The Purchase secured the young nation's ownership of the Mississippi River, the most important conduit for the goods produced by the farmers of the territories, but it also more than doubled the size of the United States and launched a new movement westward. Most settlers bypassed the Arkansas swamps, which became a haven for outlaws. Josiah Wilson's first experience in the Arkansas delta likely occurred in the summer of 1834, when a raiding party from Randolph attacked a criminal band hiding in a rough-and-tumble camp in south Mississippi County. Described as an "expedition composed mostly of our chief citizens," the posse prepared to locate and arrest the bandits who raided ships and stole "horses, negroes, cattle, and every specie of property" available. Allegedly associated with the infamous John Murrill, the outlaws preyed on the ships and barges that navigated through the numerous snags and an ever-shifting series of sandbars along the Plum Point reach, a treacherous seventeen-mile section of the Mississippi River adjacent to Randolph, defying and threatening the livelihoods of the merchants and farmers of the region.[2]

The event that finally prompted the men of Randolph to organize their posse involved an attack on a flatboat from Kentucky "laden with flour and whiskey" and "stranded on a sandbar" about twelve miles south of the town. While two of the robbers held the raftsmen at bay with leveled rifles, "the rest plundered the boat and took off in other boats," retreating west through Golden Lake and into a dense swamp shrouded in a deep forest and cane-brakes, heading to Shawnee Village, so named for a band of Indians who briefly inhabited the site in the late eighteenth century. To the south of the so-called village a sizable slough further obscured their location, and swamps abounded all around them, for they occupied a murky backwater of the Mississippi River made more complicated by damage done by the New Madrid earthquakes of 1811 and 1812. Describing the area as "so sparsely populated and wild, that justice cannot reach these freebooters" and justifying their actions by suggesting "that even the county officers participate in the profits of the plunder," the men of Randolph raised their posse and secured passage on the *Kentuckian*, a steamboat which took them free of charge to the Arkansas shore. The ship's captain, a Mr. Johnson, proved only too happy to aid in the capture of the criminals.[3]

After Captain Johnson landed them on the Arkansas shore, the men ascended a steep natural levee and took an overland approach to the outlaw den, hacking their way through the canebrakes and descending on the village, where they burned "two or three crazy cabins" and captured eight or ten of the suspects. Another steamboat, the *Tennessean*, brought them back to Randolph without charge. Within days they staged a second raid and maintained scouting parties for about a week, "returning daily with victorious trophies of fresh prisoners or recovered property."[4] The time spent "scouting" provided ample opportunity to reconnoiter the area for other purposes, such as locating parcels of property above the floodplain, something that preoccupied restless young men, like twenty-nine-year-old Josiah Wilson, who were looking for prospects. If he accompanied the posse, Josiah certainly glimpsed an unpromising terrain, but the experience likely planted an idea in his mind.

Some of the outlaws eluded capture and disappeared further into the swamps, and the reputation of the Arkansas region for harboring such criminals persisted. Ironically, the desolate landscape that so effectively shielded this criminal band had been one of the most urbanized regions in pre-Columbia North America. The trees, underbrush, and swamps concealed the remains of a vanished civilization visited almost three hundred years earlier by a different kind of desperado, Spanish conquistador Hernando De Soto. When he crossed the Mississippi River looking for gold in 1541, De Soto encountered a complex indigenous civilization growing vast fields of corn, much of which the Indians transported by river to an extended market. One particularly powerful tribe occupied a principal village called Pacaha, which was located along the river at either present-day Nodena or Pecan Point, both within a few miles of Shawnee Village and both traversed by the men from Randolph in their search for the outlaws in 1834. De Soto attacked and ransacked Pacaha, and though he located no gold, he found a "very good" town, "thoroughly well stockaded; walls were furnished with towers and a ditch around about, for the most part full of water which flows in by a canal from the river."[5] De Soto left Pacaha to search elsewhere for gold and later died of fever in Arkansas, having found no treasure.[6]

The mention of canals suggests how far back the history of human manipulation of the delta landscape extended. The very process of digging the canal worked a change on the terrain, likely reordering the unseen microbiotic world inhabiting the riverside and creating the possibility of a new one arising within the canals themselves. The attempt to utilize the opportunities presented by the river also signaled a basic understanding of hydrology, an unsurprising understanding given the Indians' dependence on the river. With agriculture as the basis of their economy, their "commerce stretched north into Illinois and south to the Gulf Coast as well as westward to Crowley's Ridge and the Ozark Highlands."[7] Thus they foreshadowed the cotton producers of the nineteenth century in their achievement of a kind of commercial agriculture, a livelihood Josiah Wilson aggressively pursued in the area in the 1850s. While Wilson occupied a relatively isolated homestead, however, thousands of neighbors surrounded the corn producers of the sixteenth century.

Their civilization likely dated to about 1250 A.D., when clashes between different groups in the area led these Indians to move from scattered hamlets into consolidated villages, and a virtual population explosion occurred. The rich delta soil supported the expansion of their agricultural pursuits as they aggressively exploited the forest resources in order to clear land for corn production but also for lumber. They utilized the lumber to build temples on the tops of large pyramidal mounds located in the center of their principal villages. They further depleted the forests to build the palisades meant to protect their villages from their enemies, to construct their homes, and to provide firewood for the larger concentrations of population. With the forest habitats declining, deer and other wildlife became scarce, and the Indians began to rely even more on agriculture. At the time of the De Soto *entrada* in 1541, several large chiefdoms vied for control over the area, and this stretch of the Mississippi Valley consisted of one of the most densely populated areas in all of North America. As many as thirty-five thousand Indians lived in northeastern Arkansas and southeastern Missouri at that time, but something shattered their civilization, and the population would not again reach the density it had achieved in the mid-sixteenth century until roughly 1900.[8]

Archeologists initially believed that epidemic disease brought by De Soto destroyed their civilization, but recent forensic examination of skeletal remains suggests instead that severe nutritional deficiencies, likely caused by massive crop failure, devastated the people of northeastern Arkansas.[9] Evidence of the nature of the disaster that undermined this civilization comes from two sources separated by more than four hundred years. De Soto's chroniclers remarked on the severe drought confronting the Indians of the region at the time of his arrival. Late twentieth-century scientist David Stahle, using tree-ring analysis, confirmed that a drought of one hundred years' duration began even before De Soto arrived. Not even twenty-first-century society could easily endure such a devastating environmental disaster. The sixteenth-century Indians, so dependent on agriculture, had contributed to the destruction of their own society by clearing fields for the production of corn and eliminating forests that might have provided alternative sources of food and fuel. Forest habitats attract deer, an important component in their diet, and support a rich abundance of wild fruits and berries which they could have harvested. Once the Indians became dependent on agricultural production, they narrowed their options. The drought dealt a fatal blow to a society reliant upon agriculture and, at the same time, drove what wildlife that remained in the area to range farther away and out of easy reach.[10]

The disappearance of the extensive Native American civilization in north-eastern Arkansas left the area largely unoccupied for most of the seventeenth and eighteenth centuries and allowed for the regrowth of forests and cane-brakes that obscured those areas once covered by the massive fields of corn. The relatively mild climate made for a long growing season, and the rich soil invited a wide variety of wild plants and trees to overgrow the area. Deer returned to graze in the forests, and as the canals collapsed the region became more swamp-like, attracting millions of wildfowl annually. In the late eighteenth century, a small group of Indians from east of the river moved into eastern Arkansas, partly to take advantage of the hunting possibilities, but they maintained only semipermanent settlements and soon moved along. Also motivated by the desire to escape the Indian wars taking place in the southeastern United States, the Cherokee, Chickasaw, Choctaw, Creek, Delaware, and Shawnee established settlements there, but they too remained only a short time.[11]

Once the United States purchased the Louisiana Territory in 1803, a demographic revolution began to take shape, one that eventually brought settlers like Josiah Wilson across the Mississippi River. The most immediate impact, however, was not in the Arkansas region. Only the poorly placed Arkansas Post, located much further to the south, presented an opportunity but one that attracted few takers. Because most early settlers moved through St. Louis or New Orleans, the Arkansas area remained largely unexplored. That might have changed sooner had it not been for an apocalyptic series of earthquakes in 1811 and 1812. For the second time in less than three centuries, a catastrophic natural disaster reshaped the landscape—earthquakes that wrought more fundamental and lasting environmental damage than had the sixteenth-century drought.

Ranking as among the twenty most devastating in world history and the worst ever to hit North America, the earthquakes occurred along the New Madrid seismic zone stretching from just above the Missouri village of New Madrid down through what would become Mississippi County and into Marked Tree in Poinsett County. It also reaches into west Tennessee to Reelfoot Lake and Dyersburg. Thousands of small and large quakes shook the region for over a year, but, according to the estimates of late twentieth-century scientists, three of the quakes probably ranged in intensity from 8.09 to 8.3 on the Richter scale, striking the Mississippi Valley on December 16, 1811, and on January 23 and February 7, 1812. Although the western United States has suffered earthquakes of greater intensity, the "geological conditions of the Mississippi Valley and the relatively low attenuation of surface-wave energy in eastern North America" resulted in far greater damage over a much wider area than would occur in the West. In fact, "no earthquakes in the recorded history of the United States have had a damage or felt area anywhere approaching" the size of the New Madrid quakes. "On a global scale, the only other seismic region which is notable for such anomalously large damage and felt areas is the subcontinent of India." The primary quake area—that is, the zone experiencing the greatest disturbance— stretched south from Cairo, Illinois, to Memphis, Tennessee, and west from Crowley's Ridge in Arkansas into west Tennessee, an area estimated to be between thirty and forty square miles, but "one million square miles or half the United States was so disturbed that the vibrations could be felt without the aid of instruments." Tremors from the quakes rang church bells in Charleston, South Carolina, and shook the cities of Boston, Washington, and New York. They were felt as far south as New Orleans and as far north as the colony of Quebec in Canada. The third quake, considered the most severe of the three, did greater damage in a wider area. "Cracks developed in brick buildings in Savannah, Georgia. A chimney top was thrown down in Richmond, Virginia."[12]

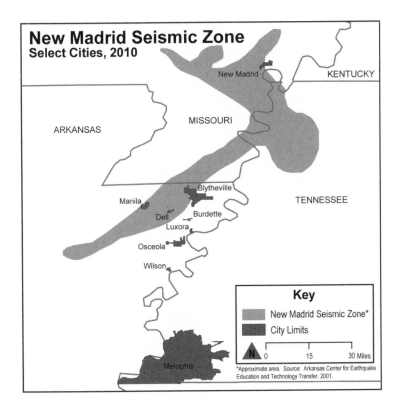

New Madrid Seismic Zone
Select Cities, 2010

KENTUCKY

New Madrid

MISSOURI

ARKANSAS

Blytheville

Manila

TENNESSEE

Dell   Burdette

Luxora

Osceola

Wilson

Key

New Madrid Seismic Zone*

City Limits

N   0   15   30 Miles

*Approximate area. Source: Arkansas Center for Earthquake Education and Technology Transfer, 2001.

Memphis

The earthquakes focused national and international attention on the sparsely settled region of northeastern Arkansas but also virtually guaranteed its continuing isolation. The Wilson family, then living in Kentucky, must have experienced both the vibrations and the horrifying sounds that accompanied the quakes. John Audubon, the famous naturalist, traveling in Kentucky at the time, reported feeling the unsettling tremors and described one of the more severe shocks later in 1812 as like "the distant rumbling of a violent tornado." Indeed, a variety of terrifying sounds accompanied the quakes, the cracking open of the earth, and the extrusions of water, sand, mud, gas, and even foreign fragments such as the skull of an extinct musk ox. Some observers reported hearing roaring, whistling, and hissing, the latter almost certainly caused by extruding water. Others characterized the quakes as sounding like thunder or cannon fire, and one observer said the quake he experienced began with "distant rumbling sounds, succeeded by discharges as if a thousand pieces of artillery were suddenly exploded." The effects made potential settlers more than a little reluctant to venture into such dangerous terrain. The earthquakes certainly must have shaped Josiah Wilson's appreciation of the place. In 1812 he was seven years old, an impressionable age, and he would have viewed the northeastern Arkansas region as frightening and mysterious. Reports that the ground rolled in the manner of a wave on the ocean would have filled so young a person with awe and dread, and published drawings of houses and people tumbling into fissures in the earth probably led many to have vivid nightmares.[13]

The extensive environmental damage caused by the earthquakes displaced settlers, most of whom were located in Missouri. The earthquake had inundated their fields and farms with water or otherwise ruined them. Prudent farmers already in the region abandoned their farms, particularly once the federal government gave them "New Madrid Certificates," awarding them land elsewhere to compensate for the damage done to the acreage they held within the earthquake zone and marking the first time that the young nation provided disaster relief. The Cherokee Indians, then living in Northeastern Arkansas, interpreted the earthquakes as a sign that they had been wrong to accommodate to white ways, abandoning their own traditions. Believing that further earthquakes would visit the area, punishing the whites who had exploited them, they departed their villages and resettled on the Arkansas River valley above Little Rock. The blighted landscape of northeastern Arkansas seemed most uninviting and attracted little more than adventurers and outlaws for decades. Fissures scarred the land and extrusions of sand across the area created three kinds of problematic phenomena: Sand blows, identified as low patches of white sand a few inches deep; sand scatters, characterized as thinner patches of sand usually located on low ground; and sand sloughs, that is linear depressions of approximately three to five feet or more that frequently fill with water. All three could be found in Mississippi County. Sand blows and sand scatters dotted the landscape across the northwestern corner of the county and stretched in an irregular pattern down the center of the county in a southwesterly direction. Farmers would later come to understand that they could work land located on sand scatters and sand blows, but sand sloughs presented the greatest challenge, and one of the most prominent sand sloughs in the entire earthquake zone was located at Big Lake in northwestern Mississippi County.[14]

Another environmental consequence of the earthquakes was the formation of the controversial "sunk lands," which created something of an engineering nightmare later in the century when pressure to drain the land led to the need for elaborate drainage designs. The sunk lands, resulting "from the local settling or warping of the alluvial deposits," existed in northeastern Arkansas and nearby portions of Missouri and Tennessee. The earthquake created three types of sunk lands: those connected with sand sloughs such as Big Lake, a phenomenon Josiah Wilson knew well, as Reelfoot Lake, also created by the earthquakes, existed near Randolph; those located in river swamps and characterized by shallow water that was not so deep as to prevent the growth of some timber; and those in lakes of standing water. All three types existed in Mississippi County and were destined to play a critical role in the struggle to develop the area in the late nineteenth and early twentieth centuries.[15]

Even had there been no earthquake, the swamps of northeastern Arkansas would have discouraged all but the stout of heart. As historian Conevery Bolton Valencius suggests, swamps both fascinated and repelled the average person in the nineteenth century. While they promised a bountiful harvest of game, they harbored fearsome dangers and were laden with troubling social meanings. One could become lost in the swamps or fall victim to a variety of deadly predators and a host of diseases. Outlaws and escaped slaves found refuge there, and their presence not only added to the mystique but also portended the possibility of disastrous encounters. Early nineteenth-century readers of literature familiar with John Audubon's "The Lost Ones" and Paul Bunyan's *Pilgrim's Progress*, which both employed allegories involving swamps, also understood the spiritual hazards of entering such forbidding environs.[16] This dubious reputation applied to all swamps, but the environmentally damaged northeastern Arkansas suffered under additional disadvantages.

Although the U.S. government began to survey the county and place lands there for entry at the Helena land office for $1.25 an acre in the mid-1820s, potential settlers remained skeptical of the area's potential. Fifteen years after the first quake awakened the residents of New Madrid, one observer described a country that "exhibited a melancholy aspect of chasms, of sand covering the earth, of trees thrown down or lying at an angle of 45 degrees, or split in the middle. The earthquakes still recurred at short intervals, so that the people had no confidence to rebuild good houses, or chimneys of brick." As late as 1869 a hunter reported, "I have trapped there for 30 years. There is a great deal of sunken land caused by the earthquake of 1811. There are large trees of Walnut, White Oak, and Mulberry, such as grow on high land, which are now seen submerged ten and twenty feet beneath the water. In some of the lakes I have seen cypresses so far beneath the surface that with a canoe I have paddled among the branches."[17]

The existence of the sunk lands and the other environmental consequences of the earthquakes inhibited settlement of the region, but changes on the horizon inspired Josiah Wilson to abandon Randolph and settle on a homestead just north of Shawnee Village. The men of Randolph, who had so assiduously attempted to protect their interests against outlaws, experienced a challenge to the integrity of their little river town from an entirely different direction, a challenge that doomed Randolph. It had been vying with the even smaller settlement of Memphis, immediately to the south, to become the most important commercial center on the Mississippi River in western Tennessee. Indeed, in the 1830s Randolph's promise seemed much greater than that of Memphis, so much so that an enterprising editor, Francis S. Latham, convinced that an important river city would emerge in the vicinity, settled there in 1834 and began publishing the *Randolph Recorder*. Situated on the second of the four Chickasaw bluffs, Randolph appeared to enjoy certain advantages over Memphis, located southward on the fourth bluff. The latter was known to have suffered outbreaks of yellow fever and cholera, and while Randolph was not entirely free of these diseases, it appeared the better choice for men concerned about the health of their families.[18]

Randolph also offered certain commercial advantages. A bustling town that handled far more cotton than did Memphis in the 1830s, Randolph became the northernmost destination of packets from New Orleans, and until 1840 it was the "the great steamboat depot of West Tennessee." An assortment of travelers and farmers from up the Hatchie River brought their produce to market from the populous counties of northwestern Tennessee, and many people disembarked from the flatboats, packets, and steamships that made daily stops at the Port of Randolph, and stayed at one of the four Randolph hotels. Some of them undoubtedly ate and drank at the Wilson tavern or slept in the family's hotel. By the mid-1830s, Randolph had a population of one thousand and "good private schools (one of them called a college), nearly fifty business houses, including a distillery, two wholesale whiskey houses and a dozen saloons." A lively center of trade and commerce, Randolph seemed to have great promise.[19]

Yet even as Latham established his newspaper in 1834, Randolph had already suffered strategic blows that gave the advantage to Memphis. Hopes dimmed for the construction of a canal between the Tennessee and Hatchie Rivers that would have drawn trade to Randolph from throughout the Tennessee Valley. Though even David Crockett advocated the canal, neither the majority in Congress nor the president then favored internal improvements, and the canal never materialized. And then in December of the very year that Latham settled in Randolph, a commission appointed by the Tennessee legislature selected the centrally located town of Covington rather than Randolph as the county seat of Tipton County, thus guaranteeing that town the important legal business of the area. Also at about that time, Memphis rather than Randolph secured a triweekly post route from Nashville, leaving Randolph with only weekly delivery by horseback. The advantage turned further toward Memphis when the federal government in 1834 purchased Chickasaw Indian lands in northern Mississippi, which assured the Chickasaws' removal and gave Memphis a vast hinterland to the south for white settlement. This encouraged certain entrepreneurs to lay plans to build a railroad from Nashville to Memphis, creating greater interest and further settlement. Eventually, Memphis secured its railroad, and Randolph faced a near fatal blow delivered by the river itself. By the 1840s it was clear that the course of the river was shifting into Randolph's environs, dooming the port, and many of the men of the town began to cast about for alternative locations.[20]

Some of Randolph's most prominent citizens, like newspaper editor Latham, simply moved to Memphis. William H. Wilson, Josiah's younger brother, eventually went to Covington, where he farmed and raised a large family. Parker Wilson, another of Josiah's younger brothers, elected to make his home in Brownsville, Tennessee, where he became a prominent doctor. Some men made more unconventional choices. In 1835 the postmaster of Randolph, Colonel William P. Miller, resigned and went to Texas to command troops under Sam Houston, taking several young men of the town with him, including Sam Wilson, another of Josiah's brothers. Young Wilson served as a first lieutenant under the command of Burr Duvall and died in the Goliad Massacre in March 1836.[21]

While his brothers, for good or ill, made their own decisions about where to cast their lot, Josiah struck out in a different direction. The choice that Josiah made to move into the Arkansas swamps might have seemed unwise, but he knew the terrain and understood the arduous labor necessary to bring timberlands into cultivation, for the area around Randolph was itself newly reclaimed. Part of an "unorganized" region until 1823, Tipton County had, after all, attracted the Wilson family and they had prospered there. When William Wilson died in November 1837, he bequeathed his entire estate to his wife and his four youngest children, but he had almost certainly settled property on his four older children prior to drawing up the will in October 1836. The document does not indicate precisely how much land he owned, nor does it specify the value of his estate. The only detail provided involved the six slaves he left to his wife, "a negro girl Susan and her three children" and a "negro girl, Nancy, and her child." Napoleon Wilson, one of Susan's children, told a Wilson descendant in 1908 that William Wilson owned a plantation near Randolph, but the incomplete deed records of Tipton County do not confirm that supposition. Nevertheless, he left a sufficient legacy to his children, all of whom would find it necessary to look for opportunity elsewhere.[22]

As the eldest son, Josiah served as co-executor of his father's estate with his mother, a trusted position that signaled his place as head of the Wilson family. A self-confident and propertied man who possessed ambition similar to that of other westward-moving planters, he migrated across the Mississippi River and onto dangerous ground in 1846. A photograph of him, probably taken in the 1840s just when he was about to make his move, captures the look of a stern man dressed in a suit and waistcoat, his shirt unstarched and his tie carelessly cinched. Josiah's right forearm rests on the arm of a chair; his large, rawboned hand extends forward. Tall and lanky with slumped shoulders, he is a plain-looking man. His close-cropped hair is swept away from his face, and a deep furrow marks the space between wary hazel eyes. He bears the grim expression common to those who had to sit for photographs in that era. Whether Wilson's stern countenance reflects the limitations of the photographic process of the period or something significant about his character is hard to tell. One can discern an almost skeptical expression reflected in his steady gaze toward the camera, an expression that he might have turned toward the Arkansas swamps.[23]

Doubts he had to overcome about the Arkansas side of the river almost certainly included very reasonable concerns about the seismological stability of the region and the environmental damage caused by the earthquakes. Government-contracted surveyors provided the first detailed descriptions of this blighted landscape as they crisscrossed Mississippi County between 1824 and 1848, and their reports would hardly have inspired a rush to purchase such seemingly inferior land. Charged with the responsibility of laying off township, range, and section lines so that federal lands could be sold to the public, the surveyors also provided descriptions of the land that, at first glance, seemed to confirm its worthlessness. To the average observer the time it took the surveyors to complete their tasks would have raised rather than lowered suspicions. To put it in perspective, most surveyors ran their range, township, and section lines in approximately two weeks, even in the worst of conditions. They expected, reasonably, to make eleven miles per day, but in only two in twenty-four surveys conducted in the county did they find that possible. One surveyor took a remarkable fifty-three days to run his lines, and another took forty-one days. Nevertheless, like the other six surveyors who literally struggled through the mire of Mississippi County to conduct their business, they dutifully followed instructions to provide detailed reports on the condition of the land, the variety of trees and vegetation, and other features they encountered. Of the 1,644 remarks made about the land, 1,164 (71 percent) of them identified periodically flooded or "wet" lands. The surveyors also reported earthquake-related damage in all but two of the twenty-four surveys, but those two were traversing the Big Lake area, widely acknowledged by contemporaries then and archeologists today to have been created by the earthquakes. Flood-prone and earthquake-damaged lands ran from one end of the county to the other. Out of the approximately 428,520 acres surveyed, 290,805.5 (67.9 percent) acres regularly flooded and were unsuited for agricultural purposes. In only four cases did surveyors report that more than 50 percent of the land lay above the floodplain. Josiah shrewdly chose one of those four townships to settle in, the one with 80.2 percent in that condition (see table 1.1).[24]

Because surveyors characterized some of the land above the floodplain as rich, a few enterprising men purchased property there regardless of the area's unsavory reputation. The earliest planters settled along the river, where a natural levee provided some safety from floods, and where ready access to river landings—and thus the market—outweighed any concerns they might have harbored about the possibility of collapsing banks. By the time Josiah made his move into Arkansas, it was too late to secure land there. Instead, he purchased a 160-acre homestead in the most favorably positioned interior township, an area with the best cultivable acreage. Remarks made by the surveyor that much of the land there was "first rate" or "good" likely inspired his confidence. In fact, of the townships surveyed, Josiah selected the one that received the majority of such remarks.[25] He laid his claim to a particularly convenient piece of property located due west of Randolph, where he maintained business interests. From the port of Randolph he had only to cross the river to the landing at Golden Lake and from there he could travel by a narrow "road" through the most waterlogged segment of real estate in the county and reach the rich agricultural lands that he homesteaded. The road from Golden Lake actually amounted to little more than a footpath, undoubtedly of pre-Columbian origins and frequently flooded. It passed between two lakes—Round Lake and Golden Lake (for which the village was named). Wilson doubtlessly counted on the seasonal low water in the late fall to render the road above the floodplain so that he could transport his harvest through to the river.[26]

For Wilson, who hoped to capitalize on the profitable cotton market, access to both the road and the river was essential, and he soon purchased other parcels along the same road and even closer to the Golden Lake settlement. His daughter Viola eventually inherited the property he came to own in this area. He also purchased parcels on another prehistoric road, leading to the village of Frenchman's Bayou, a small settlement to the south. Children by his second spouse, including his youngest son, Robert E. Lee "Lee" Wilson, inherited this property. The division of the estate between Golden Lake and Frenchman's Bayou signaled a family dispute after Josiah's death between the children borne by two successive wives, but while he remained alive, the acreage functioned together to serve his economic interests. He purchased parcels above the floodplain and those highly rated by the surveyors. Most of his property rested on prehistoric foundations, including three Indian sites, two of them large villages. Though the Pacaha had long since disappeared, their previous occupation reflected the suitability of the acreage for agricultural and settlement purposes. Although his parcels were above the floodplain, sunk lands and swamps were close at hand, as were several lakes: Golden Lake, Butler's Lake, Carson's Lake, Young's Lake, Swan Lake, and Round Lake were all very near his property. Josiah had chosen carefully and well.[27]

Although Wilson retained some interests in Tennessee, he sold property there in 1845, 1848, and 1850, the cash from which he purchased his land in Arkansas through the General Land Office. His first wife, Rachel, and their two daughters accompanied him to Arkansas, and he set to work immediately to realize his ambitions. The area was rather sparsely populated, but they were not without neighbors. Benjamin Butler, whose father had homesteaded there and served as constable in the township before his death, lived on a small farm located on a prehistoric Indian site just to the northeast of the Wilson homestead and just beyond the intersection of Cross Bayou, which ran east/west, and Frenchman's Bayou, which stretched north from the settlement by the same name. As the two bayous met, they formed Butler's Lake. Daniel Campbell, Susannah Gleason, George Evans, and Andy Hart owned farms in the vicinity positioned near the road that gave them access to the river. None of these lands, moreover, were located on the problematic sand blows, scatters, or sloughs created by the New Madrid earthquake, although low-lying lands, like that which formed Butler's Lake, existed throughout the area.[28]

An eerie cypress forest which "grew in grotesque shapes that presented weird scenes in the moonlight" made the area belong to the realm of exotica. Part of the North American or Mississippi Flyway, the delta served as an important winter habitat for a variety of waterfowl from the Great Lakes and the prairies, and the cacophony of sounds that accompanied their annual visitations seemed mysterious and foreign to those new to life in the swamps. Literally millions of ducks, geese, and other birds wintered in remote places there. Dozens of other species of wild fowl made the swamps their permanent home. An astonishing variety of four-legged animals inhabited the region on a year-round basis, and the abundance of wild game, such as white-tailed deer, beavers, opossums, and rabbits, provided for sport and a more varied diet. Other creatures, however, sometimes made life dangerous. Bears, wildcats, gray foxes, coyotes, and ferocious razorback hogs were dogged pursuers of livestock and, if riled, sometimes confronted lone individuals with serious consequences. Meanwhile, even deadlier inhabitants of the swamps lived virtually unseen—viruses and bacteria—or unappreciated as harbingers of disease—ticks and mosquitoes.[29]

The Wilson family suffered the dangers of the swamps almost immediately. Rachel Cox Wilson did not live to see the 1850 census enumerator, leaving Josiah Wilson residing in his household alone with his two daughters, Viola, age thirteen, and Missouri, age ten. What took Rachel's life is unknown, but infectious diseases, about which little of use was known in the mid-nineteenth century, were a constant source of worry and grief. If not struck by some infectious disease, she might have faced another predator: the ubiquitous mosquito, a pest which abounded in the many stagnant pools of water that were left standing when the rivers returned to their normal boundaries in the sum-mer months. As the temperatures rose, conditions ripened for breeding mos-quitoes, which then carried diseases such as malaria and yellow fever to the residents of the swamps. Malaria and yellow fever are seasonal diseases con-tracted in that region between June, when the weather has warmed enough to permit mosquitoes to breed, and October, when the first frost ends their existence. Two additional factors have to coincide, however, for the diseases to prevail. First, in order to contract malaria, the female *Anopheles* must be among the mosquitoes breeding in the area; second, a person already infected by the plasmodium parasite must be present. The female *Anopheles* must then feed from an infected person and then bite an uninfected person, thus spread-ing the disease. A different mosquito, the female *Aedes aegypti* mosquito, car-ried the yellow fever virus, but the transmission process was the same. A few scientists were beginning to suspect a connection between the mosquito and these two diseases, but they had only a vague notion of the role they played, and no one had a cure.[30]

Yellow fever typically struck more densely populated areas like cities, but malaria became endemic to the swamps. While some physicians advocated the use of quinine for malaria, which merely relieved the symptoms, the best defense was repeated infection that gradually inoculated the sufferer. A cruel and hardly satisfactory remedy, repeated infections worked only if the victim contracted the same form of malaria. Four major types of the plasmodium parasite are responsible for the disease. The most common form, *Plasmodium vivax,* is relatively mild, but *Plasmodium falciparum* frequently leads to death. *Vivax* typically causes the patient to suffer fever and anemia. In addition to those symptoms, *falciparum* can cause microvascular blockages in the brain (cerebral malaria) and the lungs (pulmonary edema), and it can damage the kidneys, leading to acute renal failure. Internal hemorrhages sometimes result in black urine; hence the term "blackwater fever" became associated with this form of the disease. While the different forms appear to cluster in different places, they can be transmitted easily if an infected person travels, and Missis-sippi County's many river landings heightened the likelihood of the introduc-tion of different strains. Thus a patient might become immune to one form only to contract and suffer from another. Although the lethal variety of malaria sometimes struck the area, most people merely suffered periodic bouts with the milder form of the disease.

People living in the county's swamps were probably more likely to die from waterborne diseases like cholera, and scientists in the early nineteenth century were nearly ignorant of its causes. They understood that contaminated water led to cholera, but they did not understand the role of the bacterium *Vibrio cholerae* nor did they have an appropriate remedy. Infection leads to severe cramps and increased bowel action, and nineteenth-century physicians, if they were present, administered generous amounts of opium to ease the cramps and then calomel, a purgative, which "made the already dreadful flux even worse." As a result of these treatments, patients frequently became dehydrated and suffered an imbalance in electrolytes, and it was from these consequences that many died. Patients who survived did so despite the remedies prescribed by physicians.[31]

Whether Rachel Cox Wilson died from yellow fever, from a lethal form of malaria, from cholera, or from some other disease cannot be known. Neither can it be ascertained what caused young Missouri Wilson to lose her hearing, but between the 1850 and 1860 censuses, she became deaf. She might have been overmedicated with quinine to fight an encounter with malaria, for excessive dosages can cause hearing loss. She might have fallen victim to meningitis, a disease prevalent in subtropical climates such as that existing in the Arkansas delta. Periodic epidemics of meningitis visited the area throughout the nineteenth and early twentieth centuries. Caused by either a virus or bacteria that infect the fluid of the spinal cord, it can enter the brain, resulting in a variety of symptoms from headaches to convulsions and hallucinations. Several kinds of bacteria are associated with the disease, and all of them thrived in the swamps of Arkansas. Meningitis could be life-threatening, but could also leave the victim with a variety of lesser injuries, including hearing loss, impaired eyesight, seizures, and cognitive disabilities.[32]

Despite the costs to his family, Josiah and a few others settled in the swamps, endured the disease environment, and raised families. Having lost his wife, Josiah began a liaison with a young woman, Martha Parson, in about 1852, fathering at least four children with her, three of whom survived to adulthood. Twenty-two at the time her relationship with Josiah began, Martha has obscure origins, but we know that she lived in a house adjacent to Josiah's cabin from the time their first child, Victoria, was born in 1853 until at least 1860. In that year she still carried the name Parson, as did her three children. While she eased his loneliness, she complicated his relationship with his daughter Viola, a complication relieved only by his daughter's marriage in 1854 to a nearby planter. But tensions remained, and the rift between the two women erupted into a full-scale legal battle in 1870 after Josiah's death. One issue arising then involved Josiah's failure to marry Martha until after their children were born, an omission which eludes satisfactory explanation. It may partly explain, however, why Josiah fails to secure even a mention by county gene-alogists and historians, who focus instead on certain other notable families. Josiah, the rough-hewn and enterprising planter, chose an unconventional existence on an interior plantation and continued to live in a modest cabin. He failed to secure the solicitations of more favorably positioned planters and did not inspire the imagination of those who later perpetuated the county's antebellum plantation myth of grand homes and happy slaves.

However much his story failed to provide rich fodder for historians, Josiah Wilson prospered in his Arkansas venture. By 1850 he had acquired a total of nine hundred acres, mostly through private transactions rather than from the General Land Office. He had one hundred acres in cultivation, twenty-one slaves, and an estate valued at five thousand dollars. His eight hundred acres of undeveloped land held thick stands of red oak, cottonwood, gum, elm, hickory, walnut, pecan, ash, and other timber, some of which he harvested. In terms of value of his real and personal property, he ranked in the top 20 percent of all slaveholders in south Mississippi County and the top 6 percent of all heads of household there. His compatriots included state senator Thomas Craighead, who, like Wilson, continued to live in a simple log structure. Other planters, however, particularly those residing along the river, built large homes. For example, John Harding McGavock built an elegant plantation house at Sans Souci landing, just south of Osceola, the county seat, in the mid-1850s. Other McGavocks—John's brothers, cousins, and in-laws—built plantation homes of substance at Nodena Landing, Frenchman's Bayou, and Pecan Point. Only one of the clan faced severe challenges in the antebellum period. Young E. J. McGavock, who established a plantation at Pecan Point, had chosen unwisely. The Point was located at a spot on the river that turned in toward the west, and over years the powerful Mississippi ate away at the bank there, eventually taking the plantation home and the remnants of several large Indian mounds as well as evidence of a sizable Mississippian-era village. The other homes and plantations along the riverfront fared better, and their owners prospered.[33]

The plantation homes on the Mississippi had a serious functional purpose, for they were positioned at "landings" where steamboats made stops that al-lowed planters to send cotton to market and receive goods and materials. Planters and their families took the opportunity to gather together at harvest to await the arrival of steamboats, a thrilling event that provided a welcome interruption to the monotony of life in the isolated swamps. It might have been at one such gathering that young Viola Wilson met the planter she married in 1854. "There was always great excitement about the arrival of the boats, and such speculation as to just how much cotton 'she' would be loading up the river. . . . As the boat would near the landing its arrival would be announced by whistles—and each line had a different whistle. Old river hands knew them all by their signals, and the word would go around that it was time to get to the river for the 'Rob' had just blown for a landing." While the "Rob" was the Robert E. Lee and thus of post–Civil War vintage, dozens of boats, with names like the *Angle Norman,* the *Crystal Palace,* the *Natchez,* and the *Queen of the West,* ran the route before the Civil War. The *Kentuckian* and the *Tennessean,* of course, had figured prominently in the county's history. For the planters, however, the arrival of the steamboats involved far more than entertainment, and a good deal of worry and bother likely accompanied any excitement they might have felt. For the black slaves required to load the 500-pound bales of cotton on board the steamships, their arrival signaled backbreaking work and endless toil. It also allowed the possibility of interaction with both slave and free black riverboat hands as well as the possibility of escape.[34]

Even as the plantation economy of Mississippi County began to expand, southeastern Arkansas remained the most dynamic growth area in the state. Planters from east of the Mississippi River who swept into the Arkansas delta between 1850 and 1860, when cotton prices soared, chose most often to locate in the more promising and less problematic counties of southeastern Arkansas. By 1860 they held the largest concentrations of slaves: Chicot (78 percent), Lafayette (64 percent), Arkansas (47 percent), Union (46 percent), and Desha (40 percent). These counties also produced the greatest amount of cotton. While Chicot County planters, for example, harvested 40,948 bales in 1860, Mississippi County, which tied with Dallas County in terms of percent of slave population (37 percent), harvested only 1,244 bales. Although the number of slaves in Mississippi County increased from 865 to 1,461, so too did the number of whites. By 1860, the county's percentage of slaves had risen only slightly to 37.5 percent of its population. It ranked fourteen out of eighteen delta counties in slave holdings.[35]

Although Mississippi County lagged behind the other delta counties, Josiah Wilson profited enormously from the expansion of his plantation. In 1850 he had only a hundred of his nine hundred acres in cultivation and grew a variety of crops, such as corn, peas, and potatoes, but he raised no cotton. By 1860, he owned 2,300 acres and harvested ninety bales of cotton. He continued to grow a variety of crops and raise livestock, but the rising price of cotton in the late 1840s and throughout the 1850s encouraged Josiah and other planters to bring their unimproved land into cotton cultivation as quickly as possible and expand their operations rapidly. The price ranged upward from a low .06 cents a pound in 1845 to a high of .14 cents a pound in 1857. Although it had dropped to .12 cents by 1860, that was still twice what it was in 1845. The ninety bales Wilson reported in 1860 probably weighed an average of five hundred pounds, and at .12 cents a pound for ninety bales, he would have netted $5,400. To put it in perspective, that would be $103.285.04 in (the year) 2000 dollars, four times the median household income in Mississippi County of $27,479 in that year.[36]

Wilson's slaveholdings also increased significantly in the 1850s. By 1860 he owned nearly twice as many slaves as he had a decade earlier—from twenty-one to forty. Twenty-one of his slaves ranged between the ages of twelve and forty, providing for a relatively large plantation labor force. They appeared to live in nuclear or near-nuclear family settings, but they were crowded into six cabins, which were likely merely smaller and less elaborate versions of the unadorned cabin the Wilson family inhabited. Wilson held only four adult male slaves and six adult female slaves, while thirty of his slaves were under the age of sixteen, suggesting that he relied more on natural increase than purchase to increase his slaveholdings. Certainly it was much cheaper to rear slaves than to buy them. The average cost of a slave in 1859 was $1,658, meaning that Wilson, with forty slaves, had $66,320 in slave property.[37]

The slaves of Mississippi County, living as they did in a frontier plantation system, necessarily endured the most extreme conditions confronting slaves in the American South. Slaves in southeastern Arkansas, a section of the state a decade ahead of Mississippi County in terms of development, likely enjoyed better conditions. They had cleared much of the land, dispensing with the backbreaking work of bringing forested lands into cultivation, and the ratio of female to male slaves there made it more likely that they could reestablish family and community relationships severed or lost when their owners moved to southeastern Arkansas in the 1830s and 1840s.[38]

One striking aspect of Mississippi County's slave population involved its concentration in the five southern townships. Of the 1,560 slaves in the county, 1,208 (77.4 percent) lived there and in relatively easy reach of one another. This made possible the existence of a much larger slave community, as large, in fact, as that of the white community in that area, which amounted to only 1,262 (tables 1.2 and 1.3). In other words, slaves represented 49 percent of the residents there. The existence of swamps and deep woods gave the slaves cover as they traveled, with or without passes, to visit with each other and to hold grape arbor (religious) meetings. Extensive research on slave family, religion, and community elsewhere suggests that the slaves of Mississippi County created similar institutions, but little is known of them because virtually nothing except the slave census schedules survives in the historical record. Even Napoleon, the slave who provided the genealogy on the Wilson family in 1908, said nothing of his own history. Born into William L. Wilson's household in 1835, he came into Josiah's possession and probably made the trip to Arkansas in 1846. He remained close to the family to the end of his life, but no one interviewed him about his own experiences.[39]

Even less is known about the slaves of north Mississippi County. Only 352 (22.5 percent) of the county's slaves resided there, and most of them were located in one river township, Canadian, or in adjacent Clear Lake. Big Lake, the township on the county's northwestern border, held no slaves, mirroring its "whiteness" later in the century. It remained white and isolated, the latter because of the existence of the impassable lake that gave the township its name. The concentration of the white population there represented an important development suggesting the existence of a racial and class divide, a divide that separated the northwestern townships of the northern section of the county into a white enclave of non-slaveholding small farmers. In the post–Civil War period, the white farmers of Big Lake Township in particular became determined to maintain Buffalo Island, as they called it, as a white enclave. They associated African Americans with plantation agriculture and that association, together with racism, fueled their determination to maintain theirs as a "white man's country."[40]

As much as they had gained during the 1850s, the planters in Mississippi County were not yet approaching the kind of intensive agriculture that the late Mississippian-era Indians had achieved prior to their disappearance. The overall population of the county, which stood at 3,898 in 1860, was still far below the levels indicated by De Soto in 1541. Although the acres in cultivation had increased from 8,111 in 1850 to 17,584 in 1859, planters and farmers cultivated only the most accessible land and refrained from engaging in massive reclamation. Planters along the river constructed some rudimentary levees to protect against floods, but they had not yet even begun to dream of any comprehensive drainage of the swamps. Still, the elimination of over ten thousand acres in forests and cane fields would have destroyed some wildlife habitats and created stresses on those areas not yet cleared as animals relocated. Ironically, as historical ecologist Linda Newston suggests, the removal of the trees, which act as barriers, might have increased the swamps, the breeding ground for mosquitoes and a host of dangerous bacteria, and thus intensified the disease environment. This was, however, still a community in its infancy, and it would be the post–Civil War population that would have to endure the consequences.[41]

The Civil War brought an abrupt end to the development of the area and reversed the upward trajectory of property holders in Mississippi County. While riverboat traffic did not end altogether, the war greatly altered its character. Now, instead of exciting the happy curiosity of those on shore and the business expectations of planters with a crop to send to market, a steamboat constituted a harbinger of war news if not war itself. The Union Army contracted with steamboats to transport troops and prisoners, and the frontier slave society of Mississippi County nearly collapsed as the federals soon captured control of most of the river. Proximity to the Mississippi, always a mixed blessing, now carried additional disadvantages. The Union Army appropriated Sans Souci Plantation for use as a hospital, and once the federals launched an offensive down Crowley's Ridge in late April 1862, the area between the river and the ridge became a no-man's-land. The swamps, forests, and thick canebrakes provided ample cover for guerillas, and, together with frequent Union cavalry raids from Missouri, made the interior of the county no safe haven for planters hoping to maintain some semblance of normalcy—and produce a cotton crop. Conditions became so chaotic that many whites fled the scene. Whether Josiah remained in the county remains unknown, but no matter his strategy for surviving the war, he undoubtedly experienced serious challenges and his family endured severe privation. Some planters took their slaves and moved to safer locations, but nothing surfaces in the record to indicate what happened to Josiah's slave population. Many slaves in the county sought freedom within Union lines, and certainly Union troops in the vicinity made that a distinct possibility.[42]

Freedmen in the county and Union troops alike faced a number of challenges, among them a homegrown resistance to emancipation and to Union activities in the area. In 1864 federal troops apprehended Captain Charles Bowen, the leader of a guerilla insurgency in the county and a man destined to play a crucial role during Reconstruction.[43] He had resigned his position as county sheriff in 1861 to organize a unit for the Confederacy, the "Osceola Hornets," and fought in several major battles east of the river. He sustained heavy losses, particularly at Shiloh, where only seven of his men left the field. He suffered a head wound in that battle, and once he recovered he returned home to recruit replacements, but instead of reporting to the Confederate command for reassignment, Bowen remained in northeastern Arkansas and led his men in guerilla warfare. His campaign against Missouri Jayhawkers, Union troops, and freed African American slaves was aided by rather than ended by his selection as county sheriff in late 1864 by the new "reconstruction" governor, Isaac Murphy.

Bowen owed his reacquisition of his old job to President Abraham Lincoln's early experiment with Reconstruction, a mild plan meant to quickly reintegrate former Confederate states once they came under federal control. Arkansas fell to Union forces in September 1863, and though most of Arkansas remained in contention, the state qualified for readmittance to the Union in 1864 under Lincoln's plan after 10 percent of the eligible voters (in 1860) had sworn an oath of future loyalty to the United States. Although Congress subsequently refused to seat congressional delegates elected to the U.S. House and Senate that year from Arkansas, Tennessee, and Louisiana, the three states that qualified under Lincoln's plan, those elected to or selected for state and county positions assumed office. Bowen had to take an oath of future loyalty in order to accept the position in 1864, but that oath did not prevent him from continuing his campaign to control the former slaves.[44]

The end of the war brought little relief from the turmoil. Bowen held the sheriff's position until 1866, when he declined to run for reelection and another former Confederate, John Long, took office. Congressional Reconstruction began in 1868, but former Confederates ultimately succeeded in maintaining control. From 1868 to 1872 Mississippi County constituted a virtual war zone. On one side stood the militia called out by the state's new Reconstruction governor, Republican Powell Clayton, a dashing former Union cavalry officer. On the other side stood local white Democrats and former Confederates determined to maintain political authority. Clayton, arguing that widespread fraud and violence occurred on Election Day, November 3, 1868, declared martial law in ten counties, including Mississippi County, the day following that election. Black Republicans made up the bulk of his militia, and they confronted angry white citizens, mostly former Confederates, who were sometimes backed by the Ku Klux Klan, an organization founded in nearby Pulaski, Tennessee. Charles Bowen and other former Confederates led the county chapter of the KKK, and ordinary citizens, whether white or black, were often caught in the middle. Freed people asserted their newfound rights and privileges only at risk of life and limb. With the Freedmen's Bureau agent located in the county unable to get a handle on the situation there, former Confederates under Bowen enjoyed a free hand in the effort to subjugate those African Americans who remained in the county. While political violence existed in almost every section of the state—northwest Arkansas being the only exception—northeast Arkansas was particularly notorious.[45]

The violence in Mississippi County culminated in a struggle for political control in 1872 called the "Black Hawk War." Governor Clayton had earlier appointed a former Union army officer, Charles Fitzpatrick, to preside over the board of registration, and Fitzpatrick proved determined to organize black voters in the county. The freedmen, who had formed "into secret societies throughout the county and often marched in armed bands to Osceola and other points, making speeches and causing a great deal of excitement," embraced the opportunity to vote. Clearly politicized, they meant to experience all that freedom and citizenship had to offer. As historian Steven Hahn has argued in his *Nation under Our Feet,* African American slaves learned certain political skills in the antebellum period that aided them in their transition to freedom and prepared them to assume the rights and responsibilities of citizens. Freedmen in Mississippi County demonstrated that they had learned these lessons and stood ready to use them. They supported Fitzpatrick when he decided to run for the state legislature that year against a popular former Confederate and one of the organizers of the county's KKK chapter, H. M. McVeigh, but they constituted a minority of the population. In fact, the black population had declined by a third, from a total of 1,461 slaves in 1860 to 971 freed people in 1870. The white population, meanwhile, had grown from 2,434 to 2,662. Only a few local whites identified themselves as Republicans, and most of these were newcomers to the county and Union veterans. With Fitzpatrick in the lead, they attempted to form a viable Republican Party, hoping to draw the bulk of their membership from the population of freedmen. Given the demographics, however, they realized that they would have to rely on some white voters as well, a coalition that that proved just as impossible to fashion in Mississippi County as it was elsewhere in the South. Although Governor Clayton is generally credited with having effectively destroyed the Klan in Arkansas by 1869, it appears that the Mississippi County Klan merely went underground. Masked groups of whites serving as the unofficial military arm of the local Democratic Party continued to surface and played the decisive role in the county in the violence that erupted in the fall of 1872.[46]

Fragmentary and contradictory reports exist concerning the events that transpired during the Blackhawk War, but Fitzpatrick's candidacy for the legislature raised the stakes in the struggle for political control and led to a heated exchange on the streets of Osceola between the candidate and the county sheriff, J. B. Murray, yet another former Confederate, in late summer 1872.

The two men drew their weapons, and Fitzpatrick shot Murray to death.[47] Fitzpatrick surrendered to authorities and was released on bail, but when he came up for trial in the fall, a large group of armed African Americans, attempting to show their support for Fitzpatrick, marched to Osceola. Former Confederates Bowen and McVeigh wrote the only accounts of what happened next, and they convey a particularly partisan view of the events. According to Bowen, a group of four hundred to five hundred African Americans threatened to destroy Osceola and all the whites within it. He almost certainly exaggerated the numbers, and the blacks who gathered in Osceola intended to support—and perhaps protect—Fitzpatrick. Bowen had at his disposal a group of armed former Confederates, who "had resumed the habits of peaceable citizens," as McVeigh put it, after the war but found it difficult to "shake off the habits of soldiers." According to Adah Roussan, wife of the editor of the local newspaper and a stalwart temperance advocate, the instigators were "renegade whites, and in nearly every instance, when trouble broke out between the races, it was traceable to a white man's whiskey dive." Regardless of the causes, Bowen and his men attacked the African Americans who had gathered in Osceola, driving them into the swamps of southern Mississippi County, probably passing through or very near the property owned by Josiah Wilson and operated by his heirs that year. They killed a "good many" of the African Americans and many of the survivors fled to other counties. Fitzpatrick, apparently recognizing his vulnerable position, left the area entirely. Bowen and his "renegade" whites had vanquished the Republican threat.[48]

Local whites had effectively maintained control of county government throughout the Reconstruction period but had done so by extra-legal and violent means. In Fitzpatrick's absence, H. M. McVeigh secured election to the state legislature that year without opposition, and in 1874 Charles Bowen served as a delegate to a new constitutional convention, one that drafted Arkansas's "Confederate" constitution, a reactionary document that imposed severe restrictions on state officials and vested considerable authority on local officials, especially county judges. Two years later Bowen secured election to the powerful county judgeship, marking the ultimate triumph of the former Confederates in the county.[49]

Although the former Confederates enjoyed political control of the county, the economic realities remained distinctly uncertain. The total wealth in real and personal property in the two southernmost (and richest) townships had dropped from the high of $464,564 in 1860 to $285,535 in 1870. Josiah Wilson, who died just two weeks after the census enumerator secured the information about his own personal holdings, had relocated to Memphis, leaving his plantation in the hands of his son-in-law, Napoleon Lafont, who had married his eldest daughter, Viola, after the death of her first husband in 1855. Lafont reported that the plantation had a cash value of $40,000, nearly half the $75,000 it was worth in 1860.[50] The fate of the McGavock family is even more telling. The combined wealth of the McGavock brothers and cousins had been reduced from $428,720 to $122,600.[51] And John Harding McGavock of Sans Souci Plantation had died—of fever—in 1861. Those at the bottom of the social structure, moreover, suffered the most devastating losses. The percentage of household heads holding property had declined to 42 percent, below the high of 70.7 percent recorded in 1860 and below, even, the figure (59.5 percent) recorded in 1850. As difficult as things were for the planters, the small farmers of northwest Mississippi County had suffered even more staggering losses. Their real and personal property wealth was reduced by 90 to 100 percent, and their anger and frustration over the situation grew only bitterer in the postwar period.[52]

While the poor whites of northwest Mississippi County withdrew into their all-white enclave, African Americans either left the county altogether or jockeyed for position in the prevailing labor market. The ruling elite had maintained their position of political dominance—over poor whites and blacks alike—but their method of doing so had worked against their economic interests. By pursuing the violence against politicized African Americans, they had driven much of their labor force out of the county. The contract labor system—something that the Union army introduced during the war and the Freedman's Bureau championed after it—had been no more popular in Mississippi County than it had been elsewhere in the South, but the cultivation of cotton required substantial labor, and most farmers and planters placed their hopes for economic recovery on the cotton crop. The labor crisis grew so acute that one planter, Felix G. McGavock, imported fifty German men from New York and eighteen Chinese men from Chicago to harvest his crop during one season, only to have the Chinese enticed away almost immediately. Later he hired over fifty Irish girls between the ages of fourteen and twenty-five years for one year from New York. Understanding they were not "seasoned" to the swamps, he dissolved a "few ounces of quinine, in a barrel of whiskey" and administered it at "the rate of three drinks of two drams each at specified intervals during the day." He also maintained an infirmary and saw to it that a "Catholic priest come to attend to their spiritual welfare." In the end, planters came to rely on African American laborers but not under the contract or wage-labor system initiated by federal officials during and after the war. The sharecropping and tenancy system eventually emerged, and the next generation of southerners, such as Josiah's son Lee Wilson, would preside over its development.[53]

It would take decades for Mississippi County planters to regain their footing and return to intensive agriculture, but as the population expanded in the late nineteenth century, it became necessary to fashion a new labor system. This new population brought greater pressure to bear on an environment with a peculiarly troubled history. Twice in less than three hundred years ecological disasters had stricken northeastern Arkansas and refashioned the landscape in fundamental ways. The late sixteenth-century drought worked a demographic revolution that left the area nearly vacant for three centuries. Just as new groups of Indians moved into the area and Anglo Americans began to consider its possibilities after the Louisiana Purchase, a series of catastrophic earthquakes wracked the region, leaving in its wake a scarred and blighted landscape. Once Anglo Americans overcame their qualms about settling in the notorious swamps of northeastern Arkansas in the 1840s and 1850s, they imported African American slaves to begin the process of carving plantations out of the canebrakes and swamps. The profitable cotton market of the 1850s helped men like Josiah Wilson create small fortunes, but the Civil War brought devastation, and the emancipation of slaves led to an intense postwar struggle over the meaning of freedom, a struggle the ex-Confederates eventually won. The postwar years witnessed a series of trials, some of them created by a significant increase in population of both blacks and whites, whose interests did not necessarily coincide. While planters adopted the sharecropping and tenancy system in place of slavery, they exploited potential laborers of both races, but their black workers became the targets of poor whites motivated not only by the extreme racism that emerged in the late nineteenth century but also by their antipathy toward the plantation model of development. The next generation of planters, like Josiah's son Lee, struggled with the postwar dilemmas as they put into place a coercive labor system and brought greater acreage into cultivation.

2

THE MAKING OF
THE MAN

D eath and controversy plagued Robert E. "Lee" Wilson early in his life. Born during the final weeks of the Civil War, he faced not only the reduced circumstances common to most southerners in that era but also endured additional burdens. When his father died in July 1870, his older half sister challenged his right to inherit. She failed in her effort to deny him a part of their father's estate, but the death of his mother of yellow fever in 1878 left him an orphan and under the power of those who had tried to disinherit him. Determined to secure his patrimony, he seized control of his inheritance in 1882 by successfully petitioning the court to recognize his right to transact his own business free of his guardian's oversight. Only seventeen at the time, he began to lay the foundation of a small empire. A member of the generation that came of age during the tumultuous decades of Reconstruction and Redemption, he embodied Henry Grady's model of a New South entrepreneur born out of chaos and destruction. He used what he had at his disposal, a small inheritance in the Arkansas swamps, and earned a reputation for a single-minded pursuit of profit and a determination to impose his will upon those who opposed him. The series of intrigues and tragedies that confronted him in his first two decades of life inculcated in him a strong desire to overcome obstacles in any manner necessary. His assumption, at a very early age, of responsibility for dependent members of his extended family imbued him with a sense of paternalism that later influenced all of his relationships, including those with his employees, both black and white, both menial and white collar.

Within days of Josiah's burial in the old cemetery in Randolph, Tennessee, a battle over his estate exposed a long-simmering discord among his children. Josiah's failure to execute a will sparked the conflict, and his tardy marriage to Martha threatened to undermine Lee's expectations. On one side stood Josiah's second wife, Martha Parson Wilson, who represented her sons, Lee, age five, and William Henry, age ten. Martha also spoke for her thirty-year-old "deaf and dumb" stepdaughter, Missouri, who lived in her household. She faced Viola Wilson Lafont and her second husband, Napoleon Lafont, who had taken over management of Josiah's Arkansas property when the elder man retired to Memphis in 1869. Martha's oldest child with Josiah, Victoria, remained strangely silent in the early phase of the court proceedings, but her husband, Dr. James Davies, later became a mediating force between the disputants.

Because Josiah held property in both Tennessee and Arkansas, separate probate courts held jurisdiction. The challenge to Lee's inheritance actually began in Tennessee when Napoleon Lafont, representing his wife, filed a petition in the Memphis Probate Court arguing that because Martha's sons were "born out of wedlock," they were not entitled to inherit from the estate of their father. Martha defended her sons' patrimony by citing an Arkansas statute, "If a man have by a woman a child or children and afterward shall intermarry with her and shall recognize such children to be his, they shall be decreed and considered as legitimate." Martha's position on this matter necessarily prevailed, and on September 12, 1870, the court awarded her the use of the house and its furnishing and $5,000 to support her household—which included Lee, William Henry, and Missouri—for one year.[1] But the battle for control over the estate was far from over, and some months of intense negotiation subsequently ensued. On March 17, 1871, Lafont designated Davies as spokesperson for his wife's interests, and three days later the contestants finally agreed to divide $16,635.38 in cash assets held in Tennessee equally between Josiah's widow and five children. One crucial concession Martha made concerned the title to the house on Poplar Street, a home Josiah had purchased for them in 1869. She relinquished any claim in the ownership of the home in exchange for the right to remain there for the rest of her life. Josiah's five children inherited equal shares to the house.[2]

More important to Lee's future, however, was the disposition of the Arkansas property. On the same day that the Shelby County Court rendered its order regarding the Tennessee estate, the Mississippi County Probate Court ruled that Josiah's children divide the 2,300 acres in Arkansas equally. The court appointed Dr. Davies as guardian of the property inherited by Lee and William Henry but listed Lafont as a secondary signatory, leaving the boys in a vulnerable position. Although Lee's inheritance there provided the stake he needed to found his plantation empire, the land held certain disadvantages. Much of the acreage that Lee and William Henry inherited was undeveloped and suffered frequent inundations during periods of overflow. Nevertheless, some of it was under cultivation, and much of the rest had potential if drained and protected from floods. Both Lee's four hundred acres and William Henry's 420 acres were located on the road to Frenchman's Bayou, a convenient arrangement for Davies, who secured the right to manage their Arkansas property. Davies, who operated a mercantile and medical practice at the village there, also farmed substantial acreage of his own in that vicinity. Viola inherited their father's prime property, including Josiah's initial homestead, and it, like Missouri's, was located on the road toward the village of Golden Lake, where Napoleon could easily manage it.[3]

Meanwhile, Martha raised her boys in the bustling river city of Memphis, a location destined to play a crucial role in Lee Wilson's later rise to wealth and prominence. The city served as the most important commercial center in a region that included not only northeastern Arkansas but also western Tennessee, southwestern Kentucky, and northwestern Mississippi. The cotton economy fueled the city's economy, although the lumber industry played an important secondary role, particularly in the late nineteenth and early twentieth centuries. Lee eventually opened both cotton factorage and wholesale lumber operations there, and though St. Louis would become more important to him as a source of investment capital, Memphis remained a focal point throughout his life.

In the 1870s, however, the income from the Arkansas land and Tennessee investments together with the annual support their mother received provided the boys with a comfortable life in a city that was in the midst of considerable turmoil.[4] Memphis had a complex social structure of prosperous merchants, professionals, and planters at the apex, a small middle class of shopkeepers and clerks, and a vast lower class of working men and domestics at the bottom. The Irish dominated the foreign-born population, making up 1,422 of the 3,284 working people in the city. Although most of them were impoverished, they occupied positions in a vast array of livelihoods. They operated grocery stores and saloons; they worked in construction and draying. The three wealthiest Irishmen in 1870 were bankers. Irish immigrants were politically astute, as well, and were able to mass their votes in key elections. They dominated the police force and the fire department, and one Irish-born orphan, John Loague, served in a number of elective offices and eventually secured election to the mayor's office (1874–75).[5]

African Americans, the city's largest minority in 1870, constituted its poorest residents. They made up 15,442 (38.3 percent) out of a population of 40,226. Only 4 percent of them owned property in 1870, and most of them worked as domestics, as drayman, as construction workers, or at loading cotton on vessels at the wharf.[6] Considerable antipathy existed between the Irish and African American populations, particularly given the fact that they competed for jobs at the lowest level. The arrival of freedmen during and immediately after the Civil War exacerbated these tensions. White attitudes, complicated by resentment of the Freedman's Bureau and the presence of black troops, hardened toward the impoverished African Americans. The city's newspapers heightened hostilities by regularly criticizing the freedmen and publishing biased accounts of racial conflicts. On May 1, 1866, a riot erupted, pitting the largely Irish American police against African Americans, and for three days rampaging crowds of whites invaded black neighborhoods, robbing, beating, and murdering any black man, woman, or child in their path. During the melee, forty-four blacks and two whites died.[7]

The city also became the home of prominent ex-Confederates like Jefferson Davis, who settled there after his imprisonment and became a director of the Carolina Insurance Company. Nathan Bedford Forrest inspired the creation of the infamous Ku Klux Klan in nearby Pulaski, Tennessee, an organization which spread across the South and sought to keep freedmen on the plantations and, ultimately, to limit their political participation. Linked by economic ties to the planter aristocracy in the vicinity, the city's fate was influenced by the health of the cotton market, a market doomed to a long, slow depression following the war. Planters in the vicinity of Memphis typically secured advances from Memphis merchants and cotton factors and then marketed their crops through the river city. The Wilson family, for example, did business with J. & J. Steele and Company, which operated out of No. 1 Exchange Building on Front Row (now Front Street). Two brothers, James and John Steele, operated the company and continued to transact business with the family after Josiah's death. Like most other such businesses, they served a clientele of planters and farmers in the region. When the Steele brothers co-signed the Guardian's Bond with Martha Wilson, establishing her right to serve as guardian of her sons and her stepdaughter, they signified how intimately connected to planter families such men could become.[8] Men like the Steeles took up residence in a fashionable neighborhood developed along Vance Street, immediately south of downtown Memphis, which also attracted some wealthy planters, who moved their families into the city. Josiah Wilson settled his family in a solidly middle-class and respectable—but not particularly fashionable—neighborhood just to the east of downtown Memphis, a neighborhood that included shopkeepers, hotel managers, and at least one policeman. Pinch, a district where most of the impoverished Irish population lived, lay just to the north.[9] In an area that was densely populated, very close to the commercial district and to the river landings, the Wilson boys grew up in an interesting and dynamically changing city.

While Memphis boasted an opera house, a theater company, a symphony orchestra, and many other refinements that suited planter families and the merchants and professionals who embraced planter values and aped their manners, it also entertained a motley assortment of visitors off the many steamships, packets, barges, and flatboats that tied up along its port. Its brothels, bars, and riverside eating establishments served a rough clientele that earned the city a well-deserved reputation for sordid behavior and criminal activity. Both before and after the Civil War, the city government only sporadically attempted to discourage these enterprises or police the individuals who frequented them. At best the police and city officials ignored them; at worst they colluded with them to reap what profits vice-ridden areas typically provide. In addition to the usual assortment of ruffians and criminals, an unruly and disreputable band of youths, known as the Mackerel Gang, caused havoc throughout the 1860s and 1870s.[10] They reached their heyday in the 1870s, just as young Lee Wilson reached boyhood. A broken arm he suffered in May 1875 was almost certainly caused by some typical childhood accident rather than as a result of an encounter with the Mackerels, but the physical courage Lee Wilson demonstrated as a man later in life probably owed something to a childhood spent in the notorious river town.

Meanwhile, the city's leaders had engaged in fiscal irregularities in the 1850s, which placed the city on dangerous ground economically. The turmoil and chaos of the Civil War and occupation by federal troops complicated the city's troubled finances, and the vast and problematic population explosion during and immediately after the Civil War taxed its decaying infrastructure. Hordes of unemployed white and black men added to the city's economic woes, and many of its middle-class and elite citizens refused to pay their taxes or were simply unable to do so. The city government's reputation for imprudence and dishonesty hardly inspired voluntary compliance with tax assessments and collections. Of course, many—like Martha Wilson—routinely paid their taxes and carefully maintained their homes and businesses, but their dedication to civic responsibility made little difference to a city poorly positioned to recover from the excesses that predated the war.[11]

By the early 1870s the city's infrastructure was in serious decay and officials seemed unwilling or unable to remedy the situation. The streets, despite the expenditure of considerable sums, badly needed repair. In poor shape before the war, they were in terrible disorder afterward. After considerable pressure from the citizenry and wrangling between the city and contractors, the city made the decision in 1867 to lay down Nicholson pavement. Composed of wooden blocks treated with creosote, Nicholson pavement hardly stood the test of time. Within a few years the pavement began to disintegrate and exude noxious fumes.[12] Meanwhile, those who secured the private contracts to collect the city's garbage failed to do their jobs. Decaying matter littered the streets and alleyways in the city and attracted "roving bands of curs, goats, and hogs. During hot weather, especially, the combined stench arising from privies and streets was enough to sicken even the most hardened resident."[13] No citywide sewer system was in place, and although privately owned sewers operated, they stretched just four and a half miles and served only the best hotels and commercial houses.[14] A privately owned water system existed, but many residents secured their water from cisterns and underground wells that, given the growing density of population, became increasingly "contaminated by seepage or vapors from nearby privies."[15] While families like the Wilsons had indoor toilet facilities, the poor continued to use poorly constructed and untended outdoor privies. Those few families able to pump water into their homes used either contaminated cisterns or wells or a water system served by the Wolf River, itself contaminated from refuse and chemical wastes.[16] The Bayou Gayosa, which ran from the Wolf River through the northeastern section of the city and passed very close to the Wilson home, had become notoriously polluted and sluggish, making it a prime breeding ground for bacterial diseases such as typhoid fever, dysentery, and cholera and for mosquitoes that transmitted malaria and yellow fever.[17]

Medical professionals understood that density of population, poor sanitation, and filth contributed to disease, but they had little or no understanding of the precise connection and were only beginning to understand the rudiments of germ theory. Consequently, little useful research had been done into adequate means of battling disease, and most populations were helpless in the face of diseases that would be brought under control in the twentieth century. Likewise, because nineteenth-century medical practitioners had little inkling of the mosquito's role in transmitting diseases such as malaria and yellow fever, they had no program for eliminating their breeding grounds. One principal means of reacting to an outbreak of acute diseases, and one that city officials and medical professionals frequently sanctioned, was simply to flee from an infected area with all deliberate dispatch. This had the tragic result of spreading many diseases like yellow fever, for example, further afield. An infected person who traveled to a distant city that was itself plagued by mosquitoes could be bitten by the female *Aëdes aegypti* mosquito, which would then bite others in the vicinity, transmitting the disease to a new and vulnerable population.[18]

From its earliest days Memphis suffered from most of the infectious, water-borne, and mosquito-borne diseases that plagued mankind in the nineteenth century. While outbreaks of infectious diseases like smallpox and waterborne diseases such as cholera were terrifying and costly, it was yellow fever that brought Memphis to its knees, and it was yellow fever that delivered the final blow to the Wilson household in 1878. The two earliest-known outbreaks occurred in 1828 and 1848, but the population was then so small and scattered that the fatality rate was relatively insignificant and attracted little notice. In 1830 only 663 people lived in Memphis, but by 1870, 40,226 people resided in the city, providing a perfect host for a major epidemic.[19] Already by the 1850s the population was significant enough that should the yellow-fever virus reach the abundant mosquito population, disaster would follow. In 1855, Memphis narrowly averted a serious outbreak of yellow fever when 250 people died. In 1867, 550 people—including the husband and four children of "Mother" Jones (Mary Harris Jones)—met their deaths when another outbreak occurred. But a more serious epidemic hit the city in 1873 when two thousand people lost their lives to the disease.[20] The 1873 event frightened the population and led to calls for improvements in sanitation. Many medical practitioners continued to believe filth and poor sanitation caused yellow fever, but city officials did not respond to these demands, and, in any case, unless they included a cleanup of Bayou Gayosa and other stagnant mosquito breeding grounds, their efforts would have been entirely in vain.

The 1878 epidemic struck a city sensitized to the dangers of yellow fever but unwilling to avail itself of what then passed for wisdom on the subject: quarantine. New Orleans was the first city on the river to be ravaged by the disease that summer, but officials there refused to acknowledge the early cases despite worried queries from Tennessee's health officer. On July 27 official word finally reached Memphis that a yellow-fever epidemic was, in fact, raging in New Orleans, nearly two months after the first suspicious death there. Two days later the Memphis board of health hastily approved a quarantine measure, but it was far too late to be effective.[21] Although quarantine was, in fact, closer to the right idea than almost anything else proposed, it is likely that had one been established earlier, it would have had little influence on the spread of the disease into Memphis, for it was far too easy to violate quarantines, especially for river towns.

When word reached Memphis on August 12 that the epidemic had broken out in the north Mississippi town of Grenada, Memphians became alarmed. Alarm turned into panic the next day, August 13, when the first confirmed cases in Memphis surfaced. The first to die included a policeman named James McConnell, who lived just a block away from the Wilson family, and a woman who operated a riverfront eating establishment on Front Row, less than a mile from the family's Poplar Street address. After nineteen more cases were publicized on August 14, a full-scale panic ensued as people rushed to the train depot to depart. According to some estimates, approximately twenty thousand Memphians fled the city between August 14 and 17, many of them finding refuge in Louisville, Kentucky, one of the few cities to remain open to the refugees. Cincinnati and Little Rock, for example, had halted all train service from Memphis, and smaller towns closer to Memphis such as Collierville also refused to receive the refugees.[22] Many of the wealthy sped out of the city in the comfort of trains, finding safe haven in resorts such as Newport and Niagara. The middle class escaped in similar comfort and found refuge in the mountains of Tennessee and Kentucky, but most of the poor straggled out of the city in whatever conveyances they could muster or remained very close or even within the environs of the stricken city.[23] The federal government provided ten thousand tents and at least one camp was established at the Shelby County fairgrounds for the poor, but this camp proved, in the end, to be no safer than the homes these people abandoned. Luckily, the camp attracted fewer than seven hundred people, with most of the poor preferring to fend for themselves in whatever makeshift shelters they could fashion. The stampede ended on the eighteenth, and "a sad, weird kind of silence" fell over the nearly abandoned city.[24] Only about twenty thousand people remained in Memphis, most of them poor Irish and African Americans. Blacks, mistakenly as it turns out, thought themselves immune to the disease. The poorest of the Irish simply had no means to flee. Only a few middle- and upper-class people remained. Martha Wilson, tragically, was among them.

Medical professionals, the clergy, and a benevolent organization of clerks known as "the Howards" attempted to deal with the mounting number of deaths, which peaked at two hundred a day within weeks of the initial outbreak. Of the approximately 14,000 blacks who remained in the city, 11,000 suffered bouts with the disease and 946 died. Nearly all of the 6,000 whites left in Memphis were stricken and more than two-thirds (4,204) died.[25] City officials aggressively pursued a number of measures considered appropriate to fight the disease. They spread carbolic acid and lime, which they thought would kill the agent causing the disease, around the area of the first cases; they burned barrels of tar on street corners around the city; and they seriously considered shooting off cannons—assuming that the sulfur from the cannons would subdue the disease—until someone pointed out that that measure had failed to work in New Orleans and only disturbed the stricken patients unnecessarily. The *Memphis Appeal* ran an article on a Creole remedy, but nothing anyone could do would stop the fever.

Why Martha chose to remain in Memphis even as most of her class departed remains a mystery. Perhaps her hold on the house was so tenuous that she feared the consequences should she abandon the home. Had she elected to take refuge with her daughter in Frenchman's Bayou, she might have survived. No cases of the disease appeared there. Who exactly remained in the house along with Martha is unknown. Missouri was staying in a school for the deaf in Knoxville, and William Henry, who was then nineteen, could have fended for himself, but Lee, who was only thirteen, would have required some attention. Had he stayed with his mother he would have witnessed the terrible disease take its toll on her. Chills signaled the onset of the disease, followed quickly by high fever and jaundice. Severe diarrhea and vomiting followed, frequently called "black vomit," and signaled the final stages of the disease. The "black vomit" was the result of internal bleeding, and few patients recovered once that symptom appeared. On September 12, 1878, just as the epidemic was hitting its peak of two hundred deaths per day, Martha died. She was buried in an unmarked grave in Elmwood Cemetery, and by that time few funeral processions accompanied the dead to their final resting places.

Whether Lee was with his mother when she died or elsewhere, her death uprooted him from the only life he had known. He was first sent to live with his father's brother William Henry Wilson in nearby Covington, Tennessee. Covington, located in Tipton County immediately to the north of Shelby County, was a small village only a few decades old at the time and primitive compared with the city of Memphis. As the county seat, however, Covington was likely the liveliest place in Tipton County. Lee attended school there, and many years later wrote a letter expressing appreciation for the schoolmaster, James Byars, who taught in a log schoolhouse. Writing to Byars's granddaughter in 1932, Lee Wilson said: "It is with great pleasure and sweet remembrances of the days when I attended school to your grandfather at Covington in the log school house. He was the most wonderful man to impart knowledge to his boys that I have ever known."[26]

Lee later regretted that he ended his formal education at age fifteen, but in 1880, in the company of forty-five-year-old freedman Napoleon Wilson, the very man who had been born in Lee's grandfather's household in 1835, he left Covington and made his way back to Arkansas. The financial calamity that struck Memphis following the yellow-fever epidemic that took his mother's life had bankrupted his Tennessee estate. His Mississippi County farmland, safely in the hands of his competent brother-in-law, James Davies, provided him with some income but most of his Arkansas property, characterized as a "cypress swamp," failed to inspire creditors to advance funds enough to provide for his needs. Like his father before him, he turned his gaze toward Mississippi County and crossed the river. He might have secured safe haven in Frenchman's Bayou with his sister Victoria but instead, "he pitched his camp on a little island that rose a few feet above the black water that was infested with moccasins and cottonmouths." He hired two more African American men, purchased some equipment and supplies, and began cutting timber. He possessed an enterprising disposition and as he "sold a few carloads of lumber," he "bought more swamp land, at 50 to 60 cents an acre" or less, remarking that "if you have guts enough to spend $500 you can make $2500 in three years."[27]

He made at least one early misstep—sometime around 1883—but even that mistake led him to an arrangement that worked to his advantage. He believed he could make quick and easy money by opening a saloon and gambling house but "after a year or so found the employees stole all the receipts and he had only $150.00 he could get his hands on." He purchased a groundhog (portable) sawmill, and placed it at the settlement of Golden Lake, which fronted the Mississippi River. "Having no money he made arrangements with the store owner to feed the laborers, and charged it against their payroll." Once he paid off his indebtedness to the merchant, he had enough money left to invest in more forested swampland. He expanded the sawmill operation at Idaho Landing at Golden Lake, and it became a centerpiece of his lumbering empire. As he cut timber, he cleared land and put it into cultivation. He grew into a tall young man, greatly resembling his father both in appearance and temperament. One admirer described him as "a long tall Ichabod Crane type of person," but unlike the fictional character by that name in Washington Irving's "Legend of Sleepy Hollow," Wilson presented a fearless demeanor, particularly toward those who attempted to thwart his ambition.[28]

It was fortuitous that he was made of strong stuff for he faced many personal difficulties over the next few years. Between 1880 and 1885, a series of further tragedies visited his extended family. These tragedies, which coincided with Lee Wilson's assumption of control over his own property, confirmed in him the sense of paternalism that was still very much a part of southern culture. He later carried this paternalism into virtually every aspect of his life and business, extending beyond his family members to his workers, from his mill hands and plantation laborers to his office personnel and his store, farm, and cotton-gin managers. But it all started with his family. By the time he reached the age of twenty in 1885, only one of his siblings, his deaf half sister, remained alive, and she was under his protection. The first to die had been Lee's fifty-five-year-old brother-in-law and the administrator of his estate, Dr. James F. Davies, in April 1881. Thirty-two-year-old William K. Harrison, a physician who had married James Davies's daughter by his first wife, took over responsibility for Lee's estate.[29] Napoleon Lafont served as one of the sureties on the bond Harrison was required to file in order to take charge of Lee's property.

Lee Wilson chafed under the restraints imposed upon him by the guardianship and resented Lafont's partial assertion of authority over his affairs. He still stood as a minor in the eyes of the law, however, and remained unable to act on his own behalf. He resolved that problem in November 1882 when he appeared with his guardian before the Circuit Court of Mississippi County. Harrison asked that court "to remove his [Lee Wilson's] disabilities as a minor and enable him to transact all of his business." The document characterized him as "a young man of fine sense and good business capacity, and fully competent to take care of himself and to transact all of his own business without the intervention of a Guardian." Just six months earlier, in fact, he had demonstrated his ability to take advantage of opportunity by paying $16.85 for 125 acres of land declared delinquent for failure to pay taxes.[30] The land was adjacent to that which he had inherited, and its purchase was but the first of many such acquisitions in which he began to build a contiguous concentration of holdings. When the judge hearing the petition declared "that he [Lee Wilson] is the owner of considerable property estate, both real and personal, in said County and is successfully managing the same," he merely recited what everyone understood. Why Harrison relinquished his control over Lee's property remains a mystery, but it likely had something to do with Lee's force of personality and the doctor's confidence in him. So eager was Harrison to agree to this new arrangement that he swore a lie to the court: that Lee was "about twenty years old," when, in fact, he was seventeen. It is likely that everyone in the courtroom that day was complicit in this fiction, as it was a relatively small community and Lee's age had long been on the record in Probate Court documents both in Shelby and Mississippi counties. Lafont's response to this event remains unrecorded, but, in any case, young Wilson had finally freed himself of his nemeses.[31]

Lee Wilson's first recorded transaction after securing control of his property occurred in May 1883 when he paid $100 to Harrison for forty acres located near the property he had inherited.[32] But that summer held even greater significance for him personally: another untimely death. His thirty-two-year-old sister, Victoria Davies, was in the last months of her life. Upon the death of her husband two years earlier, she became executor of his estate and began carrying out all the usual matters associated with such a responsibility. But she remarried in late 1882 and relinquished control to her second husband, James C. Paine, in January 1883. Once again, Victoria receded into the background and allowed her husband to administer her affairs. By the summer of 1883, physicians in New York were treating her for ovarian cysts. She died there on August 27, 1883. They packed her body in ice and shipped it back to Frenchman's Bayou for burial. Eighteen-year-old Lee Wilson served as surety on the bond filed by James C. Paine, who became executor of her estate, and when he disappeared from the scene in 1885, Wilson again served as surety for William Harrison's bond. Victoria's will, witnessed in Memphis on January 22, 1883, divided her property between Paine and the three minor children born to her and James F. Davies: Dora, age ten, Eva, age seven, and Boaz, age two. Significantly, however, Lee Wilson would later come to play the most important role in the lives of their two daughters.[33]

Victoria's death in the summer of 1883 followed the untimely—and unexplained—death of Lee's older brother, William H. Wilson, who died in January that year, leaving a wife, Bettie, and a three-year-old daughter. The widow remarried but soon died, and Bettie's parents assumed responsibility for Wilson's niece Pearl.[34] She became the only one of the children of Lee's siblings with whom he had no connection, and she suffered for the lack of better administration of her affairs. She married at thirteen, had six children, three of whom survived, and then lost her husband. She left Mississippi County and faded out of view, surfacing in the 1910 census as a "helper" in the Memphis household of her young daughter, Viola, the "adopted child" of the family with whom she lived.[35]

Eighteen months after Lee buried his two siblings, Napoleon Lafont, who had suffered a devastating injury in a sawmill accident in 1883 from which he had never fully recovered, also died. In a curious twist of fate, his death resulted in Lee Wilson's appointment as executor of his estate. Viola, though herself dangerously ill at the time, filed a bond to serve as guardian of her two children, Clarence and Ella, with twenty-year-old Lee Wilson functioning as a co-signatory on the bond. After Napoleon's death in early 1885, it became clear that he had mismanaged Missouri's affairs, and it eventually fell to Lee Wilson to clean up the mess and restore her estate. Lafont had been in control of her property for more than a decade, and both Martha Wilson and James

Davies had challenged his annual settlements during their lifetimes. In the last months of her own life in the summer of 1885, Viola empowered her half brother to take over as Missouri's guardian and signed as one of the sureties on the guardian's bond.[36]

Young Wilson aggressively pursued management of Missouri's affairs, demonstrating his competence. During the October 1885 term Lee Wilson filed papers attesting to the lack of funds "sufficient to defray her [Missouri's] expenses for support or otherwise." He stipulated that some land Lafont had purchased on behalf of Missouri in 1883 for $2,000 consisted of 320 acres of "wild and unimproved" acreage that had "been a drain upon the estate . . . instead of a source of profit." Although Lafont had cleared and cultivated a part of that land, Wilson characterized the investment as largely "ill advised and unprofitable" and asked the court to allow him to sell it in order to "apply the proceeds to her [Missouri's] support." He sold the land and by 1887 filed an additional bond as her guardian, signaling the recovery of her financial situation. Lee Wilson spent considerable sums maintaining her property, building tenant houses and outbuildings, clearing land, and hauling timber. In early 1892 Missouri's health began to fail. She died early the next year at the age of fifty-three, leaving a respectable legacy of 747.5 acres—which included all of the acreage she inherited from her father plus an additional 240 acres—to her four nieces.[37] The decision to will her property to Pearl Wilson, Eloise Lafont, and Eva and Dora Davies signaled her appreciation of the vulnerability of young women dependent upon the protection of male relatives.[38]

Lee Wilson prudently observed his responsibilities as guardian of Napoleon Lafont's property and when Lafont's son, Clarence, reached the age of twenty-one in 1888, Wilson voluntarily surrendered control over it. Although he would have little role to play in Eloise Lafont's affairs, Lee Wilson maintained a close connection with Clarence throughout his life.[39] Shortly after turning over responsibility for Napoleon's property, Wilson purchased some acreage that had been part of his father's original estate, acreage that Viola had originally inherited upon Josiah's death and left to her daughter. Clarence, apparently acting as agent for his sister, sold the 260-acre parcel for $2,160 to Lee Wilson.[40] This was one of many purchases that Wilson made that led to his acquisition of his father's original 2,300-acre estate. Clarence married and farmed the property he inherited from his parents in Golden Lake, but he suffered from malarial fevers and then fell ill with consumption in 1902. He subsequently moved to southern Texas, eventually settling in New Mexico, where he remained for the rest of his life, renting his land to Lee Wilson to operate for him.[41]

Lee Wilson played a much greater role in the lives of Victoria's two daughters, Dora and Eva, whose husbands became closely allied with Wilson. Dora Davies married John A. Merrill, a farmer who became a business associate of Wilson's, and the Merrills remained intimately connected to Lee Wilson's family and business. They even lived in the same household for some years, and after John died, Dora lived under Wilson's protection and, like her sister, became a beneficiary under his will. Eva Davies married James H. Elkins, who worked as a bookkeeper for Lee Wilson's lumber business, in 1896.[42] When Wilson incorporated as Lee Wilson & Company, in 1905, Elkins served as vice president, but, suffering periodic bouts of malaria, he died sometime between 1905 and 1910. Like her sister, Eva lived close to Lee Wilson, and he, in fact, named the first "town" he founded in her honor: Evadale, the place on Golden Lake originally called Idaho Landing, where he had his first sawmill.[43]

Throughout the period during which Lee Wilson endured these family tragedies and assumed responsibility over the affairs of his siblings and their heirs, he aggressively pursued the purchase and development of additional acreage and expanded his lumbering business. Family connections were crucial in yet another way in this regard. In December 1885 Wilson married Elizabeth "Lizzie" Beall, a "refined and well educated lady," for whom he built a three-thousand-square-foot home, described as a "handsome structure . . . beautifully and comfortably furnished inside, and . . . in the middle of a large and well-kept lawn." Wilson sealed the alliance between the two families in a business sense when he began a partnership in 1886 with his father-in-law, Socrates Beall, a local farmer and lumberman, as Wilson and Beall Lumber Company. While the older man ran the sawmill, Lee took charge of the lumbering operation and of constructing tram roads from the logging camps to the site of their mill at Golden Lake. The two men made good partners, and their enterprise expanded and thrived.

Wilson and Beall faced certain daunting challenges because of the location of their lumber operations. As one logger along the Arkansas River, where conditions were similarly problematic, complained, "Them roads in my section just ain't passable. They ain't horseable, mulable, or oxable—Hell! They ain't even jackassable." Floods easily damaged or washed away the underfunded and poorly constructed county roads, and the same thing happened routinely to the temporary tram lines that Wilson and Beall constructed to move their logs from interior locations to their mill located on the Mississippi River. Lightly fashioned affairs designed to be taken up and moved to the latest lumbering outpost, they proved especially vulnerable to high water. To make matters worse, floods threatened the mill itself. Especially high overflows caused the banks to collapse and sometimes damaged or destroyed Wilson's buildings, ruined his equipment, and carried away his milled lumber. Understandably, he became an advocate of flood control and began attending levee board meetings after the founding of the St. Francis River Levee District in 1893, purchasing land from them and even bidding for construction contracts to build the levees. Despite the obstacles he faced, his operation grew significantly larger during these years. Having begun cutting timber while still a boy in 1880, by 1889 he had expanded his sawmill to a capacity of fourteen thousand board feet per day. In 1894 he had "the best equipped sawmill in the county and perhaps in the state," valued at $12,000.[44]

Characterized as a "hustler" who worked long hours overseeing lumber workers and the construction of his tram roads, he also took charge of marketing the lumber Wilson and Beall milled. He moved easily from the swamps of Mississippi County to the offices of lumber dealers and manufacturers in St. Louis and Chicago. Driven by an insatiable desire to succeed and possessed of a commanding personality, he personified the New South entrepreneur, someone capable of working with northern capitalists in order to further his own business interests. Northerners owned and managed most other lumber companies operating in the South in this era, but Wilson represented the homegrown capitalist, the young man "on-the-make" who early envisioned an empire emerging out of the swamps. Unlike almost all other lumbermen of this era, he began to engage in plantation agriculture and rather than "cut and run," as most such operations did, he removed the timber, cleared the land more thoroughly, and put it into cultivation. In 1889, the year he boasted that his mill produced fourteen thousand board feet of lumber per day, he also put nearly five hundred acres under the plow and employed over one hundred people cultivating cotton for him. His tenants lived in the thirty tenant houses he constructed and secured advances from him at the company store he opened with his father-in-law. From the time of his first acquisition in 1882, he continued to acquire land either with Beall or individually and to expand his plantation operations. By the early twentieth century he claimed to own forty thousand acres in Mississippi County, a sizable estate.[45]

Wilson drew his work force from the rapidly increasing population, which more than doubled between 1880 and 1900, from 7,332 to 16,384.[46] He hired both black and white men in his sawmills and lumber camps and in building his railroads. While many of the newcomers came to work in the lumber industry and soon departed, many others remained and became tenant farmers, thus providing the labor necessary for the expansion of the plantation system. Land in production in the county increased by over forty-six thousand acres, from 30,111 in 1880 to 76,655 in 1900. Significantly, over half of that increase was in bringing "unimproved" acres into agricultural production as opposed to opening up new lands. Most of these unimproved acres were in forests; thus clearing and developing them was the crucial factor accompanying the increase in population and fostering the expansion of the plantation system. By 1899, planters and farmers in the county harvested cotton on 34,380 acres. Thus they continued to devote most of their energies to the production of a cash crop, and their orientation to the market became even greater than ever (see table 2.1).[47]

A steady and reliable supply of labor became an important concern of planters as they rededicated themselves to the production of cotton in the post–Civil War period. The emancipation of the slave population and the violence and uncertainty during Reconstruction and Redemption presented them with significant challenges. A demographic revolution addressed their labor needs but presented them with some additional challenges. First, a larger proportion of males to females immigrated to the region, largely young single men from the immediate region (Arkansas, Missouri, southern Illinois, Kentucky, Tennessee, and Mississippi). They landed in Mississippi County because of opportunities to work in the lumbering industry. Second, an influx of African Americans, principally from Tennessee (34.4 percent) and Mississippi (19.8 percent) migrated to south Mississippi County.[48] Some of these African American men also came to work the lumber industry—and Lee Wilson frequently hired them in his lumber camps and sawmills. The presence of so many single men, living as they did outside the confines of the family structure, led to a volatile situation, and the influx of southern whites and African Americans contributed to growing racial strife.

Settled and established communities typically enjoyed a roughly equal adult male to adult female ratio in 1870, and that was true of Mississippi County, where 49.8 percent of the adults were male. Mirroring the rise in the lumber industry was a trend toward a greater number of males in the population, a phenomenon that continued throughout the rest of the century (see table 2.2).[49] The county's proximity to the Mississippi River made it relatively easy for migrants to reach in order to work in the expanding lumber industry, but the presence of so many young single men often led to boisterous activities, particularly in the rough-and-tumble environment along the river and in the swamps of Mississippi County. Lee Wilson understood these young men and what they wanted, hence his early—and disastrous—experiment with opening his own saloon and gambling establishment. Understanding them, however, did not necessarily translate into control over their behavior. Acts of violence became commonplace, whether committed by criminals or young men who had imbibed too freely of alcohol.[50]

In late nineteenth-century Mississippi County men seemed eager for some inspiration to battle. They fought over women, over politics, over contract disputes, and over games of chance. Fights were so frequent that the newspaper often began a report of a particular incident by remarking that "the spring fights have begun" or the "fall fights have commenced." The liberal consumption of liquor was often an inducement to violence, but men did not have to be drunk to fight. Leon Roussan, the editor of the *Osceola Times* and stalwart temperance advocate, was assaulted on the streets of Osceola by a young man angry because of something said about him in Roussan's newspaper. Roussan, a former mayor and alderman who prided himself on never having engaged in such behavior, struck back at his assailant.[51] Most fights did not result in serious injury, and some served as entertainment. Roussan's encounter with the angry young man, for example, left him with a black eye and the embarrassment that the affair was considered by his personal friends as the "joke of the season." One correspondent to the newspaper wrote, "Our town is rather dull now; had several Spring fights but yours so far surpasses any of them that by comparison they pale into insignificance, therefore will not mention them."[52]

Another important aspect of the violence that pervaded the county involved racial conflict between white and black laborers, mostly over opportunities to work on plantations. While the white population increased in the period between 1880 and 1900 from 4,671 to 8,061, the black population increased from 2,654 to 8,321.[53] For the first time in the county's history, African Americans constituted the majority. Regardless of race, those who worked on the county's plantations shared certain disadvantages, which made them increasingly desperate. Whether they came to work the plantations or began in the sawmills and gravitated later toward the farming sector, they found themselves bound up in an increasingly coercive institution as planters found another means—other than slavery—to bind their laborers to them: sharecropping and tenancy, which some historians have come to regard as "another kind of slavery."[54]

The sharecropping system arose in the years immediately following the Civil War, apparently as a compromise between freedmen who wanted land and cash-starved planters who found it difficult to pay wages. With emancipation came freedom for African Americans, but planters needed them to supply their labor needs and initially employed them through contracts to work for wages. Freedmen resented the contract labor arrangements, largely because they continued to work in gangs, a reminder of their slave status, but mostly because they wanted to work their own land. The sharecropping arrangement allowed them to move out of the old slave quarters and work an allotment of land—though they did not own it—for the planter. The transition to share-cropping began in the late 1860s when planters resorted to paying a share of the crop, either a third of the crop for those who brought only their labor to the arrangement, the sharecroppers, or half the crop for those who could furnish their own mules and implements, the share tenants. In the minds of most planters, there was little difference between the two tenancies, but as disputes arose between planters and those working for shares, the courts made some fine distinctions. According to cases adjudicated in southern states in the early 1870s, the share tenant "owned" the crop and paid a share to the planter. The planter owned the crop in a sharecropping arrangement, however, and paid a third of it as a kind of wage. Because white men, some of whom owned small farms in the antebellum period but lost them as a result of the economic chaos of the Civil War and Reconstruction, had some means to purchase mules and implements, they tended to be share tenants. Black men, on the other hand, rarely had the means to supply themselves, and thus they tended to be share-croppers. The courts' fine distinctions about who "owned" the crop were not unimportant, for in cases of dispute between the share tenant and the planter, the former at least had some standing in court. The sharecropper, viewed as little more than a wage laborer, had no legal standing.[55]

Despite their apparent legal advantage, few share tenants bothered to take disputes to court. They understood that land owners influenced, if not controlled, most local courts. If they responded at all, they resorted to violence, and a number of bloody confrontations occurred between tenants and planters in the late nineteenth and early twentieth centuries. In October 1888, for example, two tenants shot planter T. B. Phillips, who operated a plantation in south Mississippi County near Lee Wilson's property. The dispute apparently arose over their "rent notes or contract." In this case, both tenants and the planter were white, but black tenants sometimes also challenged the planters for whom they worked with equal vigor.[56] Whether plantation laborers were white or black, certain common realities faced all of them. Both share tenants and sharecroppers relied on advances from merchants or planter-operated stores, advances that included as much as 25 to 50 percent interest charges.

Even if they owed their debt to a merchant independent of the planter on whose land they labored, most became indebted to the planters for whom they worked, as the planter paid off the merchant at the end of the crop year and placed the tenants' debt on their own books. The largest planters—like Lee Wilson—opened commissaries and furnished their tenants and sharecroppers directly, thus maintaining even more control over the debts accrued by their workers.[57] Some states passed laborers' contract laws, which made it illegal to leave the employment of someone to whom one owed a debt. Arkansas and most other states sanctioned the crop-lien, an instrument that bound laborers to planters because it permitted them to require tenants and sharecroppers to sign a legally binding document that required them to mortgage the year's crop, while still in the ground, to the landowners. Some economists believe that had it not been for the web of debt that ensnared the laborers, the share tenancy/sharecropping system was a credible arrangement and might have served the interests of both laborers and planters in the long run.[58]

Because cotton prices failed to recover and, in fact, continued to decline in the late nineteenth century, planters themselves were caught in a similar web of debt. Lee Wilson, for example, mortgaged 650 acres to commission merchants in Memphis, Tennessee, in 1883, shortly after securing control over his inheritance.[59] He remained in debt—and sometimes on the brink of bankruptcy—for the rest of his life. He routinely used his access to credit to acquire more land through various transactions and thus increased his collateral and his ability to borrow more funds. Like most planters, he typically made arrangements with cotton factors which required him to grow cotton, a cash crop that many believed would resume its pre–Civil War prominence. In a sense, they too operated under a kind of crop lien. While Wilson began his own cotton factorage partnership in Memphis, Tennessee, thus extending himself beyond the usual arrangement, most planters became hopelessly indebted to cotton factors. As desperate as conditions got for planters, however, they were hardly as bad as those for the landless men who labored for them. The crucial factor that separated them from their workers was their ownership of the land upon which their plantations were founded.[60] Because they owned land, planters not only had access to greater credit but also other economic and political connections that distinguished them. Wilson's familiarity with county officials made it routine for him to lease a portion of the school lands (the sixteenth-section lands) for cotton cultivation, something other large planters in Mississippi County and throughout the South were accustomed to doing.[61]

Wilson also routinely leased county convicts for use in his fields and even, on occasion, voted them in close elections. Planters like Lee Wilson became patrons to a variety of small farmers, businessmen, and county officials, extending them credit and doing them favors. After he founded the Bank of Wilson in 1908, Wilson augmented this association by providing loans to an ever-growing number of Mississippi County people. Although the size of the Wilson operation made the scale of his influence unusual, most planters enjoyed considerable standing in the communities within which they lived. They rarely had to fear the appearance of the sheriff at the door, and they never found themselves convicted for vagrancy or sentenced to work off a debt on a chain gang. Instead, they could call upon local law enforcement officials to bring recalcitrant tenants and sharecroppers to heel, and some planters, including Lee Wilson, were not above creating their own private armies of plantation thugs to bend labor to their will.

In the context of this struggle between landowner and laborer, a conflict arose between black and white plantation workers. Complicating the situation, disfranchisement and segregation statutes passed in the 1890s rendered African Americans far more vulnerable and less able to rely upon the law for protection. The populist revolt that arose in late 1880s, which for a time gave the appearance of uniting black and white farm laborers under the Populist Party, disintegrated in the 1890s, partly because Democratic Party election officials coerced black voters.[62] Mississippi County's connection to the evolution of agrarian unrest was an intimate one, with several of its prominent citizens associated with the Agricultural Wheel, an Arkansas organization of farmers that first emerged in Prairie County, Arkansas, in 1882.[63] Concerned about the declining prices for agricultural products, they pressed state officials to remedy the situation. They wanted to secure more equitable railroad rates and they hoped for state aid in forming cooperative marketing associations that would allow them to secure better prices for their agricultural products. For Mississippi County farmers, railroad rates were of little concern since no railroads had yet traversed the county and farmers there still relied heavily on the river for transporting their crops to market. Cooperative marketing associations were another matter. Wheelers in Frenchman's Bayou, the south Mississippi County township within which Wilson owned property, lobbied for the formation of a cooperative store and cotton gin in July 1888.[64]

The wide support the Wheel enjoyed among Democrats began to erode in 1888, however. county judge Logan D. Rozzell represented Mississippi County that year at the Wheel's state convention in Little Rock, a convention at which the Wheel formed an alliance with the Union Labor Party, widely believed to be a mere front for the hated Republican Party. Rozzell was almost certainly one of the many Mississippi County Wheelers who subsequently "burned their membership cards," renounced the organization, and reaffirmed their allegiance to the Democratic Party.[65] Nevertheless, the Wheel persisted, and when the organization later formed an alliance with the Populist Party, many Mississippi County men crossed over to it, abandoning the Democratic Party. Every election that occurred in the county from 1888 until the end of the century was marked by turmoil. In 1898, for example, violence erupted in Luxora, where the election pitted a Democrat who had won the Democratic Party primary election against a man who had bolted the party and was running as a Union Labor candidate. When a prominent farmer named Reed Dunavant approached the polls to vote in Luxora, "Mr. William McKinney, candidate for Justice of the Peace, hailed Dunavant, and asked him if he wouldn't vote for him? And he replied, in a badgering way, no, I can't vote for you, because you are no better than a bolter; you are supporting (C.W.) Clifton who is a bolter." Clifton heard this exchange and advanced on Dunavant threatening to "cut his throat." Dunavant warned him to keep away, but as Clifton pressed forward, "Dunavant drew his pistol and fired with fatal result."[66]

In the midst of this turmoil, African Americans faced their greatest challenge since the violence they endured during Reconstruction when first they exercised their voting rights. Calls for disfranchisement and segregation statutes began to arise in 1889 and 1890, largely as a response to the third-party threat. African Americans voted in large numbers in 1888 and 1890, electing eight and eleven black Republican legislators to the state assembly in those two years respectively.[67] The Democratic Party's strategy of using "fusion" to seduce black voters away from the Republican Party was proving less effective in the face of the Union Labor Party appeal. "Under the [fusion] agreement, the two parties ran complementary fusion tickets during local elections, in which white Democrats traded an open endorsement from leading black Republicans in exchange for a promise that all black Republicans could campaign free from Democratic harassment."[68] Although African American voters were never fully comfortable with the arrangement, it did have the effect of delivering some black votes to Democratic candidates and it provided an opportunity for black politicians to seek and hold office relatively free from the threat of white violence. The alliance of Wheelers with the Union Labor Party in 1888, however, drew many black Wheelers away from voting for fusionist tickets, and some Democrats began to seek another solution to the challenges presented by this newly invigorated black electorate. But fusion still had its adherents in Mississippi County.

As the momentum for disfranchising black voters gained ground, at least one prominent white Democrat in Mississippi County, Leon Roussan, objected. Editor of the *Osceola Times* and an ex-Confederate soldier who remained fiercely dedicated to the Democratic Party, he rejected disfranchisement, arguing in 1889 that the black man "is born and bred to love freedom, speaks our language, knows our custom and there is nothing he holds more dear, except perhaps his religion, as our national flag and his veneration for the government is proverbial."[69] But Roussan was in the minority among prominent Democrats in his sentiments, and in 1891 the state legislature passed two measures characterized as "election reforms" which severely restricted black—and poor white—voting. The election law of 1891 awarded considerable discretionary authority to the governor, secretary of state, and auditor by placing in their power the responsibility of approving the designation of election officials down to the township level. Although the law indicated that at least one of the three officials in each township should be from another party—presumably the Republican Party—it clearly placed Democrats in an enviable position. The law also required that illiterate voters have their ballots marked by election judges. No longer would they be allowed to present pre-marked ballots or to bring friends to the polls with them to mark their ballots for them. A second piece of legislation, a poll tax amendment, required that potential voters purchase a poll tax receipt and present it to the election judges in order to vote. Although the sum seems nominal by present-day standards—$1.00—it represented a significant amount in a cash-starved economy.[70]

Yet divisions existed among elite whites in the county not only over the imposition of disfranchisement, something Roussan was on record as opposed to, but also over a violent and radical racism that emerged in the 1890s. In 1897 editor Roussan's lack of timidity when it came to reporting news—or taking a position on a controversy—landed him in deep trouble and at odds with the county sheriff, W. J. Bowen, son of the former sheriff (Charles Bowen) who had led the white mob during the Black Hawk war in 1872. The dispute between Bowen and Roussan over the adoption of extreme racism had been simmering for years but broke open dramatically over a lynching that took place in Osceola in December 1897. A black man named Phillips stood accused of murdering Tom McClanahan, a white man, in a tenant house on a plantation just north of Osceola, where the victim maintained a small stock of goods for sale to other tenants. The alleged perpetrator robbed and shot McClanahan and then escaped to Memphis, but authorities there apprehended him and turned him over to Sheriff Bowen, who bragged loudly that whites in Mississippi County planned to lynch Phillips. Though the first attempt to resort to mob justice failed when deputy sheriff John Lovewell convinced a crowd to disperse, a second group convened on the jail in the dead of night, took Phillips from his cell, and hanged him. Roussan complained that he and other citizens had offered their services to Bowen to prevent such an occurrence but the sheriff assured him that he needed no assistance. The editor refrained from accusing the sheriff of conducting the lynch party, but Bowen took offense and threatened to kill Roussan. The dispute between the two men became so contentious that Roussan briefly surrendered editorial control of his newspaper to his wife and took up residence in Memphis. The *Osceola Times* continued publishing, and some of the paper's readers expressed their support for the editor. Nevertheless, he did not return to Osceola until June 1898. By that time the Democratic Central Committee (DCC) had announced its candidates for election to county offices and Bowen's name was not among them.[71]

Although Lee Wilson seemed to play no public role in this controversy, he frequently visited the editor's office in Osceola throughout the 1890s—including during the period of the controversy—and was closely allied with those who launched a challenge to the county machine, a challenge spearheaded by Roussan in 1900. In that year, Sam Bowen, yet another son of the old Confederate, secured the county DCC's nomination to stand for election to the sheriff's position. He had been serving out the unexpired term of the previous sheriff, J. L. Hearn (who had died in office), the man who replaced his brother in the position in 1898. The Bowen faction won a struggle for power within the DCC in 1900, and a significant group of loyal Democrats, including Roussan and Joseph W. Rhodes, one of Lee Wilson's oldest and closest friends, put together an independent ticket. Rhodes sought election to the circuit clerk's position against the "machine" candidate, Charles Driver, and former deputy sheriff John Lovewell ran against Bowen for the sheriff's position. In a particularly contentious election in which several irregularities occurred at polling places, the entire DCC slate won office. Rhodes and Lovewell sued and took their case to the court, eventually ousting both Driver and Bowen. Along the way, their faction seized control of the county DCC.[72]

By the time that Roussan and his cohort launched their challenge to the county machine, disfranchisement was an all-but-established fact. The 1891 election law and the poll tax amendment severely restricted black voting, but some African Americans continued to vote in Democratic Party primaries—including the controversial 1900 election in Mississippi County. Indeed, Rhodes and Lovewell drew heavily on the black vote. Although most legislators and white citizens supported disfranchisement as a means of cleaning up the political process, arguing that blacks could be too easily manipulated, politicians and planters adapted quickly and learned to exploit the new system. As early as 1894 Editor Roussan complained that "the poll tax receipt has taken the place of the quart bottle as a factor in electioneering." Indeed, the poll tax receipt became a commodity that wealthy planters purchased in order to "vote" their tenants and sharecroppers. Until the passage of the White Primary in 1906, which prohibited African Americans from voting in the Democratic Party primary election, the only election that mattered in the one-party South, planters could use the poll tax to their advantage. Even after the White Primary removed black voters from that part of the process, planters still purchased poll tax receipts for their white tenants. Certainly, Lee Wilson gained a reputation for doing so and, in fact, took it to another level in 1909 when he allegedly voted convicts he had leased from the county farm in order to secure the election of Samuel Gladish, a prodrainage candidate, as county judge.[73] On a more routine basis, however, Wilson had the advantage of having his men occupying positions on the DCC, at least on the township level, where they also served as election judges. In the town of Wilson, for example, J. C. Cullom, the cashier at the bank, served on the DCC for a decade and a half and, along with H. H. Brown, one of Wilson's farm managers, served as one of the three election judges. Jessie Greer typically worked as one of the election clerks in Wilson, where he was also a deputy sheriff. The same pattern held true for the other Wilson-dominated towns in the early twentieth century.[74]

However much some planters presided over elections and fashioned a use for the poll tax, other consequences flowed from disfranchisement that created serious problems for them. Having used racist rhetoric in the drive to disfranchise blacks, legislators unleashed a form of violent racism that worked to the disadvantage of plantation owners in the labor-scarce environment prevailing in the Arkansas delta. Fierce competition for jobs on plantations led landless whites to attack black sharecroppers in an attempt to drive them away so that they themselves could secure employment. Although no accounts of whitecapping on the Wilson plantation have surfaced, several incidents occurred in Mississippi County. As early as 1891 county officials arrested a white man for intimidating black plantation laborers, and in January 1892 fire destroyed a cabin occupied by a black sharecropper. In November 1893 "a fracas involving a negro" occurred at the cotton gin, and within months shots were fired at whitecappers. Fears of a race war, which had circulated as early as 1890, resurfaced. Such events occurred all over the South and historians have attributed different causes to them, but most of them in Arkansas involved conflict over positions on the plantations.[75] In some cases planters banded together to hunt down white night riders. Lee Wilson adopted a strategy employed by a few other prosperous planters: he attempted to attract labor by providing better working and living conditions. He meant to maintain a steady supply of workers, and regardless of his own use of force and intimidation, he tried to keep his African American laborers safe from whitecappers. He largely succeeded, but some alarming incidents occurred nearby in the 1890s and in the early twentieth century.[76]

Of the two most notorious cases arising in northeastern Arkansas, one involved a concerted effort to force black sharecroppers from plantations in Cross County in 1902 and the other, occurring in the same year, witnessed torch-bearing whites demanding that blacks be dismissed from employment at a sawmill in Poinsett County. Planters in Cross County were so concerned about the loss of black labor due to the efforts of the night riders that they hired white detectives to apprehend the marauders. Instead, the whitecappers murdered the man in charge of the detectives, and when local authorities failed to prosecute the twelve men arrested for the crime, the federal attorney charged them with violating the constitutional rights of African Americans in the effort to drive them from employment. Meanwhile, fifteen men arrested in the Poinsett County case were similarly charged and brought to trial in the federal district court in Helena. Witnesses in the Cross County case suddenly developed what appeared to be amnesia; thus charges were dismissed. But the federal prosecutor managed to secure convictions against three of the Poinsett County men, convictions based on the principle that they had violated the thirteenth-amendment right to employment of African American lumber-mill workers. The convictions were appealed to the United States Supreme Court, which ruled in *Hodges v. U.S.* in 1905 that there was no constitutional protection to employment. The State of Arkansas passed an act in 1909 outlawing night riding, but its enforcement was lax. Instead, planters and lumber-mill owners tried to keep blacks and whites apart, an effort that served the purposes of preventing labor from uniting and, at the same time, attempted, without success, to minimize conflict between blacks and whites.[77]

It was in this period, as the plantation expanded and class and racial unrest prevailed, that Lee Wilson came of age. He was no bystander, moreover, as he aggressively pursued the lumber business and began to lay the foundation of an agricultural empire. His early experience with tragedy and controversy made him into a man possessed of intense determination. The struggle his mother underwent to secure his patrimony after his father's death impressed upon him the need to aggressively protect and pursue his own interests. Barely a teenager when his mother died, he began to fend for himself and embarked on a tenacious campaign to establish his independence. By the time he was twenty years old death had taken all but one of his siblings and left him responsible for some of their heirs. It was a responsibility he readily assumed, and indeed, family constituted an intimate part of his life and livelihood. Although most of the land he acquired came in the form of swamps and tax-forfeited lands, he also paid market prices, at times, to purchase parcels his father's other children and their heirs had received, and by the year 1900, he had control over most of his father's original estate. Like other post–Civil War operators, he adopted the tenancy and sharecropping system and benefited from the lien and contract labor laws that arose to keep labor in place and under control. However, a fractious class of white workers and volatile political environment that included the imposition of disfranchisement and the emergence of a radical racism presented him with some daunting challenges. At the same time, the disease environment continued to threaten his family and his labor force, and the intractable Mississippi River, always a problem, only grew more fierce and problematic.

3

A RIVER OF WOE
Reshaping the Land

I n the year 1900 much of the property Lee Wilson owned remained mired in swamps. He demonstrated a single-minded determination to harvest the wealth of timber on it and turn it into productive farm acreage. This required attention to two interrelated factors: building levees to protect against floods and constructing both drainage ditches and a few levees to eliminate the back swamps. Although not entirely unmindful of the environmental implications of draining the natural wetlands, he set his sights on making "Mother Nature" over to his liking, and he had plenty of company. Given the wavering commitment of the federal government, however, flood control remained unachievable during the first decades of the twentieth century, and Wilson and others along the lower Mississippi River Valley suffered the consequences. He enjoyed greater success—and endured more numerous controversies—in seeing the swamps drained. The battles that ensued over drainage, some of them involving different visions of development, reflected the tensions existing within capitalist agriculture. Plantation owners, small farmers, and potential homesteaders faced off in county court proceedings and in extra-legal activities on the watery ground of Mississippi County. As the drainage efforts progressed, another contestant entered the playing field, greatly complicating life for everyone in the county. When the U.S. secretary of the interior declared in 1908 that he intended to open the "sunk lands" to homesteaders, Mississippi County became a battleground. Homeless men hoping to homestead faced planters and lumber interests determined to secure the land for themselves. Battle also ensued in state and federal courts, and in the latter Lee Wilson quickly became one of the Interior Department's chief opponents.

Wilson operated during an unusual moment in American history. By the end of the nineteenth century, settlement had spread across the country to the Pacific Ocean, and as Frederick Jackson Turner opined, the vast American frontier had disappeared. Those eager to make their way in the world searched for opportunities and began to settle in places previously considered too wild and unredeemable by those who came before. Many thousands settled along the Mississippi River and braved an environment made dangerous by floods and swamp-related diseases. Seasonal overflows that would have gone largely unnoticed earlier in the century later became natural disasters for this settler population. The decision to move into this dangerous ground seems foolhardy given the negative images nineteenth-century Americans held of swamps and given the mortality rates facing families, rich and poor, black and white, who dared to venture there. Only the desperate search for the American dream in one of the last pieces of frontier explains the demographic revolution occurring in northeastern Arkansas.[1]

The lower Mississippi Valley constituted one of the unhealthiest regions of the country in the late nineteenth century. Although assorted maladies burdened the population year-round, the summer and early fall, when mosquitoes bred in the swamps and carried their lethal cargo to human hosts, held the greatest danger to health. Lee Wilson became painfully aware of the dangers arising during this treacherous period. Five of the eight members of his family succumbed during his childhood and youth, dying between the months of June and October, the prime period for mosquito infestations. His father died on July 28, 1870, his mother on September 12, 1878, and during the 1880s, a plague of death visited his family: on August 27, 1882, his sister, Victoria, fell victim to ovarian cancer; his half sister Viola died in July 1885; and his firstborn child, Tiny, died in August 1888.[2]

Evidence of the health consequences of living in the area come from two sources. Statistics published by the U.S. Bureau of the Census in 1906 reveal that nearby Memphis, Tennessee, had the highest percentage of deaths due to malaria of any city in the survey. Only pneumonia and tuberculosis outdid malaria as a cause of death in that city. The *Osceola Times* confirms that between 1884 and 1900 malarial fevers plagued Mississippi County too. Indeed, as late as 1938, scientists identified the lower Mississippi River Valley from Cairo, Illinois, to Natchez, Mississippi, as one of the four worst malarial zones in the United States.[3] While the disease environment spared no family from tragedy, poorer people, lacking proper nutrition and medical care, and likely suffering from greater exposure to the elements in winter and less sanitary conditions year-round, were most vulnerable. The local newspaper only recorded maladies affecting more prominent white families. Recent studies have challenged the notion, meanwhile, that blacks were more immune to malarial infection.

While certain genetic traits may have offered a very small minority of African Americans some immunity, most were as likely to contract malarial infection as were whites. Thus, a close perusal of the newspaper gives only a hint of the extent of those suffering from disease.

The dangers confronting even the prominent citizens of the county signify the seasonality of certain diseases (see table 3.1). After suffering through colds, pneumonia, and influenza from December through February, the community grew healthier in the spring, with May the most disease-free month. Conditions worsened in June, July, and August and reached a peak in September. In June and July, the likely culprits were dysentery and cholera, but the greatest peril began in late July until the first freeze in October killed the mosquitoes that transmitted malaria. The peak month for disease—August, with 16.1 percent of all diseases during the year—likely combined the first of the malarial cases with the typical dysentery and cholera associated with what residents called the "heated term." Occasional epidemics of meningitis, typhoid fever, and smallpox struck additional blows to an already fragile community. Doctors rode out into the county "often and late" attending to the "considerable sickness" that prevailed in various sections. According to one correspondent in August 1898, when a typhoid fever epidemic combined with an unusually high number of malarial cases, people halted doctors "at nearly every house" as they "drove along the road," imploring them to attend to the sick and dying.[4]

Some Mississippi County planters and merchants sent their families away, particularly from late July through September, to spare them the dangers of the season and then joined them for brief vacations. Lee Wilson, for example, in April 1895, planned to take his family to Eureka Springs. "Mr. Wilson will return as soon as he gets them located, but the rest expect to spend the summer provided it agrees with Mrs. Wilson." It is hardly surprising that Wilson wished to remove his family from danger, given his past experiences and his own little family's frequent illnesses. In the year 1894 alone, Wilson's wife, daughter, and brother-in-law Dan Beall all suffered ill health. Wilson took Beall to Memphis for treatment in March 1894, only to fall ill himself. Both Dan Beall and Wilson recovered, but by mid-July Wilson was again "on the sick list," and the editor of the *Osceola Times* complained that "our town and community seems to have an epidemic of bilious fever and chills just now." Wilson's plans to take his family to Eureka Springs for the summer of 1896 fell through because Mrs. Wilson became too ill to travel and instead sought treatment in Memphis for the summer. Illness continued to visit the family over the next few years, despite sojourns in Hot Springs, Arkansas, and Glenn

Springs (near Randolph) in Tennessee. By the end of 1899, the Wilson family fled the swamps of Mississippi County to the dubious safety of Memphis. The newspaper reported in September of that year that "Mrs. R. E. Lee Wilson and family expect to make Memphis their home in future and will leave during this month."[5]

For all this trouble nothing surfaces to suggest that the efforts of planters like Wilson to control floods and eliminate the swamps were inspired by a desire to curtail disease. Rather, the economic imperative motivated them, as they had faced increasingly damaging floods in the late nineteenth century, floods made more severe by the development of river-bottom lands north of Mississippi County, which increased the volume of water moving into the lower Mississippi. Ann Vileisis in *Discovering the Unknown Landscape* aptly describes the situation: as settlers denuded their lands of trees and cut back the understory vegetation, "the soil and water previously held in place . . . washed into streams during spring freshets. The swollen stream waters accumulated in larger rivers, including the Ohio, the Illinois, and the Wabash and eventually funneled, silt laden, into the Mississippi. The enormous river flooded downstream bottomlands with growing severity." Lee Wilson and his cohorts inadvertently exacerbated the problem by their own downstream enterprises. Forests act as natural barriers, slowing floodwaters and providing some protection. By removing these natural barriers, Wilson increased the overflow in an area already prone to frequent inundations.[6]

Lee Wilson's first encounter with a disastrous flood occurred in the early years of his manhood. In 1882, the year the seventeen-year-old went to court to secure control over his property, the worst flood up to that point in time in recorded history struck the Mississippi River Valley. The river crested at 35.15 feet at Memphis, the highest level reached since record keeping began in 1828. A man calling himself "the Scotsman," who lived near Wilson's property, wrote the *Osceola Times* that the high water prevented him from leaving his farm for fifty days and reported that "all we can see is water and driftwood" after several breaks in the levee, one of them immediately adjacent to his own cabin. The disaster drove the citizens of Mississippi County to what little high ground remained, and providing for them overwhelmed the resources of the communities located there. Nearly five thousand people, roughly 62.5 percent of the population of the county, desperate for food and shelter, sought refuge in Osceola. Even if local officials inflated the refugee estimates in order to highlight the need for assistance, it is safe to assume that over half of the approximately eight thousand residents of the county endured considerable hardship as a result of the flood. Leading townspeople held a mass meeting and called on the Corps of Engineers' office in Memphis to provide relief supplies.[7]

The fact that the patchwork of poorly constructed levees failed miserably in 1882 should have surprised no one. A serious flood in 1874, when the river reached thirty-four feet at Memphis, had exposed the inadequacies of levees built by the Corps of Engineers and landowners, and in 1879 Congress, responding yet again to pressure from citizens along the Mississippi Valley, created the Mississippi River Commission (MRC) and began to erect the bureaucratic infrastructure necessary for bringing the river under man's control. Composed of seven presidential appointees, three of them officers in the Corps of Engineers, the MRC had as its chief responsibility work that involved "the task of preparing surveys, examinations and investigations to improve the river channel." Although Congress also charged the commission with the responsibility for flood control, public opposition to the use of federal funds to protect private landowners caused legislators to refrain from allocating resources for that purpose. Congress rejected any responsibility for constructing levees to prevent flooding and continued to tie appropriations to improvements in the river's navigability.[8]

In order to further its primary mission, the Memphis District of the corps focused its efforts on the Plum Point "reach," that seventeen-mile stretch along Mississippi County that had long created problems for those navigating the river. John Murrill and his gang of freebooters had found ships negotiating the reach easy targets in the 1830s. The Corps came to regard their work along the reach as a testing ground for various techniques for maintaining a reliable channel. The engineers knew and understood the larger concern facing landowners and hoped that their efforts would incidentally reduce the severity of floods. They certainly selected a challenging stretch of river to expend their resources on. There the river changed course on almost an annual basis, and there even a slight modification of its path could doom ships captained by men unused to the capricious Mississippi. Corps engineers believed that "if they could solve the problems there then they could handle problems anywhere else on the river."[9]

The Corps expended great effort on the Plum Point reach to remove the snags, stabilize the channel, and keep Osceola's landing open to navigation. Mark Twain, in *Life on the Mississippi*, recalled the Corps efforts along the Plum Point reach: "The military engineers have taken upon their shoulders the job of making the Mississippi over again—a job transcended in size by only the original job of creating it." Twain had his doubts about the prospects for success. "One who knows the Mississippi will promptly aver . . . that ten thousand River Commissions, with the mines of the world at their back, cannot tame that lawless stream, cannot curb it or confine it, cannot say to it, Go here, or Go there, and make it obey; cannot save a shore which it has sentenced; cannot bar its path with an obstruction which it will not tear down, dance over, and laugh at." As Twain predicted, Corps engineers failed. The lack of political will in Congress and the inadequacies of existing technology doomed their efforts.[10]

The floods in Mississippi County only worsened. In January 1894 Lee Wilson watched anxiously as a portion of the bank at his sawmill on Golden Lake began to collapse. While Socrates Beall supervised the relocation of "a large lot of lumber" there, Wilson hastened to Cairo where he found buyers for the quartered oak and ash.[11] Just over a year later, Ed Williams of Elmot, a vulnerable community along the river's edge just above Osceola, engaged a force of hands in "moving his stock of goods," as "his storehouse was in danger of a caving bank."[12] Less than a week after Williams's narrow escape, a portion of the road along the river near Golden Lake collapsed as the mayor of Osceola and the county's circuit-court clerk were passing. Both men managed to jump out while their buggy fell into the river, but their horses barely escaped drowning. "Luckily, as the terrified animals struggled, their harnesses broke, freeing them from the wagon" and the two men were able to rescue them "from their perilous condition."[13]

Since 1828, the year the Corps of Engineers began keeping records, the Mississippi River's "highest stages" had varied considerably, but the general tendency by the late nineteenth century was higher. For example, prior to the Civil War, the high mark at Memphis, Tennessee, was 34.16; after the Civil War the high mark reached 37.66 in 1897. The situation worsened in the twentieth century, as evidenced by the floods of 1912 and 1913 when the waters crested at 45.23 and 46.55 respectively. This tendency brought personal tragedies and economic losses, but it also demonstrated the folly of the Corps' half-hearted efforts at flood control, a responsibility some members of that agency wanted to assume but which Congress was reluctant to sanction. Mississippi County residents endured the consequences. During the1898 flood, the old problem of collapsing banks proved a costly handicap to those living along the river's edge. A blacksmith shop near Wilson's lumberyard at Golden Lake collapsed into the river, and two merchants at Barfield had to move their warehouses. Even after the floodwaters had receded, the weakened banks continued to present problems for Wilson. In November a collapsing bank, again at Idaho

Landing at Golden Lake, "carried off several hundred sacks of seed." That same year, despite an army of men, flooding destroyed one of Wilson's tramways. Wilson had to rebuild it in order to move timber to his mill at the river.[14]

With Congress insisting that the responsibility for flood control rested with landowners or local government, the state of Arkansas moved into the breach. Responding to agitation from prominent landowners and politicians in northeastern Arkansas, the state legislature voted on February 15, 1893, to create the St. Francis Levee District (SFLD) for the express purpose of limiting floods along the Mississippi and St. Francis rivers in northeastern Arkansas. Levee construction partially funded by the state had a long history in Arkansas, extending back into the antebellum era, but the SFLD was a far more ambitious endeavor. The legislature endowed the district with responsibility for flood control in eight counties—Mississippi, Crittenden, Craighead, Poinsett, Cross, St. Francis, Lee, and Phillips—encompassing 2,500 square miles, or 1.6 million acres. The district comprised some of the most fertile land in the United States, but over 90 percent of it was in swamps or subject to periodic overflows. Recognizing the need for expensive flood-control efforts, the landowners and legislators pressed for the power to tap a potentially valuable source of revenue. Just six weeks after passing legislation creating the new district, the legislature awarded to the SFLD all state swamplands lying within the boundaries of the district and authorized it to sell the swamplands and thus provide the funds to construct the levees. Lee Wilson was among the earliest purchasers of such lands, acquiring sixty acres at Round Lake with his friend J. W. Rhodes in or about 1895. A few years later Wilson acquired 1,100 acres at Golden Lake, partly from the SFLD, and in several transactions between 1898 and 1902 he purchased a thousand acres at Moon Lake. All of these acres were in Township 11 North, Range 10 East, thus in the heart of the Wilson estate of southern Mississippi County. Not only did he acquire title to these lands, he benefited from the district's drainage projects and accepted another feature of the legislation: that allowing the district to tax landowners to help pay for the improvements (the levees). He also hired engineers to drain some of the lands left untouched by the district's projects.[15]

The district's main function in the early years involved constructing levees to prevent flooding, something that the floods in 1912 and 1913 challenged. The floods both undermined the district's efforts by destroying levees and demonstrated certain flaws in its approach. With drainage enterprises along the upper Mississippi and the Ohio rivers in the previous decade dumping a greater volume of water into those rivers, earlier levees and new ones built to earlier standards proved inadequate. The rising river stages at Memphis during the late nineteenth and early twentieth centuries dramatically illustrated the challenges confronting the district.[16]

Construction defects, the result of too little supervision and inspection of the materials going into the levees, also contributed to a bad situation. Woody material incorporated in levees rotted, leaving weak pockets, and as floodwaters pressed against the levee, they found those weak spots and slowly ate away at them, often eroding the levee at its base. Another problem arose because levee builders used the soil at hand to construct the levees, and in Mississippi County that usually meant a combination of clay, silt, loam, and sandy soils, some of which were more permeable, hence more likely to saturate than others. The uneven distribution of these soils in the levees together with the woody debris embedded within them created the potential for disaster. And even if the levee held, water sometimes seeped under the levee, showing up on the land side as much as a hundred feet inland. There it would appear as a "boil," warning onlookers to the danger and necessitating the construction of a cylindrical levee around it, a kind of well, lest the water burst through the boil and flood the area behind the levee, weakening the levee on the land side, causing it to collapse.[17]

Wilson and the levee district got their comeuppance in 1912, when an April flood easily overran the existing levee structures. The most serious flood ever to strike the lower Mississippi Valley, it ran the length of the Mississippi River and extended far up the Missouri and Ohio rivers, creating havoc for communities and farmers in its path. Along the lower Mississippi River alone, the flood inundated 10,812 square miles of land. One contributing factor to the severity of the 1912 and 1913 floods was the weakening resolve of Congress. By 1910, the expansion of the railroad system had caused some political leaders to consider rivers as a secondary means of transportation. Consequently, Congress allotted less money in 1911 and 1912 for river improvement, leaving the Corps of Engineers with no choice but to abandon systematic revetment. In 1912 twelve breaks in the levee occurred in the Memphis District of the Corp of Engineers, one of them in Mississippi County, where the river eventually ran ten feet above flood stage. The levee between Golden Lake and Random Shot, immediately adjacent to thousands of acres of Lee Wilson's land, failed to withstand the onslaught. Wilson employed "all the labor he had" in sacking the levee, which "consisted of filling 'gunny' sacks with earth, usually mud, carrying them to the top of the levee and laying them side by side lengthwise across it." Such measures raised the levee by two or three feet. But the rain continued to come down, the river rose steadily, and late in the evening of April 9, 1912, the water finally rolled "over the top of the levee carrying logs and driftwood as if they were straws. Men lost their footing, and in the darkness rolled down the thirty or forty feet of embankment." The river "burst through" near the town of Wilson, and the break soon expanded to "about a quarter mile wide" leaving "much of the south end of the county" submerged. The flood washed away houses, outbuildings, and fences and drowned livestock or left them stranded in hard-to-reach places, where they starved.[18]

The flood hit the town of Wilson particularly hard and completely covered the Wilson lumberyard, ruining equipment and scattering his logs and lumber. Wilson employee Robert Robinson wrote, "Our company has received as severe losses as anyone I suppose, the town of Wilson being three to six feet in water; all of the lumber flooded and covered with sand, making it hard to sell. . . . Our Mississippi [River] Mill is ten feet in water at the present time and our damage there is going to be quite heavy as we had about twenty five thousand dollars worth of lumber there." The Frisco Railroad sustained severe damage but not as much as Wilson's Jonesboro, Lake City, and Eastern Railroad. "The J.L.C. & E. road is probably the greatest loser," wrote the editor of the *Osceola Times*, "for practically all of its lines are built through the lowest levels of the St. Francis valley where the water for miles is of great depth. Much of it will have to be practically rebuilt." Wilson had purchased the railroad in late 1900 and had invested tens of thousands of dollars in upgrading it. He would have to start over. But the loss of the road was also "hard on the settlers in the western part of the county, as they depended on this road for transportation more than on the county roads, and its various branches are the only connecting link between many of the most important settlements of the county." No lives were lost, however, and by the end of the month, the waters began to recede, and confidence rebounded. After all, there would still be time to get some kind of crop in the ground.[19]

Farmers and lumbermen picked up the pieces and began to rebuild, but their efforts came to nothing once the flood of the following year hit. Again, the flood stretched the length of the Mississippi, Missouri, and Ohio rivers and their tributaries, virtually laying waste to cities, towns, and farms. The city of Dayton, Ohio, for example, lay in ruins. Wilson once again put a large force of men to work desperately trying to secure the levee near Wilson, but it broke in the same spot precisely "a year to the day from the break of 1912."[20] The newly repaired levee "had no covering of Bermuda grass to help protect it. The dirt in it was loose and soon became so saturated with water that no fight could save it and it was soon gone." Luckily for Wilson, his north Mississippi County plantation, Armorel, remained above the flood, and he moved men and equipment there, where he had time enough to plant what was already cleared and just enough time to put a force of men harvesting timber there. He also secured a timely loan of $100,000 from some St. Louis financiers.[21]

Still Congress refused to accept a greater role in facilitating flood control. Wilson and other landowners chafed under the constraints imposed on their enterprises and continued to press the Mississippi River Commission to address the situation. Meanwhile, they focused on the other main preoccupation of planters and farmers in the region: drainage of the swamps. In this respect, they were better served by the SFLD. From its inception, district officials planned levees that not only protected land from flooding but also effectively drained thousands of acres. Scientists who study river and wetlands ecology lament the success of the drainage efforts precisely because they eliminated the pools of standing water in which a variety of plants, fish, and microorganisms thrived. Although too little is known about the precise nature of this ecosystem since drainage destroyed it before scientists could subject it to study, examinations of wetlands elsewhere on the Mississippi River establish that many forms of biotic and microbiotic life depend upon the seasonal ebb and flow of the river. As it washes into the floodplain, it not only brings nutrients that enrich the soil; it provides the conditions that make life possible for many organisms. Together with the denuding of the area's forests, the drainage of these wetlands also destroyed the habitat of wild fowl, not only those that used the area as a feeding and resting place during their seasonal passage from Canada to Mexico and back but also those that lived and died in the dense and isolated wetland forests.[22]

The destruction of the northeastern Arkansas wetlands occurred even as the twentieth-century conservation movement began to question such activities, including the draining of the Florida Everglades that accelerated at approximately the same time.[23] Nevertheless, by the end of the twentieth century, drainage efforts and levee construction had eliminated fully 90 percent of the wetlands along the Mississippi River. What was unique about Mississippi County was a certain complicating factor shared by a few other counties in northeastern Arkansas and southeastern Missouri: the sunk lands that more rapidly filled with water when the Mississippi River rose and its tributaries overflowed and then drained, but not entirely, leaving their status as swamps or lakes unclear. In some cases the sunk lands also contained problematic sand blows that complicated drainage efforts even further.

The initial drainage districts created in the first two decades of the twentieth century aimed at reclaiming land along the eastern side of the county. Lee Wilson and a cohort of planters orchestrated the formation of these districts and contended with howls of protests from some small farmers who feared the imposition of taxes necessary to fund construction. They also knew that higher land prices would accompany development, and many men would find themselves unable to purchase land there, a fact they fiercely resented. These fears were well founded. Between 1900 and 1930, the average price per acre of improved land in Mississippi County rose from $23 to $97. No other county in the state had land as costly as that in Mississippi County.[24] Clearly, something phenomenal was going on, and it worked to the benefit of plantation agriculture along the county's eastern edge. The number of acres in farms in the county more than doubled (from 124,684 to 335,034) between 1900 and 1930, but most of those working the farms were tenant farmers. While there were twice as many owners operating farms in 1930 than there were in 1900 (from 500 to 986), there were eight times as many tenant farmers (from 1,204 to 9,561). The need for plantation labor no doubt explains the dramatic increase in population.[25]

Impatient with the failure of the SFLD's meager efforts at drainage, Lee Wilson became one of the loudest voices speaking out for an entity that had as its primary focus the drainage of overflowed lands. He and other landowners began to lobby for the passage of legislation making it possible for county governments to create drainage districts. Appointed by the county judge in 1901 to a committee to explore how the county might augment its drainage efforts, Wilson and others aggressively lobbied the state government, and in 1902 Arkansas passed legislation, modeled after a Missouri law and similar to one implemented in Illinois, that made it possible for landowners to petition the county judge to form districts within their counties. The legislation authorized counties to sell bonds to fund the improvements and tax the landowners to pay back the bond holders. The districts would have as their express and sole purpose the drainage of land, leaving flood control to the SFLD and to the reluctant Corps of Engineers.[26]

Wilson clearly benefited by the first drainage district created by the county in 1902. Drainage District 1, also known as the Grassy Lake and Tyronza Drainage District, made possible the construction of a drainage canal, which promoters promised would redeem "a hundred thousand acres of Mississippi County's best land . . . from ponds and sloughs, now the homes of frogs and mosquitoes." That was only a slight exaggeration. The twenty-five-mile canal, which cost $72,585 to construct, drained 98,644 acres. It ran from west of Osceola to the Tyronza River and drained some of Wilson's south Mississippi County land. A second district quickly followed, this one draining land in north Mississippi County. Minimal controversy accompanied the creation of these, but in April 1904 the third district, also covering a portion of north Mississippi County, "brought a large delegation of land owners" to the courthouse to protest. Opponents believed that the project had as "its object the drainage of a large body of timbered lands owned by saw mill proprietors and land speculators who would be individually benefitted for which the farm land owners who would not receive any, or but little benefit, were to be taxed to pay for the ditching." The county judge approved some changes but passed favorably on the creation of the district.[27]

The contention that land speculators and lumber companies orchestrated the creation of drainage districts speaks to a popular conception that was only partially accurate, for many speculators and lumber companies opposed drainage as they had no intention of incurring the taxes required to fund construction of levees and canals. Other such entities, like Lee Wilson & Company, aggressively pursued them. Small farmers too entered the debate on both sides, but they typically made up the most numerous set of individual opponents. Many of them had lived for years in the area and had accustomed themselves to periodic overflows. Nevertheless, they raised little objection to the creation of four small districts approved by the county between 1904 and 1908. When two larger districts were proposed in 1908, however, some small farmers threatened violence. Drainage District 9 took over both the responsibility for and the name of the old Grassy Lake and Tyronza Drainage District 1. Stretching in a southwesterly direction from the north end of the county to the southwestern corner bordering on Poinsett and Crittenden counties, it eventually provided drainage protection for 168,156 acres. It effectively drained Armorel, Wilson's north Mississippi County plantation, and much of the rest of his land in south Mississippi County. A second enterprise, Drainage District 8, known as Carson Lake Drainage District, promised to drain 56,942 acres, including three of Wilson's towns: Marie, Bassett, and the all-important Wilson, Arkansas. It ran 128 miles and was essential to Wilson's interests. There was some overlap along the edges of the proposed districts, however, and that contributed to the circumstances leading to a dramatic courthouse confrontation.[28]

Tensions arose long before the case for creating the districts came up for hearing at the county court. At a picnic Wilson sponsored to promote them, two fights broke out, one of which he brought to an end by the sheer force of his personality. When a prominent planter at the picnic who opposed the proposition threatened Wilson's friend S. E. Simonson, one of the enterprise's chief organizers, Wilson stepped between them and declared "Sir, I am astonished at your attitude. Mr. Simonson is my invited guest and he has done nothing to deserve such treatment. He has only properly upheld his point of view." Wilson's attorney, J. T. Coston, later showed him a raft of threatening letters and asked him candidly whether he wanted to continue pursuing the creation of the districts. Wilson responded, "I have never let a mob tell me yet what I can't do and what I can do and I will come out of the court house a corpse before I will surrender." [29]

Wilson had plenty of opportunity to reconsider his boast. When county judge Logan D. Rozzell held session to rule on the creation of the districts on July 8, 1908, the scene at the Mississippi County Courthouse portended disaster. Armed men from both sides of the issue stationed themselves outside the courthouse, and opponents hung a noose at the entryway, a not-so-veiled threat aimed at the judge. Inside the courthouse Lee Wilson stationed his armed men "sitting in the windows, standing in the doors and . . . in the rear." One supporter of drainage walked "back and forth among the Lee Wilson crowd telling them not to shoot at the feet, but to shoot at the belly and to shoot to kill." Given the legacy of violence in the county, it is a small miracle that gunfire did not break out. The county sheriff, John A. Lovewell, angry over having lost his bid for reelection in the Democratic primary just a few months earlier, offered little assistance. Once an ally of Wilson's in a struggle for control over county government in the mid-1890s, Lovewell had become an implacable foe. Some accused the sheriff of having spread false rumors in order to create an issue and position himself for election as an independent candidate. According to these accusations, he had circulated a pamphlet accusing Lee Wilson of planning to secure the contracts to build levees and ditches in order to enrich himself and, in the process, burdening landowners with confiscatory taxes. According to Lovewell, Wilson would then buy their lands at the courthouse door for a pittance.[30]

Lovewell's accusations might have had resonance for many in the room who knew that Wilson had purchased thousands of acres of land at the courthouse door over the years, paying no more than the taxes owed on them. Some might have remembered that in 1882, when he was but seventeen years old, he had made his very first purchase of land, securing 125 acres for $16.85. Wilson's defenders scoffed at the notion that he was out to confiscate the property of other landowners and tried to explain away Wilson's involvement in constructing previous levees, an involvement that critics alleged doubly enriched him. He had, Wilson's friends argued, merely completed the construction of levees left unfinished when a contractor departed the county without fulfilling his contract. This explanation failed to satisfy everyone, and although Sheriff Lovewell had deputies in the audience, no one could predict what orders they were operating under. Had one man raised and fired a weapon, the room might well have exploded like a powder keg.[31]

As the restive crowd heard testimony from the three men appointed as "viewers" by the court, a more routine matter came to the forefront. The viewers had the responsibility of assessing the value of the proposed drainage districts to the specific landowners and then apportioning taxes accordingly. It was typical for landowners to question their assessments—even Lee Wilson would do so on occasion—and their petitions to reduce the amount did not signal opposition to the creation of the district. In this particular case, some landowners had a legitimate complaint about excessive assessment, as their lands were situated where the two districts overlapped, and they faced assessments from both of them. As they began their testimony, voices in the crowd shouted their protests against the creation of drainage districts while others interrupted with cries of support. Feelings ran so high that Judge Rozzell retired to his chambers with the attorneys representing the two sides of the issue.

The attorneys for those seeking reassessment were themselves prominent men and landowners, including John B. Driver, Jr., whose father was a former director of the SFLD (1897–1901) and a longtime friend of Lee Wilson's. In the judge's chambers that day, Driver met with the attorneys representing Wilson and the other organizers and encouraged the judge to recess court for a long lunch, allowing them the opportunity to restore calm. But when court resumed after a three-hour recess, one of the leaders of the opposition to the creation of the district, said to have been handpicked by Lovewell to inflame the crowd, "insisted upon making a speech" and accused Wilson and Coston "of having but one purpose, that of stealing the homes of the poor people." According to one source, he then produced a rope and urged the crowd to use it to "hang the d——scoundrels." The crowd surged forward, and the judge abruptly adjourned court and departed by a back door. As his carriage carried him away from his courtroom, a man ripped the noose from the courthouse door and, accompanied by a small crowd of angry men, followed the judge down the road and out of town. A delegation of night riders later visited his home and threatened violence if he approved the district.[32]

Most of the crowd remained in the courtroom after the judge's departure, including Lee Wilson. When Lovewell's handpicked rabble-rouser attempted to harangue the crowd further, Wilson told him to "shut your G——D——

mouth and let me tell these people the truth." The crowd, though still some-what unruly, listened as Wilson attempted to explain his position. "He was often interrupted with great demonstrations of approval to the chagrin of the adversary." The crowd eventually dispersed, and the issue languished. Judge Rozzell placed the case on the docket to be heard in September, but as that date drew near, he continued it further, preferring to allow his successor to rule on the explosive issue. The old judge, who at seventy-two years old had a distinguished career in public service behind him, was ready to retire. He had declined to seek reelection in the Democratic primary, which had taken place March 25, 1908, and a younger man, thirty-two-year-old W. C. Armstrong, secured the nomination. Armstrong won the general election in the fall, but died—of natural causes—within six months of taking office and before he had an opportunity to issue a final ruling on the Drainage District cases. His death in May 1909 presented a unique opportunity for prodrainage forces. The newly elected county sheriff, Coleman B. Hall, along with a small group of men, including Wilson's attorney, J. T. Coston, "took the early morning train to Little Rock" just hours after Armstrong died, with the intention of asking the newly elected, prodrainage governor George Donaghey to appoint Samuel L. Gladish as Armstrong's replacement. Coston "had stumped the state for Dona-ghey" during the Democratic Party primary the previous year and helped bring Donaghey to Mississippi County, where he appeared at a picnic sponsored by Lee Wilson to promote drainage. But when the Mississippi County delegation reached Little Rock, they found the governor was in El Dorado at a barbecue. Brought to the telephone by way of horse and buggy, the governor listened to the request and agreed to appoint Gladish to hold the position of county judge until a special election could be held.[33]

With the election scheduled for July 1, 1909, the county Democratic Cen-tral Committee, which now included key Wilson allies, had no time to orga-nize a primary so they selected Gladish as their candidate by a vote of 21 to 10. One disappointed contestant, however, L. W. Gosnell, refused to accept the committee's decision and rallied supporters at a "mass meeting" in Osceola behind his independent candidacy, a candidacy based on opposition to the creation of the controversial drainage districts. While Gosnell traversed the county seeking support and lambasting the drainage efforts, Gladish himself "took a horse and buggy or rode a horse to the different sections of the county," and, emphasizing the state of the roads and the need for drainage in water-logged Mississippi County, wore "two pairs of trousers and when he arrived at the place where he was to speak, he would take the muddy ones off." As the date of the election neared, all indications were that it was going to be very close. Lee Wilson asked his attorney for advice on whether he could "vote the convicts" working in his operation. The county convict farm was located on his property near the town of Wilson, and he used the convicts as plantation laborers in return for paying the cost of their support. Although his attorney declined to render a legal opinion on the issue, Wilson "voted" them nonetheless, and Gladish won the election. He secured 1,596 votes to Gosnell's 1,480 votes, a 116-vote majority. D. F. Taylor, the Republican candidate, received 713 votes. Wilson's convicts, who numbered far less than the 116-vote difference between the two leading candidates, could not have made a difference in the outcome, but the fact that he "voted" them left a lasting impression, earning him a place in the political folklore of the county. It took another year and a half to gain approval of the controversial drainage districts, mostly because of intense negotiation with landowners over their assessments. Once the county court approved the districts, the bonds to fund construction had to be brokered and an engineering firm hired to begin construction.[34]

If the establishment of drainage districts stirred controversy, so did the successful completion of their work—and these later controversies included a new array of antagonists. Together with similar projects in Poinsett County, the success of the drainage efforts brought the Arkansas sunk lands to the attention of the federal government. Ironically, the state of Arkansas inadvertently set the stage for controversy over ownership of some of the sunk lands when it created the SFLD in 1893. The newly appointed board of directors hired engineer Harry N. Pharr to oversee the construction of the levees, and he immediately focused on a troubling situation. He knew he needed to construct a levee that would cut through the sunk lands in Mississippi and Poinsett counties, but he suspected that the sunk lands belonged to the federal government, not to the state of Arkansas or to the SFLD. Certain prospective purchasers of district lands, like Lee Wilson, encouraged Pharr to seek clarification from the General Land Office.[35]

By virtue of the Swamplands Act of 1850, Congress had awarded to Arkansas, as to other states, the swamplands within its borders. The act stipulated, however, that the states had to present surveys of their swamplands to the federal government for approval. The problem in northeastern Arkansas was that the surveyors in Mississippi, Poinsett, Craighead, and Green counties failed to survey all of the swamps, particularly the notorious sunk lands. The surveyors characterized some of them as lakes, and to further complicate matters, they made numerous mistakes in drawing the meander lines, mistakes that would later figure prominently in the question of ownership. The fact is, as part of a natural watershed, the sunk lands overflowed when the Mississippi and St. Francis rivers rose in the spring every year. In the late summer and winter, they receded although they never fully drained. Thus their boundaries—their meander lines—shifted according to the season. If the surveyors reached there in spring, they outlined one set of boundaries; if they arrived in late fall, as they typically preferred, they designated quite another.[36]

Pharr's request for clarification from the federal government as to the ownership of the sunk lands resulted in a statement from Secretary of Interior Hoke Smith in 1895 that the federal government had no interest in them. The SFLD, in order to secure the funds to construct levees, sold some of the lands in question to Lee Wilson and others, but in the early twentieth century a dispute arose between some of these purchasers and landowners who had land adjacent to the drained acres. The dispute revolved around whether the sunk lands were swamps or lakes. According to riparian rights, if a lake was drained, a landowner having property adjacent to the lake could claim the land exposed. If there were several landowners around the lake, they were each entitled to an appropriate share of the reclaimed land according to the frontage on the lake they owned. If, however, the sunk lands were swamps, then riparian rights did not apply.[37] These landowners argued that the SFLD had no right to sell the land underneath the lakes. In 1902 the SFLD—with Lee Wilson's backing—once again appealed to the General Land Office for clarification, and the Interior Department once again asserted that the land belonged to the state and thus the SFLD had the right to sell it. This ruling encouraged Wilson to make further purchases at Moon Lake later that year.[38]

However, after a series of confrontations and court proceedings between lumber companies and small farmers in the sunk lands, the Department of Interior reversed itself. Secretary of Interior James Rudolph Garfield, a dedicated conservationist appointed by Theodore Roosevelt to the position in 1907, signaled the arrival of an environmentally conscious federal perspective. On December 12, 1908, Garfield issued a statement declaring that all the sunk lands, including those "erroneously" characterized as lakes, belonged to the federal government. As unsurveyed swamplands, they had not been patented back to the state according to the 1850 Swamp Land Act. He further ordered them surveyed for the purpose of opening them up for homesteading. The declaration was bad news for claimants who had been arguing that the lands were lakes and thus subject to riparian rights; it was bad news for SFLD, which had been acting, in good faith, as though the state had the authority to grant them the right to the land; and it was bad news for those who had legally, or so they thought, purchased some of the disputed land from the district or from third parties who had secured their titles through purchases from the district. Lee Wilson's claim to over two thousand acres was now contested. After having played a role in petitioning the General Land Office twice, once in 1894 and a second time in 1902, Wilson must have been particularly offended by the new ruling. He was so fed up, in fact, that he began to assert title as a riparian claimant to land in Moon Lake adjacent to land that he had purchased in 1902. He must have figured that if that was the score, he could play that game too.[39]

It is entirely possible that Wilson had been harvesting timber from backwoods swamplands he did not own all along, but he faced a reckoning over his activities at Moon Lake in September 1909 when the government seized 684,856 feet of saw logs and 700 ranks of ash bolts he had cut on his riparian claim there. The government further indicated that it intended to pursue reimbursement for an additional 1,519,278 ft B.M. logs and 1,409 ranks of ash stave bolts Wilson had already harvested and sold. Special agent W. B. Johnson, dispatched to Mississippi County by the Interior Department to investigate Lee Wilson & Company's timber trespass, found the company's president to be defiant and uncooperative. "On September 25, 1909, I interviewed R. E. Lee Wilson . . . and he then stated to me that he would make a statement with reference to this trespass by his company when I had opportunity to see him a few days later. I have since made several unsuccessful efforts to see him, and am this day advised by T. J. [sic] Coston, of Osceola, Arkansas, one of his attorneys, that he will now decline to make a signed statement, but that he will not at any time deny having cut the said timber, and will defend the title to that now under seizure, and stand suit for that removed." In other words, Wilson said "so sue me."[40]

The opinion of the secretary as to the ownership of the sunk lands led to congressional hearings in 1910 that provoked considerable controversy. Senator Jeff Davis had convinced Representative William A. Oldfield to present the bill in the House, but by the time it came up for hearing, Oldfield distanced himself from the procedures, stating that he had introduced the measure only as a favor to Davis. Arkansas senators played competing roles during the hearings. While Senator James Clarke carefully refrained from criticizing any of the claimants, Senator Jeff Davis revealed considerable antipathy to those he characterized as "timber thieves" in his testimony before the committee. He declared that "Chapman & Dewey, a large milling concern, and Mr. Lee Wilson, and a dozen others I could mention have . . . simply stolen timber until it has made them immensely rich." Davis suffered some embarrassment when it was later revealed that he represented the St. Francis Levee Board in a case it had against Chapman and Dewey, a case that would be strengthened if his testimony convinced the House committee that the board's claim to owner-ship of the sunk lands was superior to all other claims. So eager was he to prevail that he attacked a group of citizens he had been cultivating since his days as governor of Arkansas (1900–1906) when he championed the "little man." Referring to the five to six hundred "squatters" there as agents of the lumber companies rather than bona fide homesteaders, he testified that "these people have what they call 'donacks' there, which is a little mound like a bea-ver's hill, and they will put a little doll pen on that, or build them a little river boat, in which to live, or else they will find a little high elevation somewhere, and they will put up a small shack, or they will go there in a dugout, and that is what they call actual residence." He dismissed them further by alleging that they were squatting on land that could not possibly be farmed and suggested that Wilson and Chapman/Dewey were sending men into the disputed land to claim it as homesteaders in order to facilitate the harvesting of timber there.[41]

Attorney J. A. Tellier, who represented the homesteaders at the hearing, painted quite a different picture of the battle taking place on the ground in Arkansas. The homesteaders had powerful forces arrayed against them, not only the SFLD and those who had purchased land from the district; they also faced harassment from the timber companies who were themselves at odds with the district and claiming the same land.[42] As a claimant, Lee Wilson fit two of these categories: first, based on purchases from the district and second, based on his riparian rights to land. Tellier, in order to correct Miller and Davis's characterization of the homesteaders, introduced hundreds of tele-grams and affidavits filed by homesteaders in Mississippi County.[43] Most in-teresting are the 109 notarized affidavits.[44] The men asserted that they had lived on their homesteads for twelve months or less and over half of them had their families with them.[45] The implication was that they had responded to Secretary Garfield's 1908 announcement that he intended to open the land for homesteading. Seventy-five claimed to have built houses, and five more had houses under construction. Although only a few had outbuildings like hog pens, chicken houses, and corncribs, fifteen had barns.[46] Twenty-five men had gone to the trouble to dig wells, a costly and labor-intensive endeavor. While only twenty-three claimed to have cultivated a few acres or put in a garden, thirty-three testified to having cleared at least a few acres. Significantly, most of them said that all of the land they each claimed could be cultivated every year if the drainage projects then pending at the county court were approved.[47] Finally, of the 109 affidavits, seventy-nine had been filed between February 19 and 29, 1910, and were clearly meant to counter allegations made by Jeff Davis and W. B. Miller at the hearings. To counter the allegation that they were not bona fide homesteaders but placed there by Wilson and Chapman/Dewey to undermine the legislation, all of those who submitted affidavits in February swore they had no lumber contracts and that they were there to homestead.[48]

Thirty-five of the homesteaders resided on homesteads in south Mississippi County, some of them very likely on land claimed by Lee Wilson. They, like those elsewhere in the county, were family men in the prime of their lives who had settled on their homesteads within the previous twelve months. Most of them had constructed houses, nine of them had sunk wells, and a few of them had put up an outbuilding or two. Only a third of them had cultivated or cleared land, but all of them swore that some of their land could be cultivated under present conditions and that most of it could be cultivated if the drainage projects pending at the county courthouse were approved. All of them swore they had "no contract with any individual or corporation, either expressed or implied, for the sale of timber or division or sale of said land" and that they were there to homestead for themselves and their families.[49]

Of the thirty-five south Mississippi County men, only thirteen could be located on the manuscript census of population, likely because they lived in such inaccessible locations that the census taker simply overlooked them. Those thirteen men ranged in age from twenty-two to sixty, and all but two of them were married men with families. All but two of them had been born in Arkansas, Missouri, or Tennessee. Most of them listed themselves as farmers on the census, although there was one engineer and one public schoolteacher. Only one was black, thirty-nine-year-old Henry Still, who lived with his wife, Mattie, and their five children near Wilson, Arkansas. Because night riders actively discouraged blacks from settling on homesteads, Still was particularly unusual. But, aside from being black, Henry Still was like the other homesteaders in every respect. In the final analysis, most of the homesteaders located in south Mississippi County were family men from the South who were hoping to achieve landownership and provide security for their families.[50]

In the end, the hearings accomplished nothing but embarrassment for their sponsors and everyone else concerned: the SFLD, purchasers of land from the district, and riparian claimants. Neither did the potential homesteaders gain anything. No bill was reported out of committee, and the issue was batted back to the courts, which merely prolonged the struggle and gave the riparian claimants time to address the problem on the ground in Arkansas. There they had the upper hand in local courts. Wilson and other planters and lumbermen, in a striking defiance of federal authority, secured injunctions in the Mississippi and Poinsett counties' courts prohibiting homesteaders from cutting timber on the lands they claimed, a prohibition that prevented them from making the improvements they needed to make in order to certify their claims. One homesteader complained to the Interior Department in February 1911 that "these [injunction] cases are set for some stated term of the court, but then the time arrives, the case is continued for cause." Unable to cut and market the timber on their homesteads, and therefore, unable to sustain themselves, the homesteaders moved off the land to secure employment elsewhere. This "creates an abandonment of the land," giving claim jumpers "usually in the employ of Lee Wilson," the opportunity to move in and take possession.[51] To make matters worse, the Interior Department issued orders in 1914 prohibiting settlers from cutting timber and delaying final certification of homesteader claims until the courts could decide the issue of riparian rights.[52]

The year 1914 held other significance as suits filed by the government in 1912 to "quiet title," in other words to settle the issue of ownership once and for all, finally reached fruition. Interestingly, the government was now referring to all the defendants as riparian claimants, whether they had purchased land from the district or held land on the periphery of lakes and were merely claiming riparian rights. The government's language may well have been calculated and meant to support the position that these were not lakes but swamps and thus not subject to riparian rights. In fact, the four suits involved both men who had purchased sunk lands from the SFLD and those who were claiming riparian right. While Lee Wilson fit into both categories, the government's suit against Wilson identified only the property he had purchased from the district at Moon Lake. After the federal government filed its suit to quiet title in the Wilson case, another decision by the Supreme Court would have caused Wilson concern. On December 1, 1913, the United States Supreme Court ruled in *Little v. Williams* that riparian rights did not apply for the parties involved in that particular case, either Johanna Little, who had purchased Walker Lake from the SFLD, or J. J. Williams, who was claiming it by virtue of owning land adjacent to it in north Mississippi County. The court was very clear in its decision, arguing that the SFLD did not have title, did not itself have riparian right, when it sold the "lake" to Johanna Little. The lake, in fact, was not lake, but swamps and had never passed out of the hands of the federal government. By this time the federal government's suit to "quiet title" against Lee Wilson in his claim to Moon Lake was pending in the federal district court. On February 20, 1914, that court ruled in favor of the government, essentially divesting Wilson of his claim to Moon Lake. The federal court made identical rulings in the three other suits it had filed to "quiet title."[53]

By now the government's argument was all too familiar to Wilson. He appealed the decision, and when the eighth circuit in St. Louis ruled in favor of the government's position, he appealed to the United States Supreme Court. On November 5, 1917, in *Lee Wilson & Company v. United States*, the court ruled that the title to Moon Lake resided with the government. Although this seemed to settle the matter from the judicial standpoint, the court itself suggested in its decision in the *Wilson v. U.S.* case that the proper remedy for riparian claimants like Wilson was to be found in Congress. In the last paragraph of the opinion, Chief Justice Edward D. White wrote that if riparian claimants deserved "equitable consideration . . . it is manifest that the prayer for their enforcement is, in the nature of things, beyond the sphere of judicial authority, however much relief on the subject may be appropriately sought from the legislative department of the government."[54]

Congressmen and senators from Arkansas hardly needed reminding by the Supreme Court that legislation could address the problem. Although the 1910 Oldfield Bill had been controversial and unsuccessful, beginning in 1916 several bills were introduced in order to assist not homesteaders but riparian claimants. None of these bills involved property Wilson claimed, but that changed after the Supreme Court's decision against him in 1917. Senator Joseph T. Robinson, a former governor (1909–13) and now senior senator from Arkansas, presented a bill in the Senate in early 1918. Senate Bill 5566, which was accompanied by House Bill 13400, named Wilson specifically and identified other acreage he claimed in the sunk lands of south Mississippi County: eleven hundred acres at Golden Lake; Senate Bill 746 addressed the sixty acres he claimed at Round Lake. Neither bill, significantly, referenced his property at Moon Lake, property Wilson apparently lost interest in after the Supreme Court ruled against him in 1917. According to the two bills, he had cleared, drained, and was cultivating the land in Golden Lake and Round Lake. The commissioner of public lands and the secretary of the interior both supported the bills, signaling a remarkable departure for the Department of the Interior. Even J. A. Tellier, the champion of the homesteaders, stipulated that he had no objection to the passage of the bills, which were designed to give Wilson the preferential right to "purchase" the land he claimed at a price of $6 an acre at Golden Lake and $12.35 an acre at Round Lake. This was at a time when the average price per acre in Mississippi County stood at nearly a hundred and fifty dollars per acre. But the bills' originators argued that the price represented "approximately what the lands would be reasonably worth if they had not been drained and otherwise improved at the expense" of Lee Wilson. They further argued that homesteaders had no standing in the matter, that the courts had determined that only the federal government or the riparian claimants had a legitimate claim. The Sixty-fifth Congress passed the bills on March 1, 1919.[55]

Other riparian claimants also found relief in Congress, particularly by a measure introduced in June 1921 by Representative William J. Driver, House Bill 6863. Senator Robinson introduced its companion in the Senate. These bills, like their predecessors, were designed to give preferential rights to riparian claimants to purchase the land they claimed. Again, "riparian claimant" included not only those who had purchased the land from the district but also those who owned land adjacent to the swamps or lakes and were claiming riparian right. The tenor of the discussion in the House suggested that the district favored claimants who had drained and developed their acreage, a description that perfectly characterized Lee Wilson's activities. The homesteaders there were left without an advocate. House Bill 6863 was amended to give the secretary of the interior the authority to decide on a case-by-case basis the merits of other riparian claimants not named in the bill. It passed on September 15, 1922, and was signed by the president a few days later.[56]

At the same time that the Interior Department abandoned the homesteaders in favor of development-minded individuals like Lee Wilson who destroyed an important wildlife habitat, its secretary, Franklin Knight Lane (1913–20), was championing the creation of the National Park Service, something conservationists supported. Although Lane has a mixed record on the environment, having pushed for the construction of a controversial dam in Yosemite Park in 1913, he supported other conservation efforts, such as the designation of Big Lake as a National Wildlife Refuge in 1915. This occurred in the context of a violent struggle in the area between members of the Big Lake Shooting Club, who claimed exclusive rights to hunt in certain areas, and locals who wanted to continue to augment their own diets and incomes by hunting there. Further complicating the situation was the presence of out-of-state market hunters who vied with both local groups. In the midst of this ongoing struggle, a group of small landowning farmers west of Big Lake—in the Buffalo Island region— petitioned to create a new drainage district that would drain not only their lands but a significant portion of Big Lake.

Lee Wilson's role in this development is impossible to discern, but he was far from disinterested as he was said to have constructed a large clubhouse of his own at Big Lake. Of course, he may have been a member of the Big Lake Shooting Club himself, and the reports of his having built his own commodious clubhouse may have an exaggeration. While Wilson had pursued an aggressive strategy to drain and develop the land he claimed, he also took pleasure in the swamps and backwoods of Arkansas and appreciated the opportunity to hunt and fish, particularly at Big Lake where he "entertained lavishly visitors from New York, Chicago, San Francisco, Cleveland and other cities." Considerable business and politicking took place at the clubhouse, but personal relationships were also cemented there. In 1906, he entertained his prospective son-in-law, Frank Wesson, of the Smith and Wesson Company in Springfield, Massachusetts. Wesson hoped to marry young Victoria, but Wilson, who wanted to discourage the Harvard-educated "dandy," took Wesson by mule wagon, to Big Lake, a daylong trip over rough roads. He had misjudged Wesson, however, who had hunted throughout the West and in Canada and turned out to be the "life of the party."[57]

When the petition to form Drainage District 16 reached the court in 1914, the Buffalo Island farmers and everyone else concerned arrived at a crucial moment, a point of no return. The "duck hunter's paradise," which had been under assault for more than a decade, particularly because of the aggressive market hunters, was facing annihilation. The calamity facing Big Lake's wild fowl particularly began with the expansion of the market, an expansion augmented by the construction of railroads through the region in the early twentieth century. Railroad promoters pitched advertisements to sportsmen, writing that "a sportsman residing in St. Louis can leave his office at the close of business hours, and after a leisurely discussed supper, seek refreshing slumber in the luxurious berth of a Pullman coach, to awaken next morning in the hunting grounds of Arkansas." Playing both sides of the fence, the railroads also provided "refrigerated cars for the transport of commercial game and fish," something that spoke to the interests of both resident and commercial hunters. The hunters, whether local or otherwise, were attracted to the area by the abundant native population of animals and birds, but also by the millions of ducks and geese who traversed the famous North American flyway from Canada to Mexico. During the winter of 1893–94 alone, hunters marketed 120,000 mallards, and one gunner was said to have killed eight thousand of them. By 1910, a game warden estimated that five thousand ducks a day were shipped out of Big Lake by market hunters. Given that "ducks were bringing about 50 cents each in 1910," while "common laborers were paid only a few dollars a day," many local men found employment and sided with the out-of-state interests.[58]

When the Big Lake Shooting Club asserted exclusive rights to hunt in the area in 1901, the stage was set for a dramatic confrontation. They "employed a resident staff of ten, including gamekeepers who patrolled the lake to search for poachers" and even "issued their own permits for fishing or hunting on their preserve." Club members attempted to court local businessmen and law enforcement officials but were not entirely successful, particularly after they invoked their exclusive rights to hunt the area they claimed. In 1901 they purchased a ten-foot strip of land circling Big Lake for the sum of $5 from two lumber companies and claimed riparian right to the lake. Local hunters hired by out-of-state interests challenged the purchase, arguing that since the Little River passed through Big Lake, it was a navigable stream and therefore not subject to riparian rights. The market hunters lost the argument in court in 1903 and then resorted to violence, burning the club's "20-room, $8,000 clubhouse" in July 1904. Club members hired Pinkerton guards and built an even grander structure, worth approximately $50,000, only to see it burned to the ground in 1910. By 1915, one game warden hired by the club had been shot and wounded no fewer than twelve times.[59]

While this was a complicated struggle in which local men objected to the club's exclusive use of what had been a common ground, the intrusion of the market had introduced an out-of-state player with little interest in the well-being of the local population and even less interest in the protection of the environment. Fortunately, William Hornaday, director of the New York Zoological Park, had turned the spotlight on Arkansas in 1913, ranking it as "next to Florida, the rearmost of all our states in wild-life protection." Characterizing the sunk lands as "the greatest wild-fowl refuge anywhere in the Mississippi valley," he enjoined, "awake, Arkansas!" Hornaday declared, "You have permitted hired market gunners from outside your borders to slaughter the wild-fowl of your Sunk Lands literally by the millions, and ship them to northern markets with very little benefit to your peoples."[60]

In the end, it was a strange convergence of interests that led to the designation of Big Lake as a National Wildlife Refuge in 1915. The Drainage District 16 proposition was the final straw for Joseph H. Acklen, a member of the club who had spearheaded the fight against the out-of-state market hunters. Acklen had served a stint as the state game commissioner of Tennessee and another as the head of the National Association of Game Commissioners. When

President Woodrow Wilson created the new position of chief game inspector upon taking office in 1913, he had appointed Acklen to the post. By that time both federal and state game laws were lending support to the club's fight against the market hunters, but the violence against the club showed no signs of abating and then came the drainage proposition. Acklen used his influence in Washington and in 1915 President Woodrow Wilson signed an executive order placing 3,500 acres into the Big Lake Reserve. The refuge was later expanded to include an additional 7,538 acres. The order "made it unlawful for any person to hunt, trap, capture, disturb or kill any bird of any kind within the reservation." Together with an international treaty signed with Canada the next year designed to protect migratory birds, the creation of the Big Lake Wildlife Refuge "signaled . . . that the problem [of the slaughter of wild fowl] had been recognized and efforts would be made to do something about it."[61] Meanwhile, back in Mississippi County, the new drainage district's boundaries were redrawn to comply with the establishment of the refuge.[62]

Lee Wilson's perspective on the establishment of the refuge is unknown, but as much as he intended to develop his land and make it profitable, he understood the importance of wetlands and hunting preserves. While hunting wildfowl became illegal in Big Lake after 1915, club members continued to fish in the abundant lake and hunt other prey. Besides, plenty of wilderness areas existed elsewhere in Arkansas and surrounding states, and Wilson could use them to entertain important guests. As late as October 1932, he wrote to his attorney and several businessmen that he was "planning on a hunt on Peters Island, beginning the opening of the deer season in Arkansas. . . . I am trying to kill two birds with one stone. I want to familiarize myself with conditions on Peters Island, as I will be sober, and at the same time do a little hunting." Wilson owned half of the island, located between Tunica County, Mississippi, and Lee County, Arkansas, and was negotiating for the purchase of the rest of the island. He intended to maintain part of it as a private hunting preserve but also had his eye on the timber. He later declined to make the purchase, characterizing the timber there as insufficient considering the price. Wilson also maintained a 320-acre "preserve" in south Mississippi County and willed "the use of it" to the Boy Scouts of America upon his death in 1933. They had to pay the taxes and restrain from disturbing the acreage by removing the trees; nor could they transfer title to it. In January 1934, however, the Boy Scouts declined the bequest, writing that "not only are the stipulations of the bequest hampering, but there seems to be grave doubt as to the suitability of the property, as situated and described to us, for Boy Scout purposes at present or for

a considerable time ahead." Wilson had apparently willed them one of the last
pieces of swampland left in the county, something that he probably regarded
as a preserve, a hunter's paradise, but which the Boy Scouts may have rightly
concluded was a bastion of mosquitoes and malaria.[63]

The creation of Big Lake National Wildlife Refuge stands as a contradiction
to the otherwise unbridled exploitation of Mississippi County's natural resources. The last word on the direction of the county's development came
with the signing of House Bill 6863 in 1921, marking the culmination of two
decades of struggle over the sunk lands, and it is perhaps no surprise that
plantation and lumber interests, like Lee Wilson & Company, prevailed. With
greater resources at his command—both economic and political—Wilson
could influence the outcome of the struggle even as he faced setbacks along
the way. The battles he waged with homesteaders and, for a time, with the
federal government over the sunk lands were matched in intensity by those
fought over the issue of creating the drainage districts. Of equal ferocity was
his battle to overcome the natural landscape and turn it to agricultural purposes. The process of making Mother Nature over to his liking was still very
much a "work in progress" in 1920, but in the previous two decades, the total
acreage "in farms" had doubled, increasing from 124,684 to 277,670 acres.
Homesteaders, however, failed to secure a substantial share of the reclaimed
land. The number of owner-operators had doubled, but the number of tenants and sharecroppers had more than quadrupled.[64] Plantation operators like
Lee Wilson brought acreage into cultivation by employing the labor of landless men and their families. His little empire stretched "eight miles wide and
twenty-seven miles long." Like the "feudal baron of old," one journalist lik-
ened him to, he took extraordinary measures to make certain that he exercised
maximum power over it and over the people who worked for him.[65] He would
harness labor as artfully—and as ruthlessly—as he harnessed nature. In the
meantime, however, he faced the daunting challenge of rationally organizing
his sprawling enterprise, a challenge he addressed by using a combination of
New South ideas and Progressive Era strategies.

4

A NEW SOUTH
ENTREPRENEUR IN THE
PROGRESSIVE ERA

While Wilson spent the first two decades of the twentieth century contending with the forces of nature, he also focused on consolidating his control over his operation and organizing his lumber and agricultural businesses into profitable enterprises. He developed a decentralized management structure, but he ran the company like a sole proprietor. Although he incorporated in 1905 as Lee Wilson & Company, "in his own mind he disregarded the corporate entity and treated the property as his own." When he grew weary of the "burdens placed by law on corporations" in 1917, he dissolved the corporation and reorganized as a "trust" company, with members of his family holding shares of the trust.[1] Even then he had "no thought whatever of restricting his power over his property," and no family member ever challenged his authority. Characterized by one observer as a "feudal baron," by another as a "farmer prince," and by a third as a "benevolent dictator," he understood the power of the larger-than-life image he projected and never corrected misrepresentations that worked to his advantage.[2] For example, although published reports of the acreage he owned ran from forty thousand to as high as sixty-five thousand acres, he actually had title to approximately forty-seven thousand acres. He also operated thousands of acres for family members and out-of-state entities. However much he *owned* outright, he held substantial collateral in land, livestock, business inventory, buildings, and credit accounts, which, together with his forceful personality, enabled him to secure additional operating funds from the elusive northern investors so energetically promoted by New South entrepreneur Henry Grady in the late 1880s. Wilson's innovations, however, occurred during the Progressive Era, and certain developments—and challenges—in that era influenced him to adapt progressive measures to a southern, rural setting.[3]

Models of the kind of organization Wilson developed emerged in the Arkansas and Mississippi deltas in the late nineteenth and early twentieth centuries, but the Delta Pine and Land Company (D&PL) in the state of Mississippi became the only one that rivaled him. Chicago investors launched D&PL in 1886 as a land speculation company involved in selling tracts of land to local and out-of-state buyers. When a consortium of British textile manufacturers purchased the company and its holdings in 1911, they had grand plans to grow long-staple cotton to supply their need for fine cloth, but the land had to be cleared, drained, and readied for cultivation. By that time, Wilson had been cotton farming for more than twenty years. Like other big operators in the delta, he utilized the tenancy and sharecropping system, opened furnishing stores, and operated cotton gins. Long before D&PL planted a row of cotton, Wilson divided his land holdings into four (later fourteen) separate plantations, each with its own farm manager. Delta Pine and Land Company, which came late to plantation agriculture, also adopted the tenancy and sharecropping system and divided its holdings into separate entities and pumped millions of dollars into their venture only to find that long-staple cotton would not grow well in the wet delta landscape. They instead supplied short-staple cotton to mills in Manchester, and though they endured some difficult years during the 1920s agricultural depression, the venture usually paid sufficient dividends to stockholders in England. Delta Pine and Land developed a showplace plantation operating on thirty-eight thousand acres, and by the 1920s attracted visitors from around the country and abroad. Like Wilson and a few other large producers, they employed an industrial model, emphasizing efficiency and productivity. Apart from the fact that Wilson owned greater acreage, he differed in at least one important respect from D&PL. His operation bore the stamp of his large and forceful personality, and he remained intensely involved in every aspect of his company throughout his life.[4]

Although Wilson enjoyed almost unquestioned authority to operate his company on the ground in Arkansas, he faced challenges in three key areas: the development of a rational organizational structure for his sprawling enterprise, the difficulty of maintaining sufficient working capital, and the ongoing struggle to secure steady access to the market. In all three areas he faced difficulties arising out of new state or federal progressive regulations. First, federal laws imposed on corporations prompted him to dissolve the entity and adopt the unusual "trust" arrangement. Second, state banking laws exposed him to unwelcome scrutiny and the need to adjust his banking operations. Third, his purchase of an intrastate railroad led to bothersome "inspections" by state officials. His experience with ownership of the Jonesboro, Lake City, and Eastern Railroad (JLC&E) also illustrates his domineering management style and his aggressive pursuit of working capital. Within a few years of purchasing the entity, he survived a confrontation with its old board over a refunding strategy he wanted to employ and prevailed in a three-year strike by railroad workers. He also engaged in a decade-long battle with the St. Louis and San Francisco ("Frisco") Railroad, which ran through the county on its way to Memphis from St. Louis. One story dramatizes his tendency to resort to extreme measures to secure what he wanted. The Frisco's administrators initially refused to schedule a stop by its fast train to St. Louis at the town of Wilson, a matter of great inconvenience to its proprietor. "One morning, as the fast Frisco hove in sight from Memphis, one of Mr. Wilson's [JCL&E] trains 'stalled' across the Frisco lines" and forced the fast train to stop. "A railroad official noted the coincidence of the stalled train with the boarding of the Frisco train by Mr. Wilson himself. But thereafter all trains stopped at Wilson."[5] This possibly apocryphal story represents only a humorous episode in the battle he waged with the Frisco over matters of considerably greater importance.

Before he purchased the JLC&E in 1911 and began his battle with the Frisco, Wilson engaged in another struggle with someone much closer to home. At the beginning of the twentieth century, he still operated the Wilson and Beall Lumber Company in a partnership with his father-in-law, but Wilson began to dream of bigger things. When they started their business in 1886, the two men made good partners for one another, with Wilson, the energetic and younger man, building short-line railroads into the deep woods and aggressively marketing their lumber, and Socrates Beall, the older man, taking charge of the sawmill and keeping the operation running. In early 1904, however, Wilson wanted to take over his father-in-law's interest in the company and sought to buy him out. Beall, who was nearly seventy years old and in failing health, refused to sell and sued when Wilson "dissolved" the partnership. In April 1904, the two men and their local attorneys traveled to Little Rock for a hearing. The court decided the case in Wilson's favor, making it possible for him to strike out on his own.[6]

Despite his dispute with Beall, Wilson remained on cordial terms with his wife's siblings, and, indeed, they formed part of a large extended family that essentially sheltered under Wilson's expanding economic umbrella. Among them was Elizabeth's brother, Daniel, who worked for Wilson as a lumber dealer. Her sister, Mercy, married a prominent planter, O. T. Craig, who had business connections with Wilson. Wilson also took a keen interest in the children of his sister, Victoria Davies: Boaz, Eva, and little Dora. As though to memorialize the family relationship and, at the same time, to establish the economic link, he named not only a daughter but a town after Victoria. He named the small hamlet of Evadale after one of his nieces and the town of Marie after his youngest daughter. A fourth town he named Wilson, and a fifth town he named Armorel (standing for Arkansas, Missouri, and R.E.L.). Wilson and Elizabeth actually raised Dora, and even after she married a prominent planter, she and her husband, John Merrill, sometimes resided in the Wilson household. By the early twentieth century, Wilson provided housing for his extended family members and for all his employees.[7]

In April 1905, shortly after the court allowed him to sever connections to his father-in-law, Wilson incorporated and declared a paid-up capital stock of $1 million, almost certainly an exaggeration. Again, Wilson understood the importance of appearances. The papers filed with the secretary of state's office listed Wilson as president, J. H. Elkins (his niece Eva's husband) as vice president, and Michael J. Blackwell, Wilson's longtime bookkeeper, as secretary and treasurer.[8] According to his attorney, Charles T. Coleman, he individually owned the stock in the company and "conveyed all of his property, including his interests in the various partnerships" to the new corporation. "The corporation took his place in each of the partnerships, itself becoming the partner." His attorney later suggested that the corporation might not have had the "legal power to enter into these partnership agreements" but that "no one ever questioned it."[9]

The "partnerships" referred to by Coleman demonstrated Wilson's strategy for decentralizing his growing organization. Some of these partners had previously worked for him as salesmen, including Oscar M. Hill, who eventually ran the Hill and Wilson store in Marie, Arkansas, and H. S. Portis, who operated the Idaho Grocery Company in Bassett, Arkansas. Both these young men ran their stores at the center of one of Wilson's plantations. The town of Marie, located south of Wilson, served as the focal point of one of his plantation units; the town of Bassett, located southwest of Wilson, became the center of another of his plantations. At Keiser, Arkansas, ten miles northwest of Wilson, he formed a partnership with John P. Keiser, a young man in his early thirties who operated a general merchandise business at the heart of yet another Wilson farm. A fourth partnership was located in north Mississippi County and served Wilson's Armorel operation. Over the years, these plantation "farm" units expanded from four to fourteen, and Wilson found that having mercantile establishments attached to them provided his tenants and sharecroppers with easier access to supplies, but it also allowed Wilson to shift much of the administrative and financial burden of managing his growing plantation empire to his store managers.

Wilson's partners formed a cadre of young men loyal to Wilson, and while they treated their portion of the partnership property as their own, they coordinated their business affairs through the Wilson office. Over the years, as the company expanded, the arrangement evolved in a manner that closely bound the managers to the central office, requiring them to buy their implements, food, feedstuffs, and supplies from the company's head office in Wilson, borrow money and pay interest to a Wilson bank, and pay rent to the company for the premises they operated in. They also bought electric current from the Wilson electric plant, ice from the Wilson ice house, lumber from the Wilson sawmill, and flour from the company gristmill. "At the close of the season," the managers settled "up with all the Wilson concerns" to which they were indebted and paid each a profit. Their books went "to the executive offices at Wilson where a corps of expert accountants" inspected them.[10]

Although he established a decentralized structure and had accountants to inspect the books, Wilson maintained an almost dictatorial interest in his various constituent businesses and encouraged his employees and "partners" to master efficient business practices. In 1918, for example, he launched a Business Science Club, paid a Professor Parrish to preside over bimonthly meetings, and required his employees to attend the sessions along with him, "mastering each lesson as we go along." Professor Parrish taught a course modeled after the techniques of Arthur Sheldon, which emphasized the science of efficient service and profit making, a method that combined the concept of service and profit in a way that appealed to Lee Wilson, the plantation paternalist. Sheldon, who ran a publishing house and a business school in Chicago, was a famous Rotarian who coined the association's motto in 1908, "He profits most who serves best." He believed that certain natural laws governed ethical business practices and the course he sponsored emphasized both efficiency and ethical behavior. Wilson may have become aware of Sheldon's course through an advertisement, such as one published in the *Rotarian* in 1917, a "course of study for busy businessmen" on the "science of business building."[11] In a letter to one of his employees, bank cashier Kelley Cullom, Wilson announced that "your presence will be appreciated by me personally and will show your interest and intention to cooperate with us. I feel a keen personal interest in this work now because I know great personal benefit will be gained by you and this benefit will incidentally redound to the good of our organization, of which you are an integral part. So, I ask that you bring with you lesson book one and a pencil and note book. We are going after this thing now and intend to get all there is in it." Wilson's message to Cullom was clear. He was to attend the sessions with his pencils sharpened. The message also reveals that Lee Wilson, at age fifty-three, was still schooling himself in business techniques, still searching for the best method of insuring his own success and that of his various enterprises.[12]

At the same time that Wilson engaged Professor Parrish to bring the Sheldon method to his employees, he was in the midst of restructuring his company in a much more fundamental way. In February 1917, Wilson dissolved the corporation he launched in 1905.[13] Wilson resented the implications of the Sixteenth Amendment, which authorized a graduated income tax in 1913, and other taxes imposed after the United States entered World War I, and he devised a novel solution to avoid having to pay corporate taxes. After dissolving the corporation, the former stockholders of Lee Wilson & Company signed a quitclaim deed conveying all the property in the company to Lee Wilson.[14] He then began to lay the foundation for creating a trust company. Attorney Charles Coleman's evaluation of the trust reveals that resentment about taxes and restrictions on corporations were not the only motivating factors for the creation of the trust. Coleman put it bluntly: "Mr. Wilson's main purpose in executing the declaration of trust was to guarantee the preservation of the unity of his estate . . . so that it could not be divided and dissipated by his children."

By the time that Wilson began restructuring his company, he had become painfully aware of the inadequacies of at least two of his three children. Victoria, safely married to businessman Frank Wesson, caused him no concern, but his son, R. E. Lee Wilson Jr. (known as Roy), and younger daughter, Marie, inspired little confidence. Marie first triggered the apprehension of her parents when, at the age of fourteen, she eloped with the family chauffeur and gave birth to a son—before a polite time had elapsed. They named the son John G. Nelson Jr., and for a while the couple and their baby resided in the Wilson household in Memphis.[15] Marie filed for divorce in early 1910 and eventually left her son in the care of her parents, who renamed him Joe *Wilson* Nelson, obliterating all but the most obvious trace of the unsuitable marriage and marking the child as a "Wilson." Mississippi County lore has it that Wilson gave the senior Nelson a horse and $5,000 to disappear. Marie began traveling and visiting fashionable friends in New York, Florida, and Paris, France. She gained a reputation for restlessness that troubled and sometimes inconvenienced her family. In the spring of 1915, for example, in leaving the Saint Paul Hotel in Minneapolis, she forgot to remove her jewels from the hotel safe. The hotel management corresponded with the Bank of Wilson, saying they had received "a letter from J. G. Harmon from the Hotel Jefferson, St. Louis, enclosing the key to the safety deposit box and directing us to express the package to Miss Wilson at Wilson, Arkansas, and to place a value of $3,000 [or $63,723 in 2009 dollars] thereon." The Saint Paul Hotel charged the Bank of Wilson for shipping and insuring the jewels. Marie eventually married—and later divorced—John Harmon, her traveling companion. The episode provides a glimpse into the lifestyle Marie led.[16]

As anxious as Marie made her parents, Roy's behavior had more serious consequences. Lee Wilson saw himself building a legacy to pass along to his children but particularly to his son, who he imagined would lead the company into even greater things. Roy, who graduated Yale in 1912, came back to Wilson to work in the family business, occupying a desk directly facing his father's desk in the Wilson office, but he had already showing signs of the afflictions that haunted him his entire life. He had a serious drinking problem and suffered from debilitating bouts of malaria. He served as vice president and general manager, but he cared more for hunting and playing polo, a sport he acquired a taste for at Yale. He styled himself the country gentlemen, and his frequent absences caused his father to rely on other employees, particularly his principal farm manager, Jim Crain. Roy's most noteworthy "accomplishment" was his marriage to Natalie Armstrong, a descendant of the Welsh socialist Robert Owen, who in 1825 founded New Harmony, a utopian community in Indiana. Natalie eventually inherited part of her great-grandfather's books, books which found a home in Roy Wilson's library in Wilson in the grand mansion he built with his father's money. Styled after the Tudor mansions Roy and Natalie visited on an extended vacation in England in 1922–23, it represented everything the old socialist/utopian hated most. Lee Wilson himself had little more use for such extravagance. Once asked why he continued to reside in his handsome but more modest home rather than build a mansion like that his son occupied, Lee Wilson responded, "because my son has a rich man for a father and I don't."[17]

Whatever inspired Lee Wilson to dissolve the corporation—his dissolute children or the burdensome corporate taxes and regulations—he intended to have "the power to use and dispose" of all of his property "just as if it had not been converted into a trust estate." The trust document made clear that "the legal title to said property is vested in me absolutely, with full and absolute power to manage, control, sell, assign, pledge, hypothecate, mortgage, convey and otherwise dispose of the same, and to execute any and all instruments necessary for that purpose, including assignments, bills of sale, mortgages, deeds of trust, bonds of title, deeds and conveyances, as fully and completely as if I were the sole and absolute owner of the legal and equitable title thereto." Instead of stockholders, his family members held certificates of interest, but the certificates gave them no authority to interfere with Wilson's management of the business. They received a share of the profits, and Wilson himself decided what share of the profits they deserved. Each of the five certificate-of-interest holders had two thousand shares: Lee Wilson, Elizabeth Beall Wilson, Roy Wilson, Victoria Wilson Wesson, and Marie Wilson. The only possible power that the holders of certificates could exercise was that of electing the trustee. It is hard to imagine that any of them would have launched an attempt to unseat Lee Wilson as trustee, and apparently no such attempt was ever made.

As Wilson began imposing a systematic organizational structure on his sprawling plantation, he developed a multifaceted approach to financing his operation. Even here he relied on a decentralized system. His managers, for example, secured their own credit accounts with wholesalers and extended credit to their customers in the lean months between harvests. They negotiated advances with cotton factors, such as Wilson-Ward and Company in Memphis, another Lee Wilson business he founded with John M. Ward in 1897.[18] The company advertised as cotton factors and commission merchants, but like other factors, it secured its own funds from selling the staple, and its lean months coincided with those of the cotton growers themselves. It was in this context that Wilson devised a strategy that demonstrated both his genius for innovation and his penchant for taking chances.

In late 1908 Wilson opened his own bank, an enterprise that provided a new world of capital opportunities to him and became an integral part of his overall operation. Together with his plantations, mercantile establishments, cotton gins, lumber businesses, and cotton factorage, the bank made it possible for him to leverage credit and accumulate capital. The Bank of Wilson developed formal relationships with banks in Memphis, St. Louis, Chicago, and New York City—called "correspondent banks" in banking parlance—and this allowed it to circulate its checks and provided Lee Wilson with introductions to an expanding number of potential lenders.[19] Always operating close to the margin, Wilson would find himself in great need of extensive credit in the coming decades, and while the Bank of Wilson served a number of prominent planters, small farmers, and ordinary people in Mississippi County, it functioned principally as Lee Wilson & Company's guarantor and as Lee Wilson's personal piggy bank. As a guarantor, the bank routinely provided credit references for various Wilson enterprises to wholesalers unaware of the bank's connection to Wilson. In 1919, for example, the First National Bank of St. Louis asked the bank about the creditworthiness of Hill and Wilson in Evadale and received a very positive response from the bank cashier, who refrained from acknowledging his operation's connection to Hill and Wilson.[20]

Wilson's founding of the Bank of Wilson occurred just when the question of inadequate farm credit facing farmers nationwide began to concern federal officials. Groups of farm representatives began visiting Europe to study how countries there addressed the problem of funding agricultural enterprises, and even William Howard Taft sponsored one such commission in the last year of his presidency. The Farm Credit Act of 1916 was one result of this period of intense study, an act that created twelve Farm Loan Banks across the country, including one in St. Louis. This legislation made it possible for farmers to form cooperatives—made up of local farmers, businessmen, and planters—and secure federal funds to provide loans of up to 50 percent of the value of their farms and improvements. Lee Wilson became a director in one such cooperative and accessed this avenue of credit.[21]

Wilson's struggle to secure adequate capital to maintain and expand his operations has to be understood in terms of the historical context of capital and credit in the United States. The South had suffered under serious handicaps in the period after the Civil War, in part because of disadvantageous legislation passed in 1863 establishing a national banking system. Southern representatives who would have blocked such legislation had resigned their congressional seats two years earlier in order to join the Confederate States of America, and the 1863 act, together with acts that refined and strengthened it in 1864 and 1865, naturally favored the North and the West over the South. The need to rebuild the South after the Civil War absorbed available capital, and Arkansas's problems were compounded by its struggle with a prewar debt caused by the collapse of its banks in the early 1840s. Expenditures during Reconstruction to fund education, expand the railroads, and rebuild the levee system further complicated the situation. These expenditures were predicated on a projected growth of the economy that never materialized. The agricultural economy failed to recover in the postwar period, and Arkansas's New South advocates found it difficult to accumulate enough capital to expand economic opportunities. State banking regulations were weak and largely ignored, and thus banks were often chartered with too little capital to guarantee their success. In any case, with no federal deposit banks in the South, it was difficult to insure the adequate circulation of what money there was available. The Bank of Wilson was a product of these conditions and reflected the aspirations of entrepreneurs like Lee Wilson. While it suffered from all the weaknesses inherent in an essentially unregulated banking system, it provided Wilson with a crucial advantage in the quest for capital opportunities.[22]

Wilson began to organize the bank in the summer of 1908 and hired William Jett, a southerner living in Haner, Pennsylvania, to work with certain key Wilson executives in soliciting the interest of potential shareholders. Jett was happy to have the chance to return to the South and anticipated taking over as the cashier, an executive position at the time. To sell shares of bank stock and raise the funds necessary to open for business, he worked with Michael J. Blackwell and Wright H. Smith, both of whom were executives of Lee Wilson & Company. They planned to sell a total of four thousand shares at $6.25 per share, raising $25,000 in capital, but when the bank opened, the fifty-four initial subscribers had purchased only 971 shares, raising only $6,312.50. Nevertheless, they advertised a capital stock of $25,000. Lee Wilson & Company was the single largest shareholder, purchasing 160 shares for $1,000, and Lee Wilson himself owned forty shares. Those two hundred shares represented nearly 20 percent of the shares sold, and Wilson employees subscribed to another 16 percent.[23] Together, then, Wilson and his employees commanded 36 percent of the bank shares. A variety of individuals, such as planters and professional men located in south Mississippi County, many of whom had a personal or business connection with Wilson, owned the remaining shares. Joseph Rhodes, a merchant from Golden Lake, and Wilson's oldest friend in Mississippi County, took shares in the bank and became its first vice president.[24] The remaining shareholders were an assortment of planters, farmers, and clerks, and the only interesting departure from this pattern was Townsend Davis, an African American farmworker in Golden Lake Township who purchased four shares of stock for $25, a substantial sum for a man of his situation.[25]

Aside from holding over a third of the shares in the bank, Wilson and his employees dominated the board of directors, taking up six of the ten positions.[26] Wilson also exercised influence in less direct ways. After months of preparation, the Bank of Wilson opened its doors on October 7, 1908, in the same complex of offices as Lee Wilson & Company, the Wilson Drug Store, and the post office. So located, it was well positioned to attract customers, but it was also subject to close observation by Wilson executives. Wilson employee Michael Blackwell served as the bank's first president, Joseph Rhodes, a former county clerk, became vice president, and William Jett took charge as cashier. Blackwell and Rhodes had little to do with the day-to-day activities of the bank, however, and Jett assumed a variety of responsibilities, everything from organizing the books to establishing the all-important relationships with correspondent banks. Yet he was not his own man at the bank. Not only did he have Blackwell and other Wilson executives there in town to account to, he submitted requisitions for stationery, office supplies, and equipment to Wright Smith, a director of the bank who worked for Lee Wilson & Company, Memphis, a company Wilson founded dedicated to marketing lumber products. Smith ordered and paid for the items and became more involved in the day-to-day running of the bank than any other Wilson employee. Soon Smith and Jett became embroiled in dispute over the accounting system Jett had put into place, and Jett complained to a friend that he had returned South to take the job organizing the bank only to find that "the whole crowd had got together and tried to freeze me out." On November 30, 1908, seven weeks after the bank opened, he submitted his resignation, effective December 31, 1908. It was a rocky start for the young bank.[27]

Over the next few years the Bank of Wilson went through a series of cashiers, young men in their twenties, some of whom were on their way to better opportunities elsewhere.[28] One thing that remained constant, however, was Lee Wilson's hand in the affairs of the bank. Trusted employees and friends served as presidents and vice presidents, while Wilson himself remained an omnipresent member of the board.[29] From the beginning the shareholders, including Wilson and Lee Wilson & Company, borrowed "liberally" from the bank, which provided them with the necessary funds to carry on their business affairs but also gave the bank sufficient security to "sell" the loans to other banks.[30]

There was nothing illegal in what the Bank of Wilson was doing in providing a source of credit and capital for Wilson, his businesses, and his friends, but when the state of Arkansas passed a new banking law on March 3, 1913, the newly appointed commissioner took issue with the extent to which Wilson used the bank's credit. "An Act for the Organization and Control of Banks, Trust Companies and Savings Banks" passed the legislature during the heyday of progressive reform in Arkansas, and legislators intended to subject the state's banks to regulation, insure their solvency, and maintain the public's confidence in them.[31] In early 1914 the Bank of Wilson filed a "declaration of intention to continue business as a bank under the provisions of the law," a declaration required under section four of the act. A few months later inspectors fanned out across the state to "exam" banks, and the Bank of Wilson hosted one such inspector on May 22, 1914. A week later John M. Davis, the state bank commissioner, dispatched a letter to the bank's directors suggesting a number of "corrections concerning the affairs of your bank." He listed ten matters that required attention, three of them addressing the use Wilson and the other shareholders were making of the bank's credit. "Your active Officers, Employees and Directors appear to be quite liberal borrowers from your bank," wrote the commissioner, who suggested they should be required to give good security for all their loans. He cautioned that "many banks get into serious trouble by loaning their funds in excessive amounts to their active Officers and Directors." But Commissioner Davis aimed his harshest words at Lee Wilson. "You have one loan exceeding the limit . . . that of R. E. Lee Wilson . . . caused by Mr. Wilson endorsing various other notes. This should be reduced immediately and Mr. Wilson should not be permitted under any circumstances to increase his indebtedness to the bank, but on the other hand should be required to reduce it." As if that criticism was insufficient to make the point, the bank commissioner, in a separately numbered item, elaborated further: "From the report of the examiner, we further find that R. E. Lee Wilson owes directly, $8,000.00 with no security, and that he is endorser upon notes of various amounts, many of which he is personally interested in the sum of $76,000.00. This is indeed a very bad showing for your bank, and should be corrected." A scribbled note in the margin beside this entry in the bank commissioner's letter says "ask RELW to comply."[32] The bank cashier, E. A. Mitchell, probably had the unenviable task of speaking to Wilson about the matter.

It is not difficult to imagine why the bank commissioner was concerned at the amount Wilson owed or was "personally interested in." According to the "State of the Condition of the Bank of Wilson," filed with the bank commissioner on September 30, 1914, the bank held a total of $100,335.32 in "loans and discounts,"[33] and Wilson's share of that amount, $84,000, meant that he had received or was "interested in" 84 percent of the bank's credit, identical to the percentage he had received or was "interested in" in 1912. If Commissioner Davis had learned of Wilson's economic circumstances in 1914, his concern might have turned to alarm. The floods of 1912 and 1913 had nearly ruined Wilson, and had it not been for a timely loan from a St. Louis financier, his house of cards might have come tumbling down. The Bank of Wilson depended upon Wilson for its own solvency, and his was a shaky foundation upon which to build an edifice. Yet Wilson made an effort to comply with the law and escaped the criticism of the bank commissioner after audits in 1915 and 1916. The records for 1917 are not available, but in 1918 Wilson once again came under fire. Commissioner Davis wrote to the bank's board of directors in September 1918, "We note that your directors personally and enterprises in which they are directly interested have borrowed about $80,000 of the $143,000 loans in your bank. This is certainly a pretty good showing for your inside officers."[34] That represented 55.9 percent of the bank's loans, and the acceptable percentage, according to Davis, was no more than 30 percent.

The loans Wilson had through the Bank of Wilson told only part of the story of the debt he was carrying. At the end of 1913, even as Wilson was reeling from the second flood to strike in two years, he was called upon to pay the obligations of a bankrupt corporation which he had personally guaranteed. His involvement with the National Car Advertising Company dated back to 1907 when, together with his new son-in-law, Frank H. Wesson, he purchased stock in the company, a Chicago-based corporation "engaged in the business of street car advertising." Precisely how Lee Wilson and Frank Wesson became involved in the National Car Advertising Company remains unknown, but the venture proved to be a disastrous investment. Very little information about the business survives in the historical record, but Wilson and Wesson soon "acquired control" of the company when it began to experience financial difficulties. They both "personally endorsed the company's notes and guaranteed its obligations." They made a gentleman's agreement to share the losses should the company fail. When the company went bankrupt in late 1913, they lost their entire investment and were called upon to pay off the company's debts. Wilson was in no position to assume responsibility for any additional debts at that point in time so "Wesson advanced the funds necessary to make good" on their "endorsements and guarantees." Wesson paid $80,000, half of which Wilson was responsible for according to their gentleman's agreement. In 1916 Wilson gave Wesson "a small amount of cash" and a $40,000 note at the rate of 6 percent interest per annum. Over the next three years, Wilson paid no more than the interest on the note, but in 1919 he had his Memphis cotton factorage company, Wilson-Ward and Company, write a check to Wesson for $40,000. The debt was charged back to Wilson on the Lee Wilson & Company books, making Wilson's total debt on those books that year $106,886.94.[35]

Wilson's economic situation during this period would likely have been much worse had it not been for a windfall resulting from the Great War, a windfall that worked to his son-in-law's advantage too. Frank H. Wesson served as an executive with the Smith and Wesson Company, makers of the famous revolver. As the conflict in Europe expanded in 1915 and 1916, the need for armaments increased and Wesson's fortunes grew in tandem. But, more important to Lee Wilson, cotton prices reached heights not seen since immediately after the Civil War. Certain cotton growers across the world became engulfed in war and ceased producing just as demand for cotton products increased, a demand fueled by the need to clothe the armies of the various belligerents. Between 1916 and 1917, the price of cotton rose from 16.36 to 27.09 cents per pound. At five hundred pounds per bale, that meant an increase of from $86.80 to $13.45 per bale, a handsome additional return. The price reached a high in 1919 of $35.34 ($176.70 per bale) but the next year dropped suddenly to prewar levels as other cotton producers came back into production and demand for the staple decreased. Before the collapse in prices, Wilson had been able to recover from some of the losses he sustained prior to the wartime boom, paying off the St. Louis creditor that had rescued him from bankruptcy in 1913. His obligation to his son-in-law was the last one he was able to dispatch before the cotton economy soured once again.

Throughout this period the Bank of Wilson had been a useful ally in Wilson's struggle to keep his economic ship afloat. His various entities borrowed money from the bank, sometimes in ninety-day notes, and when it came time to redeem them, the manager paid the interest and secured another note in the same amount in its place. The bank typically sold the notes for a similar period to a banking institution, usually in St. Louis, Chicago, or New York, paid the interest when it came due, and extended the note further. This was not an uncommon practice in banking circles. Bankers in the South typically looked to the end of the agricultural cycle—September, October, November, and December—to receive payment and were usually willing to wait on the promise of a good crop year. Lee Wilson and the Bank of Wilson had been good customers, and the shell game they were all playing depended upon everyone continuing the game.

Despite the admonitions of the bank commissioner in 1914 and 1918, the Bank of Wilson continued to extend generous loans to Lee Wilson and his various enterprises. However, it avoided further criticisms from the commissioner, perhaps because it increased the amounts loaned to individuals and entities not connected to Wilson.[36] Indeed, between 1914 and 1920, the amount of loans and discounts listed by the bank nearly doubled (from $100,335.32 to $196,281.20) while the bank's total resources increased by two-and-a-half times (from $116,189.38 to $273,631.62). The latter was, in part, a result of an aggressive campaign on the part of the bank to secure deposit accounts from three public entities: the county government, the two largest drainage districts, and the St. Francis Levee District. Wilson helped found and lead a county taxpayers' association, which secured certain reforms, including a requirement that the county government spread its funds throughout the various banks in the county. Wilson's influence with those in charge of the largest drainage districts in the county secured their deposits. The St. Francis Levee Board adopted a policy of depositing its funds in the various banks in its eight-county region and, unsurprisingly, the Bank of Wilson received a share of that entity's deposits. In 1923, the bank increased its capital stock from $25,000 to $50,000. Whatever liberties Wilson was taking with the bank, they appear to have escaped the attention of the bank commissioner in the 1920s, and they did little damage to the bank's economic condition.[37]

The Bank of Wilson remained viable even as other banks in the state and around the country began to collapse in the twenties and the early thirties. Its success where others failed speaks to Wilson's ability to manipulate the assets at his command, but it also owed something to careful management practices. The directors and the bank's cashier watched deposits and aggressively pursued payments on notes and liquidation of overdrafts, even when it meant harassing people of influence. For example, S. E. Simonson, a close friend of Wilson's and director of one of the largest drainage districts in Mississippi County, owed a $5,000 unsecured obligation that he could not pay when it came due on January 1, 1920. The bank's cashier, Kelley Cullom, first kindly reminded him of the obligation, and when Simonson asked to be allowed to simply renew the note for the full amount, folding the interest owed into the new note, Cullom responded that "on account of a ruling made by the Board of Directors" he would have to insist upon at least a 10 percent payment plus interest due. Simonson and the bank subsequently played a game of "cat and mouse" over the next two years, with Simonson reminding them of favors he had done the bank, and the bank needling him relentlessly for payment. As director of the Grassy Lake and Tyronza Drainage District No. 9, Simonson had placed at least $10,000 of the district's money on deposit in the Bank of Wilson. In a transaction brokered by Wilson himself, the money was held as two $5,000 certificates of deposit at a 5 percent interest rate.[38] When the Bank of Wilson began to pressure Simonson to repay his own personal note in January and February 1921, Simonson reminded the cashier "I think you have not less than $10,000.00 of funds principally thru my assistance at rate of 5 percent and this . . . is contributing to your advantage in time of need."[39] Simonson's reminder worked. The bank allowed him to recycle the debt for ninety days, and when Cullom pressured him once that period had elapsed to pay the loan, Simonson responded, directing his letter to Wilson: "You will recall also Mr. Wilson that this loan of mine was a kind of reciprocal arrangement and I think that all things considered you are not being burdened and at the same time are being profited. It would not be convenient for me to pay this note now."[40] Cullom responded that the "directors insist on maturing obligations being taken care of promptly," and, almost certainly under Wilson's command, he went on to say "if you wish you may call the Certificate of Deposit of $5,000.00 payable to the Grassy Lake & Tyronza Drainage District No. 9 referred to in your letter and take up your note to this bank."[41] In other words, Wilson called his bluff. The next month Simonson paid the bank 5 percent on the principal and all of the accrued interest, but by the end of 1922, the bank was still carrying his $4,500 note.[42] Aside from the small payment he had made and an increase in the interest rate to 9 percent, Simonson also eventually provided security in the form of a life insurance policy to secure the debt.[43]

One of the reasons that Wilson and the bank played this game of cat and mouse with Simonson was simply that the bank commissioner frowned on the practice of recycling debt without payment of interest owing and at least some payment on the principal. Unsecured loans were particularly problematic, and Simonson's willingness to sign over his insurance policy to the bank was an important concession. A more pressing motivation, however, involved the agricultural depression following World War I, a depression that struck the Wilson enterprise with considerable force.

Wilson's response to the disastrous drop in cotton prices in early 1920 was vintage Wilson. He started another company: a cotton factorage called Lee Wilson & Company, Blytheville (Wilson-Blytheville). This might have followed naturally from his association with the Sheldon course of study. Having an office on the ground in Mississippi County, close to the source of production, spoke to the issue of efficiency. Wilson had moved his family back to Wilson from Memphis, Tennessee, and wanted to exercise close supervision over those charged with marketing his cotton. In fact, Wilson likely thought he had little choice. The cotton factor industry itself was in crisis, with the number of cotton factors operating in eastern Arkansas dropping from sixty-four to two.[44] He hired two experienced cotton brokers, Arthur J. Haaga and Sam L. Thomas, likely thrown out of work by the bankruptcy of the firms for which they worked, to comb the country in search of cotton factors and textile mills willing to purchase Wilson cotton.[45] Wilson-Ward and Company, Wilson's cotton factorage business located in Memphis, continued to do business, but it had been hit hard by the economic crisis. The Wilson-Blytheville office secured contracts with them as well as with other cotton factors with offices in Memphis, marketing bales with Anderson-Clayton and Company, a cotton factorage with a branch in Memphis and main offices in New York and New Orleans. Wilson-Blytheville also marketed through Beane Brothers of New Orleans and Leathers and Matthews and Company of Spartanburg, South Carolina, and they sold cotton directly to textile mills in both New England and the South.[46]

Cotton factors not only served the purpose of securing buyers for the staple; they also provided credit to planters while they awaited the sale of the cotton. The collapse of the cotton market after World War I seriously compromised the ability of the factors that remained in business to advance funds. Under these circumstances, Wilson had to be more aggressive in his pursuit of credit, and his St. Louis connections once again served him well. Wilson-Blytheville entered into agreements with the Liberty Central Trust Company and the National Bank of Commerce, both of St. Louis, to advance funds on cotton held in a warehouse in Blytheville. The agreement that Wilson himself signed with the Liberty Central Trust Company in September 1921 carefully spelled out the terms.[47] The company would extend loans at the rate of $60 per bale, an amount considered "safe" by the lender since cotton, if it could be sold at all, was selling for about 15 cents per pound, equaling $75 per bale of cotton. Cotton held in the Blytheville warehouse would stand as collateral for the loan, and warehouse receipts identifying the cotton by bale numbers and grade were to be placed on deposit in a local bank. By allowing the Farmers Bank and Trust Company of Blytheville to serve as custodian of the warehouse receipts, the Liberty Central was accommodating Wilson. As Haaga and Thomas scoured the country looking for buyers, the cotton market fell flat, and then in early 1922, the cotton textile industry was hit by a strike.[48] Wilson was able to give his creditors something toward the balance owing, but he had to convince them to allow him to pay the interest and carry most of the principal forward.[49] The patience of his creditors was rewarded in the end. On May 9, 1922, Arthur Haaga of the Wilson-Blytheville office sent a check for $5,000 to St. Louis, a check which "paid off the last of the obligations" due Liberty Central. Over the course of the previous eight months, that office had borrowed a total of $62,000 from the Liberty Central, but the sums owed by Wilson's Blytheville office were not the only debts he incurred.

The Wilson-Blytheville office also secured a line of credit from the National Bank of Commerce in the fall of 1921, ultimately borrowing $26,500 from them, again using cotton warehouse receipts as collateral. The relationship between them was not quite as cordial as that between Wilson and the Liberty Central. A problem emerged which signaled concern over the "grade" of cotton represented by the warehouse receipts. "You will recall that it was understood at the time that we made the advance against this cotton that we were to permit you to make any necessary substitution in the cotton receipts, provided you substituted other receipts . . . of equal or better grade than that withdrawn." When Wilson-Blytheville failed to comply with this requirement by substituting class "A" with class "B" cotton receipts on September 28, 1921, the bank objected.[50] Wilson and the National Bank of Commerce maintained a running battle of words and wits over the next month, with Wilson-Blytheville ultimately conceding the point in order to secure an advance on its line of credit.[51] The bank held up delivery of funds in late October, releasing them only after receiving a telegram from the custodian bank in Blytheville that Grade A collateral had been received, reiterating, "we do not want to advance the money on other than middling or better grade of cotton."[52] The quibble was of no small moment to the bank. In late October more cotton was coming on to the market, driving down the price. The difference between what they could expect to receive and what they had loaned per bale was shrinking. Nevertheless, Lee Wilson & Company proved able to pay off its indebtedness to them by June of 1922.[53]

Wilson took advantage of every avenue of credit available, but his continuing success depended upon a recovery of cotton prices. A marginal recovery occurred in the mid-1920s, but many farmers and planters went bankrupt before that came about.[54] Wilson's substantial assets which he could leverage for credit and his willingness to engage in aggressive marketing saved him from a similar fate, but a transaction in the spring of 1922 illustrates the extent to which Wilson himself was feeling the pinch and likely accounts for his ability to satisfy the loans he had outstanding with both the Liberty Central and the National Bank of Commerce. On April 1, 1922, he contracted with William R. Compton Company to market $500,000 ($6,385.029 in 2009 dollars) worth of Lee Wilson & Company bonds. The American Trust Company, which actually served as trustee of the bonds and which probably held most of them itself, extended the funds to Wilson, who pledged the town of Wilson and all its buildings as collateral. If Wilson defaulted on his payments, American Trust could seize the collateral Wilson pledged. It is doubtful that the town and its buildings were worth $500,000, but American Trust Company officials certainly understood that the town was the heart and soul of the enterprise and that Wilson would devote his considerable energy and business acumen to preserving it. They also had reason to believe in Wilson. One of the officers of the American Trust Company, Thomas Dysart, was also connected to the Liberty Central Trust Company, a connection that had only enhanced Wilson's reputation as a businessman worth taking a risk on. Dysart also wanted to insure that Wilson could repay the debts he owed to the Liberty Central.[55]

Wilson's efforts to secure sufficient credit and working capital required considerable attention and innovation. His imposition of a rational organizational structure over his sprawling enterprise necessitated the same kind of focus and experimentation. Aside from funding and managing his company, Lee Wilson became preoccupied with the necessity to secure continuing access to the market, particularly in an era when a new mode of transportation emerged for Mississippi County planters and farmers. Lee Wilson's acquisition, manner of operating, and sale of the Jonesboro, Lake City, and Eastern Railroad speaks to his own particular brand of entrepreneurship. He had been building short-line railroads into the Mississippi County's hinterlands since the 1880s for the sole purpose of transporting the valuable timber he was harvesting to market and thus realizing a profit. In the earliest years he laid tracks from the point of harvest to his first mill on Idaho Landing (Golden Lake) at the Mississippi River's edge. Once the lumber was milled, he barged it to Memphis and marketed it through Lee Wilson & Company, Memphis. Laying track and running locomotives into the swamps had always been a risky proposition, but given the right circumstances, a very lucrative one. The varied and unpredictable condition of the soil, which made farming so uncertain and drainage so problematic, also made maintaining a rail line a formidable challenge. The swamps of northeastern Arkansas delayed railroad construction into the area by a few decades, with the major lines initially choosing to avoid the treacherous terrain despite the promise of profits to be had by having access to the timber and cotton there. The outcome of a battle between railroad magnates Jay Gould and James Paramour inadvertently set up the conditions which brought the Jonesboro, Lake City, and Eastern Railroad into Mississippi County.[56]

While Jay Gould built railroads into Arkansas in order to exploit the state's timber, James Paramour was at least partially motivated by the desire to increase the supply of cotton to his compress in St. Louis. In the late nineteenth century Gould entered into an agreement with Paramour, with the latter agreeing to purchase a Texas line and begin extending it north in order to connect with Gould's St. Louis and Iron Mountain (known as the Iron Mountain). But Gould reneged on the agreement, apparently hoping to force Paramour to sell. Paramour, angry and bitter over what he considered a betrayal, refused to sell and determined to carry his Texas line through Arkansas to St. Louis. Ultimately, both Gould's Iron Mountain and Paramour's St. Louis and Southwestern Railroad (the "Cotton Belt") angled through Little Rock on their way from Missouri to Texas.[57] A third railroad, the Kansas City, Fort Scott, and Memphis, which was later acquired by the St. Louis-San Francisco (the "Frisco"), also established a route through Arkansas, but its trajectory was south and east toward Memphis.[58] The Frisco and a branch of Gould's Iron Mountain intersected three miles to the east of Jonesboro, where the little settlement of Nettleton sprang up, right on the edge of the Craighead County swamps. Paramour's Cotton Belt, meanwhile, ran parallel and a few miles east of the Iron Mountain, passing through Blytheville in northern Mississippi County.

The intersection of the Frisco and Iron Mountain railroads at Nettleton was attractive to lumbermen, and in 1892 the F. Kiech Manufacturing Company began operating there, accessing the wealth of timber located immediately to the southeast in the sunk lands. Although most of the sunk lands were located in Poinsett and Mississippi counties, some few thousand acres lay in southeast Craighead County. Over the next few years Kiech sent timber crews into these swamps, and as they moved further and further into the sunk lands, he needed a more reliable method of moving the timber to his mill. Lumbermen working the swamps typically used "broad-wheeled wagons pulled by oxen over the muddy trails of the bottoms," but Kiech began to lay plans to build a railroad. He contacted "prominent men in Jonesboro who were also interested in opening the [south]eastern part of the county for trade and exploitation" and in 1897 a group of investors formed the Jonesboro, Lake City, and Eastern Railroad. The first track was laid from Nettleton east to Lake City (close to the Mississippi County line), completed in October 1898, and almost immediately brought prosperity to the latter, and by August 1899, a three-mile extension from Nettleton to Jonesboro was completed, linking the Craighead County seat to the sunk lands.[59]

The desire to connect to Paramour's Cotton Belt at Blytheville, forming an important interchange, provided the incentive to move further into Mississippi County.[60] The JLC&E began extending into northern Mississippi County in the summer of 1900, passing through Leachville and then on to Manila in the fall. It was early spring of 1901 before construction crews began to lay track toward Big Lake, but once it was reached, difficulties associated with building a bridge to span the lake slowed progress, and it was the summer of 1902 before the JCL&E reached Blytheville. The railroad's directors then formed another company, the Chickasawba Railroad, largely to avoid distance tariffs imposed by the Arkansas Railroad Commission. After extending its line to Blytheville, it had fifty-five miles of main line, which still permitted it to pose as a short-line railroad and charge a higher rate. The new Chickasawba line stretched east from Blytheville through Wilson's Armorel and then to Barfield Point on the Mississippi River. In 1905 the two roads merged and rates were reduced to correspond to the Arkansas Railroad Commission's rate schedule.

The year 1905 was important for another reason. Two Mississippi County brothers, John B. and William J. Driver, purchased the JLC&E from the Jonesboro promoters. Their father, John B. Driver Sr., had only recently died, but he had been prominent in promoting the construction of railroads and the drainage of the swamps. His sons shared his enthusiasm for both, as did Lee Wilson, but they were sometimes at odds with Wilson. It was John B. Driver Jr., who had challenged Wilson over assessments proposed by the creation of Drainage District 8 in 1908, for example. The Driver brothers would run the JLC&E profitably for five years before selling it to Lee Wilson in late 1910.

By the time he purchased the JLC&E, Wilson had been railroading for more than two decades. He had run various short-line railroads, designed specifically to reach timber he wanted to harvest, but most of them were crudely constructed and were not identified as anything more than tram roads. At least three of them, however, provided both freight and passenger service, the Wilson and Beall, the Quinine, and the Wilson Northern. The Wilson Northern, founded in 1905, was the most important of these and stretched from Wilson north and west to Keiser, another of his towns. Unlike the JLC&E, the Wilson Northern had never been a profitable enterprise, at least not on paper. Of course, Wilson's goal was likely not so much to run a profitable railroad as to access the timber, and that was where he made his money. After acquiring the JLC&E in 1910, Wilson "sold" the Wilson Northern Railroad to the JLC&E in 1912, and soon the latter was running at a deficit. But the purchase of the unprofitable smaller railroad was not the only reason for the poor showing of the JLC&E. Soon after purchasing the JLC&E, Wilson spent $30,000 on a new depot in Jonesboro and upgraded the track there. He also purchased more modern locomotives and new rolling stock. "The JLC&E had begun operations with antique locomotives, cheap equipment, and with generally as little outlay of expenditures as possible. As a result their locomotive and rolling stock were in a dilapidated condition, resulting in frequent breakdowns and costly delays."[61] Wilson changed all that with the purchase of several pieces of equipment. He also spent considerable sums on improving the roadbed, particularly along the Wilson Northern. Initially there had been "no effort . . .

made to build a subgrade or to provide ballast." As a result, a heavily laden log train could cause the track to "sink into the mud and water, and the train crews would sometimes travel for several miles without ever seeing the rails." A joke circulated among the people of lower Mississippi County that you paid for the ticket and got the "rocking free."[62]

Many of the improvements Wilson made to the roadbed and tracks of the JLC&E and the Wilson Northern were literally washed away in the 1912 flood, and even as he was rebuilding, the 1913 flood dealt the railroad another blow. By this time neither the JLC&E nor the Wilson Northern were showing a profit, and the shareholders were beginning to grow weary of the roads' poor economic performance. The first signs of serious disaffection revealed themselves at a special meeting of the board of directors of the JLC&E, called in June 1913, specifically to accept the resignation of Wright Smith and to designate his replacement. Smith had been serving as second vice president and general manager since 1911, and before that he had worked for Wilson as vice president of Lee Wilson & Company, Memphis. It had been Smith who had criticized Jett's performance at the Bank of Wilson in 1908, and now Smith found himself the object of criticism. In a long letter read at the June 9 meeting, Smith defended his actions as general manager of the JLC&E, indicating that he had accepted the position of general manager two years earlier with the understanding that the JLC&E's debt was to be refinanced in order to fund a number of significant improvements to the road. But the debt was not refinanced as he had been promised, and he argued that he had done the best he could under difficult circumstances, using "the earnings of the road" to make a number of improvements. He pointed out, however, that although the earnings had increased from "an average of $14,750.00 per month in 1911 to an average of $23,400.00 per month" in 1913, they were insufficient to meet the needs of the road. Although Wilson had indicated to him "that my services had been entirely satisfactory to all of the stock holders," Smith felt compelled to resign his position after Wilson told him that there was "no hope in the immediate future of any re-financing plan being put through." A subsequent resolution on the part of the shareholders lauding Smith's performance was probably the routine blandishment that typically accompanied the removal of top management.[63]

John Burns replaced Smith as general manager but within a year was on the hot seat himself. Burns's dismissal came as the culmination of a power struggle among the shareholders of the JLC&E, a struggle Wilson eventually won. It all began when Wilson called a stockholders' meeting in early 1914 to approve $1,800,000 in bonds in order to provide the funds necessary to finally finance the improvements that former general manager Smith had been promised three years earlier. "Of that sum, $1,000,000 was earmarked for road improvements and $724,000 to pay off the debts then owed by the company." To Wilson's shock, some stockholders not present at the meeting later balked at the proposition and actually filed suit to "block the approval of the bond issue, and asked that a temporary receiver be appointed to manage the road and investigate its financial structure." They had no faith in the management of the railroad, particularly that of John Burns. Wilson responded to the criticism of the stockholders by forcing Burns to resign and firing "the entire management staff." Wilson replaced Burns with L. C. Gaty, one of the original stockholders in the Bank of Wilson and "a former manager of the Wilson Northern," as the new manager of the JLC&E. In fact, Gaty was very much the profile of the Wilson executive. He was thirty years old and an associate "of Wilson's in a number of enterprises." Wilson also began to buy up shares of JLC&E stock belonging to some of the small stockholders, determined never again to have his will questioned by those with interests enough in what he considered his own enterprise to confront him.[64]

Having prevailed over the revolt of the shareholders, Wilson next faced a challenge from quite a different front. Discontent among railroad employees in Jonesboro had been brewing since early 1914; in 1915 they went out on strike, a strike that lasted for three years. Wilson refused to accede to the demands of the union, however, and hired other workers to take the place of those on strike. By January 1917 the strike had petered out and the prosperity resulting from World War I enabled Wilson to reward his "loyal" workers with 10 percent raises.[65] The JLC&E emerged from these problems—the floods, the stockholders revolt, the strike—in fairly good shape. Although it rarely showed anything resembling a profit, its revenues grew and the number of passengers and the amount of freight it hauled increased. With the price of cotton rising to new heights in 1916, Wilson determined to demonstrate the agricultural promise of his region by sending a thirty-eight-car trainload of cotton to market. Because the JLC&E was oriented more toward transporting lumber, however, he had too few boxcars to carry the load. Due to the wartime emergency, the Interstate Commerce Commission released the JLC&E from having to comply with a requirement that cotton be shipped in boxcars. The thirty-eight flat cars carried an average of sixty bales each (thus 2,280 bales), worth $197,904 at the 1916 price.

Transporting timber, however, remained a critically important function of the JLC&E, and in 1917, Wilson renewed efforts to harvest and market timber. An appeal to the Arkansas Railroad Commission in 1918 to allow him to end passenger service from Luxora to Osceola, billed in part as a cost-cutting service, resulted in an adverse decision. In fact, Wilson's motivations were two-fold. He argued that the service between the two towns was unprofitable, that the railroad "had sold only one passenger fare out of Osceola in the year 1917" and he wanted to tear up the tracks on that line and use them in extending the JLC&E to the Mississippi River from the town of Wilson, an extension which would "run through a stretch of virgin timber" he owned and planned to harvest. His assertion that he had sold only one passenger fare was almost certainly disingenuous, as Osceola citizens opposed the cessation of passenger service, characterizing it as a blow to the town's economy and to the country folk who depended on the train for transportation. When the Arkansas Railroad Commission ruled against him, he simply ignored the decision, assuming that no one in Mississippi County would dare challenge him. "Gathering a large gang of Negroes from his sawmills and plantations, together with wagons and mules, he set out early one night to eliminate the Osceola branch. By morning the entire rail had been pulled up, loaded on wagons, and hauled away. The branch was effectively abandoned. . . . The rail was later used to build the branch at Wilson . . . to the Mississippi River." The Arkansas Railroad Commission took no notice, and Wilson suffered no consequences from his act of defiance.[66]

Wartime conditions resulted in unique challenges and opportunities, but the JLC&E's deficit continued to grow, largely because of the need for repairs to the tracks and the purchase of new equipment. Wilson bought two engines in 1916 from the Baldwin Locomotive Works, for example, and ordered two more in 1918. With the "spot price of cotton" down "to a low of 6 cents per pound" by September 1920, a recession in the construction industry resulting in decreased demand for lumber, and the falling off of both freight and passenger service, Lee Wilson and the JLC&E faced several years of declining revenues.[67] The prices for cotton and lumber would recover to some extent in the mid-1920s, but the coming of the automobile doomed passenger train service. Automobiles were becoming far more affordable and roads more passable in northeast Arkansas. "By 1924 a Ford touring car could be purchased for "$295, f.o.b. Detroit" and the Gregory Bus line provided cheaper and more frequent service between the area's towns.[68] Wilson may well have seen the handwriting on the wall. Certainly he made a shrewd and timely decision in 1925 when he sold the JLC&E to the Frisco Railroad. Because of several outstanding financial responsibilities, including a fine against the JLC&E by the Interstate Commerce Commission, the origins of which are unknown, the selling price of $200,000 went into an escrow account, and it is impossible to determine from the sources available precisely how much Wilson himself actually collected, but at the very least he had unloaded an unprofitable and costly enterprise.

The sale of the JLC&E did not end Wilson's involvement in the railroad business. The next several years witnessed a series of battles between Wilson and the Frisco over freight rates. In 1929 Wilson created the Mississippi River Western Railway Company, almost certainly as a strategy on his part to force the Frisco to his way of thinking about freight rates. He appealed to the Interstate Commerce Commission (ICC) to allow him to turn his railroad into an interstate rather than simply an intrastate railroad, probably as another maneuver in his battle with the Frisco. Suddenly, in 1931 he sold the new company to the Frisco, earning himself a place on the Frisco's board of directors.[69] As part of his arrangement with the Frisco, he gained the privilege of running his little railroad over the Frisco lines. Although he had used an appeal to the ICC for an intrastate road, he resented the ICC's interference in his operations. In late 1932 he wrote to L. W. Baldwin of the Missouri Pacific Railroad that the ICC should be done away with and that every railroad should "paddle its own canoe." He complained that the ICC was "composed of broken down politicians and professors" who "don't know any more about railroads than I do about preaching."[70] In order to run his engines over the Frisco line, Wilson had to subject them to ICC inspection, and remarks made by one of Wilson's employees are most revealing of the boss's attitude. The exact nature of the ICC inspector's complaint is not made clear in the correspondence, but inspector H. H. Kane had reported unfavorably on the condition of some of Wilson's engines. Wilson apparently suggested to the employee charged with bringing them up to specifications, W. M. Wallace, that perhaps the matter should be taken up with Kane's superior. Though Wallace himself had bridled under the "indignities" he had endured at the hands of the inspector, he counseled caution, telling the boss that after giving it "considerable thought," he advised against taking the matter over Kane's head, as a full-on inquiry might bring to light "several other defects . . . and a heavy fine." He admitted "I do not mind confessing to you that we have violated the Interstate law in the past, in fact, we never fire up the engine that we do not violate the law. And in as much as inspector Kane has threatened to haul us up in court, this move [going over his head] might provoke him to action, and probably would cost us more than to spend the money complying with the law."[71] The matter went no further, and the engines were repaired, at least enough so that they passed Kane's inspection.

* * *

By the time of Wilson's tirade against the ICC, his serious involvement in running a railroad company was at an end. The era of the railroad had lasted only a very brief time in Mississippi County. The swamps had delayed the laying of the JLC&E's track until the early twentieth century, and the coming of automobile and bus service had undermined it as a means of conveying passengers by the mid-1920s. Wilson's experiences with the JLC&E reflected his management style: his use of capital and credit, his determination to run the road on his own terms, and his willingness to engage in lawsuits to secure his interests. The strike of railroad workers, however, presented him with a unique challenge. A threat from below—from his laborers—was not something he was accustomed to. Like other plantation owners and railroad barons, he operated in a labor-scarce environment and employed several strategies to keep labor in place. Wilson was unique in providing better housing, health care, and schools for both black and white labor, but he also employed more draconian methods when all else failed. His need for labor in order to maintain profitable operations ranked second only to his hunger for capital, and it was in this period that he perfected his strategies for maintaining his labor supply and faced his greatest challenges from below.

COTTON
Upper Arkansas Delta Counties,
Early 20th Century

Areas of predominant cotton cultivation

*Crowley's Ridge. Source: United States Environmental Protection Agency, 2003; Dr. Jeannie Whayne, 2010.

5

BUILDING IT OF BRICK
AND HOLLOW TILE

hen Lee Wilson crossed the Mississippi River in 1880 to establish a lumber camp on his Arkansas property, Napoleon Wilson, a forty-five-year-old black freedman born as a slave into the Wilson household in Randolph, Tennessee, accompanied him. Napoleon's decision to join fifteen-year-old Lee Wilson in his Arkansas venture suggests an attachment that some freed people retained for the families of their former masters, but it more importantly reflects the lack of opportunities existing elsewhere for blacks after Reconstruction ended. The older black man remained closely tied to Lee Wilson throughout his life, and the "farmer prince" acquired a reputation—though sometimes contradicted by the facts—of treating blacks more generously than other employers. Certainly Wilson's construction of a state-of-the-art industrial school for African Americans in 1924 reveals a benevolence rare among planters, and his rage when white arsonists burned the school to the ground demonstrates not only that he understood it as a personal affront but that he saw it also as a threat to the African Americans on his plantation. As he stood viewing the smoldering ashes of the school on the morning of its scheduled dedication, he silently smoked his cigar and said nothing. When he returned to his office, however, he "gave vent to his feelings" and vowed to begin again, "except that I may build it of brick and hollow tile" to prevent the arsonists from repeating the offense.[1]

"They have helped me to make what I have," Wilson declared, "and I wanted to do something to help them in a substantial way." S. L. Smith of the Julius Rosenwald school building program, who had traveled from Nashville to attend the dedication, recorded these remarks, accepting them on face value. However, far more than benevolence motivated Wilson. Just a little over two years earlier—in January 1921—a black man, Henry Lowery, was burned to death on a neighboring plantation in front of a crowd of approximately five hundred men and a few women. Lowery, who stood accused of murdering Lee

Wilson's brother-in-law and niece, endured an agonizing forty minutes before expiring. Burning a criminal at the stake was an ancient form of capital punishment, usually reserved for such crimes as heresy, witchcraft, and treason. Regarded as "cruel and unusual," it was no longer employed as a method of execution in the United States, but it served as a particularly savage alternative to hanging blacks in the American South in this period. It functioned as both retribution and warning to the black community, but the burning of Henry Lowery occurred at a particularly problematic time for Arkansas planters. Just fifteen months earlier, an exodus of black labor from Phillips County followed a massacre of blacks, accelerating a trend begun earlier in the decade and mirrored in other southern rural areas: a migration of black southerners to northern industrial cities. The $52,500 Wilson spent of his own money to build the Wilson Industrial School for his African American laborers probably seemed a small price to pay to reward black labor for remaining in place, but he also likely hoped the school would encourage other African American laborers to migrate to the area. The plantation system in Mississippi was continuing to expand, and the demand for labor constituted a crucial component of planter operations. The burning of the black school, however, revealed the resentment some whites continued to feel for blacks and exposed, once again, the vulnerability of African Americans in a racially charged environment, a vulnerability against which Wilson could or—in the case of Lowery—would offer only limited protection.

Two other very pragmatic reasons also influenced Wilson to construct a school for the children of his black laborers. One key feature of the Wilson Industrial School involved the limited range of classes it offered, reflecting Wilson's point of view regarding the proper education for African Americans. While the white school offered a college preparatory track, the black school only prepared blacks to work in one of the Wilson enterprises.[2] Building the school also made good sense for Wilson from another perspective. A shrewd businessman to the core, Wilson hedged his investments according to a lifelong practice of leveraging every asset he owned to provide working capital and expand operations. Required by lenders to insure buildings against fire and other damages, Wilson likely collected a substantial portion of his original investment in the structure and used insurance money to construct a new building and then pledged it as collateral for some future purpose. It cost $60,000 to rebuild the school, $58,454 paid for by Wilson himself, the rest covered by the Rosenwald fund and the black community. What the Rosenwald people saw as sheer benevolence was far more self-interested. In other

words, Wilson knew how to invest in infrastructure—like the black school—
and yet mortgage the property and have use of the funds.[3]

Another black institution that Wilson patronized, the black church, functioned more independently from his oversight, however much he might have
liked to think otherwise. Fortunately, the WPA Church Records Project carried
out in 1940–41 sheds some light on the semi-secret world of the black church
community. Although the questionnaires on only twenty-six of the eighty-nine
black churches in Mississippi County have survived, they provide a wealth of
information about the nature of these institutions and reveal a rich, active,
and vibrant black community in the county.[4] Aside from regular Sunday services, the parishioners associated in missionary societies, Bible study groups,
and young people's clubs, among other organizations. One hundred black auxiliary organizations connected to the twenty-six churches, almost four for each
church, existed.[5] Through these organizations ordinary African Americans
exercised control over their own institution, the black church, and exerted
influence over the larger black community.[6]

African Americans attended these churches for spiritual sustenance, but
they also found in them a social space typically free of white interference.
Their auxiliary organizations were "central to the mission" of elevating and assisting the rural African American population and included some outreach beyond their own communicants, particularly to the poor and imprisoned. Sunday school services provided biblical instructions, burial associations insured
congregants for death benefits, and missionary societies specialized in aiding
sick or destitute members. Perhaps most important were the usher boards, the
"small but powerful" organizations that essentially governed the church. They
were particularly important in rural churches, which were often too poor to
keep a preacher fully employed. In such cases, they might share an itinerant
preacher who visited them twice a month. Under these circumstances, the
usher boards became even more powerful, as those who served on them were
more permanently moored to the community than was the preacher.[7]

The interest that Wilson and other planters took in the formation of black
churches failed to translate into control over them. Planters believed "that the
existence of a church would make tenants more satisfied and that hearing sermons against vice would keep them more honest and less rebellious. Workers
who accepted the status quo would probably work for smaller wages and accept poorer housing conditions."[8] Lee Wilson may have embraced this notion,
given his "support" of at least two of the black churches on his plantation. The
Wilson Church of God in Christ began as a mission in 1915 or 1916 and first

held meetings in a dilapidated cabin on the outskirts of the town of Wilson, but Lee Wilson & Company eventually allowed the church use of the old theater building and when it burned, the congregants rented the ground floor of the Colored Masonic Hall, where it was still meeting at the time of the WPA interview.[9] Lee Wilson also figured in the history of Norman Chapel, a Colored Methodist Episcopal (CME) church at Evadale. He sold a one-acre parcel to the founders of Norman Chapel in about 1898, who then constructed a frame building.[10] African Americans themselves, however, typically founded and built their own churches. Rev. Jessie P. Payne, a skilled carpenter, constructed Nehemiah's Chapel in Blytheville with his own hands and the help of his congregation. African American Methodists (AME Church) founded and constructed Walker Chapel in the town of Burdette, located north and west of Wilson.

In spite of the *assistance* of white planters and regardless of the considerable effort on the part of the African American community, the condition of the buildings and their meager accouterments speaks to the economic vulnerability of the congregations. The economic circumstances confronting African Americans made it difficult to do more than construct humble and lightly adorned buildings. Twenty-two of the black churches in the sample were housed in frame structures, and one each was brick, stone, or stucco (the materials of the remaining churches are unknown). Fully half of the frame buildings, like the AME church in Wilson, were unpainted, and none of them had congregations wealthy enough to decorate their churches with stained-glass or memorial windows. Most of the choirs and singing groups performed without pianos or organs for accompaniment, for only eleven of the churches on which we have records had pianos and only the Wilson AME church had an organ.[11] Many of the congregations struggled to support their ministers and, in fact, most ministers worked at some other occupation, at least part-time, in order to support themselves. Reverend Payne of Nehemiah's Chapel in Blytheville was a carpenter, for example. Sixteen of the preachers had only common school educations, nine had at least some high school, and only one had attended college. Rev. Prince Albert Goldsberry, who preached at Carter's Temple, the CME Church in Blytheville, had attended Southland College, a Quaker institution near Helena, Arkansas, and Lane College in Jackson, Tennessee.[12] Each of the two black preachers in Wilson, Arkansas, had only a few years of common school education.[13]

The educations, though usually meager, of the preachers enhanced their credibility with those who worshiped in their churches. The preacher's place in the black community was of central importance, and he owed his prominence in large part to the fact that he was sometimes the only one among them who was not directly, or, at least, not completely, economically dependent upon the white community. Instead, he served as an intermediary between his congregation and powerful whites, able to secure funds or other favors for the church and sometimes even able to intercede on behalf of those of his flock who had transgressed in some fashion. The African American clergy has come under some criticism for their apparent cooperation with white landowners in encouraging a meek acceptance of their earthly condition, and this criticism of black preachers was extended to the black church generally. As Lois E. Myers and Rebecca Sharpless have observed, "The tacit approval of white landowners, created an apparent collusion between the church and the landowner." But Myers and Sharpless cite Evelyn Brooks Higginbotham's criticism of this bleak portrayal of the black churches, suggesting that their autonomous worship services and their assumption of equality in Christ represented an implicit denial of their inferiority. So subtle and unassuming was this "resistance" on the part of black preachers and their congregations that most in the white community, particularly planters, remained unaware of it.[14]

Within their churches African Americans attempted to draw a curtain between themselves and whites who would interfere with them. As the congregants of Nehemiah's Temple in Blytheville learned, however, the white world sometimes intruded upon that sacred domain. In the spring of 1917 a group of whites, who objected to Rev. Payne's sermons, abducted the minister with every intention of doing him serious harm. Payne, like other Church of God in Christ preachers who criticized America's participation in World War I, urged his listeners to claim conscientious objector status. As a result, the white vigilantes tarred and feathered the minister.[15] Just as the preacher survived this brutal attack, so too did the church, which by 1941 had enough seating capacity for five hundred worshipers.[16]

The boldness of the white posse in attacking Rev. Payne reflected Mississippi County's tortured history of race relations. The African Americans who came to northeastern Arkansas to work in the expanding plantation system there hoped to better their situations, but most found only limited opportunities and further disappointment. Henry Lowery, for example, a mature black man from Magnolia, Mississippi, rode the crest of black migration into Mississippi County in early 1919, obviously hoping that the need for labor would translate into greater economic benefit. He was far from alone. Between 1900 and 1930, the number of tenants and sharecroppers in the county increased from 1,204 to 9,566, and the number of acres in cultivation rose from 124,684 to 335,034, a phenomenon that ran far ahead of a modest trend upward for the state as a whole.[17] The massive reclamation efforts in northeast Arkansas largely account for the expansion in Mississippi County, but planters also benefited by a misfortune facing their counterparts elsewhere. In the late teens southeast Arkansas faced a devastating encounter with the boll weevil. The trajectory of the weevil infestation moved from Texas north to the Arkansas River and then east through Mississippi, reaching Alabama by 1915. The voracious pest did not cross the Arkansas River until the mid-1920s, so northeastern Arkansas planters had some time to prepare for the onslaught and, in the meantime, benefited from the movement of labor out of infested areas.[18]

Whatever reasons Henry Lowery had for coming to northeast Arkansas, he found a place with Owen Craig (sometimes referred to as Osben or Oscar), who operated a smaller plantation of only 689 acres, called "Stonewall," adjacent to Wilson's and next to the Mississippi River. Although its proximity to the Mississippi made it more vulnerable to occasional overflows, Craig's tenants and sharecroppers acquired slightly more personal property than did those on Lee Wilson's plantation, a fact that may have influenced Lowery to take his chances there. Personal property tax registrations in the black communities where Wilson and Craig operated their plantations—School Districts 25 and 16, respectively—demonstrate that another difference existed. Blacks in the Craig area had a greater chance of holding property of some sort. School District 16 was one of the smallest in the county, taking up about half of Troy Township and holding a total of only 343 adults registered on the personal property rolls. The larger Wilson district, which encompassed all of Golden Lake Township and parts of Troy and Carson Lake, held 1,124 taxpayers. In the Wilson area, adult black registrants constituted 74.8 percent of the total but owned only 19.3 percent of the personal property. Adult blacks in School District 16, who represented 90 percent of the registrants, outperformed the Wilson tenants and sharecroppers, claiming nearly 47.6 percent of the personal property wealth. In the final analysis, however, individual blacks in School District 16 acquired only slightly more personal property on average: $152.12 to School District 25's $141.63. Henry Lowery's acquisition of $250 of personal property—most of it tied up in his three mules—put him well ahead of blacks in both districts but still behind whites (in District 25, whites held an average of $376.57; in District 16, whites held $419.06).[19]

Unfortunately, Lowery's tenure on Craig's Stonewall Plantation coincided with the post–World War I drop in cotton prices and a deepening agricultural depression. Many planters absorbed their losses by shorting their tenants and sharecroppers at settlement time, but even those who shared the misfortune

with their plantation laborers faced disappointed and sometimes angry workers. Craig enjoyed the position of a planter, but he operated on a much smaller
scale than Lee Wilson and did not have the assets to easily negotiate the down-
turn in the economy. For example, in 1920, Craig claimed $2,825 in personal
property; Lee Wilson registered $175,415.[20] Craig was heavily in debt to F. G.
Barton Cotton Company, cotton factors operating out of Memphis, and to several other creditors, Lee Wilson among them.[21] Craig owned only 689 acres
so had less collateral with which to secure his loans. By 1920, he had turned
the management of his plantation over to his son, Richard "Dick" Craig, and
under these circumstances, the younger Craig may have found it impossible to
resist the impulse to shortchange his tenants and sharecroppers at settlement
time.[22] If that was the case, he had little fear that he would face repercussions.
Tenants and sharecroppers, both black and white, typically understood the
consequences of confronting a planter and knew that the law almost always
sided with the landowner in disputes over settlement of the crop. Henry Lowery, however, not only violated the customary acceptance of planter settlements but also engaged in a gun battle that left two prominent whites dead
and two others wounded.

At first, Lowery seemed to thrive in northeast Arkansas. He gained a reputation as a hardworking and efficient farmer and quickly integrated himself
into the black community in the vicinity. He had earned "a diploma from some
correspondence detective school" and commanded the respect of his peers
in the three black lodges of which he became a member: the Masons, the
Knights of Pythias, and the Odd Fellows. Hundreds of other black men in
northeast Arkansas were affiliated with these three organizations or with the
Mosaic Templars (headquartered in Little Rock), or with the United Brotherhood of Friendship. In 1905, approximately 21,867 black men in the state belonged to three of these organizations—Masons, Odd Fellows, or Knights of
Pythias.[23] They were simply a fact of life in the African American community
and were, in many cases, closely affiliated, if not exactly connected to, black
churches. Although there was a period in the late nineteenth century when
black churchwomen were suspicious of fraternal organizations, that had long
since been overcome, and many preachers sponsored and held positions as officers in them. Even if the preacher himself was not personally affiliated with
one of them, because of the poverty of the black community, black churches
had to stand in as meeting places for fraternal organizations.[24]

However much he integrated himself into the black fraternal organizations
in the area, Henry Lowery might have thought twice about moving to Mississippi County had he known about the area's reputation for lawless attacks on

black farmers. Night riders represented a destabilizing element throughout the Arkansas delta; thus planters like Lee Wilson supported a 1909 law that made it a crime to engage in such activities. Night riders in Mississippi County focused on those areas with the heaviest black populations, along the eastern and southern portion of the county where African Americans constituted up to 84 percent of the population in 1920.[25] Blacks made up 63 percent in Wilson's Golden Lake Township and 81.6 percent in Craig's Troy Township.[26] Although theirs was a criminal offense, night riders, who often obscured their identities, proved difficult to apprehend and prosecute. Sometimes, though, authorities succeeded in capturing and convicting them. Such was the case in March 1915 when night riders threatened African Americans working near O'Donnell's Bend—in Swain Township about twenty-five miles north of the town of Wilson—and destroyed property belonging to planter J. D. Spann. Swain Township fit the demographic for such attacks with 63 percent of its population African American. The episode began when unidentified men posted notices warning blacks to leave and threatening one white man that "if he continued to employ Negroes his property would be burned." On March 17 they torched one of Spann's tenant houses, prompting the county sheriff to dispatch deputies to hunt down the perpetrators. Using bloodhounds, the deputies apprehended seven men, although other "marauders escaped to the Tennessee shore." One of the first men treed by the hounds, Jesse Swafford, "turned state's evidence," leading to the indictment of four of those in custody. The court convicted two of them, Mart Rogers, the alleged ringleader, and Giles Simpson; Judge W. J. Driver sentenced them to prison for seven and four years respectively. Adah Roussan, who had succeeded her husband as editor of the *Osceola Times* upon his death in 1904, believed that "this probably finishes the night riding business in the Osceola District for some time to come," but she lamented that "there has been considerable activity along the same line in other parts of the county." Circuit Court judge Driver warned that "parties found guilty of this offense in the future will not get off as light as these three men did."[27] Judge Driver moved on to the U.S. House of Representatives, however, and Adah Roussan's optimism proved unwarranted.[28]

Many whites riding against blacks in northeastern Arkansas—and elsewhere along the lower Mississippi River Valley—harbored an intense hatred of the plantation barons who secured the best land and sponsored expensive improvements that poorer farmers found difficult to bear. As the barons developed the acreage, land prices increased, making it more difficult for landless farmers to climb the agricultural ladder from tenancy to land ownership.

But to even have a hope of climbing the ladder, they had to have a foot on the first rung, and this motivated many other night riders who hoped to secure the tenant-farming positions on the plantations. The presence of the lower-paid black sharecroppers simply worked against their economic interests. Most blacks, impoverished and without livestock and implements that could secure them a tenant-farming position, remained as sharecroppers only and thus commanded a smaller wage. From a purely economic standpoint, their presence drove the cost of labor down, something efficacious to planters but lamentable to white labor. And black farmers like Henry Lowery, who had managed to acquire livestock, thus moving to the second rung of the agricultural ladder, would have enraged some whites who saw that as impudence and viewed black farmers like Lowery as obstacles to their own advancement.

What happened to Lowery has to be understood in terms of the postwar economic environment but also the acceleration of the black migration out of the South. While many blacks, to the alarm of white planters, joined the "great migration" north, others chose to remain in place despite certain vicious acts of violence that arose during the Red Scare and labor strife that followed the Great War. Urban race riots in Atlanta and Tulsa are best known, but rural violence also occurred. The most notorious episode of racial conflict in the rural South in this period occurred in Phillips County, Arkansas, just over a hundred miles south of Mississippi County in early October 1919. The events leading to the Elaine Race Riot began in that summer when blacks in Phillips County organized the Progressive Farmers and Household Union of America and hired a white attorney in Little Rock to file suit against planters for a fair settlement of the crop. Returning black soldiers believed they had proven their worth to America and expected reward, but white veterans believed *their* reward was a resumption of the racial status quo. Phillips County—like Mississippi County—had a history of night-riding activities aimed at black labor. An unbroken campaign of terror against African Americans prevailed there from Reconstruction onward, and authorities, like those in Mississippi County, characterized some of the episodes as efforts to drive black labor off so that white laborers could secure the tenancies. White night riders also destroyed property belonging to planters, revealing a class divide that presented challenges to authorities. The events occurring in Phillips County in the fall of 1919, however, demonstrate that planters could bridge that class divide when necessary. In the end, however, they would pay a steep price themselves.[29]

On the night of September 30 two white deputies disrupted a meeting of the Progressive Farmers and Household Union of America in Hoop Spur

Church near the town of Elaine. Although no one knows who fired the first shot, gunfire broke out, leaving one of the whites dead and the other wounded. A black "trustee" traveling with the deputies alerted authorities in Helena, the county seat, and those officials soon interpreted the encounter as the beginning of an "insurrection" of black sharecroppers. The union members had, in fact, hired the white attorney in Little Rock to represent them. But whites spread the alarm that the blacks had murder in mind instead, and the white community reacted hysterically, leading to a wholesale massacre of African Americans, one in which not all of the blacks behaved passively. At least some took up arms against the whites who attacked them and their families, but white officials prevailed upon Governor Charles Brough to bring in troops, newly returned from the trenches of World War I, stationed at Camp Pike. Troops restored order in the end, perhaps participating in the massacre of unarmed black men, women, and children, but before peace arrived, whites from surrounding counties—and even some from Mississippi and Tennessee—flooded into Phillips County and ran rampant, killing blacks indiscriminately. The army reported an official death toll of twenty-five blacks and five whites, but some unofficial estimates suggest that hundreds of African Americans lost their lives and many more left Phillips County in fear, leading to a serious labor shortage just when hands were needed to harvest the 1919 crop.[30]

Mississippi County planters, like their counterparts elsewhere, reacted with fear of a black insurrection. Within weeks of the massacre, rumors circulated that Sheriff Dwight Blackwood had dispatched a deputy to travel to Memphis "to purchase arms with which to suppress an expected uprising in this county." Although the sheriff emphatically denied the rumor and the newspaper insisted, despite all evidence to the contrary, that "there has never been any trouble between the races in this county," officials called a mass meeting of the county's "colored citizens together with a number of our best white citizens" to quell fears.[31] The meeting, which took place at the courthouse in Osceola in early November 1919, featured speeches by two prominent African Americans, the preacher at the Osceola AME church and the principal of Osceola's black school. A black preacher from Blytheville gave the invocation.[32]

The participation of the preachers in the mass meeting speaks to the reputation for collusion between planters and prominent blacks, but it also represents what Gunnar Myrdal regarded as the crucial role such persons played in mediating between the white and black worlds. Circuit judge J. T. Coston—Lee Wilson's longtime attorney—served as the principal speaker and explained "conditions in the south and north, so far as the Negro is concerned." Coston's remarks reflected the fear that African Americans would flee Mississippi County, just as many in the Phillips County area had done, exactly when most planters needed them to harvest their crops. Once Coston finished speaking, J. W. Lowe, a black businessman in Osceola, presented a resolution that deplored the unrest and violence occurring in Arkansas and elsewhere, speaking not only to race riots but to labor strikes, and pledged that the county's black community would "stand in a body as one man to discourage any loose talk on the part of our people and to inform them that there is no just cause for any of us to allow any agency or agencies among us to sow discord, to destroy the things we hold so dear, law and order." The resolution concluded with a promise and a request: "We pledge our friendship to our white friends that they can depend on us to do our duty as negro citizens of this county, that no disgrace shall be brought upon the fair name of this great county, and in turn we ask the friendship of our white friends."[33]

These expressions of confidence in the friendship existing between blacks and whites in Mississippi County stand in stark contrast to the facts. A "class initiation" of members of the Ku Klux Klan, then enjoying a revival in the South as a whole, at Wilson, Arkansas, in 1923, demonstrates the true nature of that relationship. An unidentified observer reported that a crowd of hundreds of men gathered "at the well known 'Blue Hole,' which is between Wilson and the River," very near where Lowery had been burned to death eighteen months earlier. "At the gate each visitor was seen to put his mouth to the ear of one the guards as though whispering some secret word." They began to arrive at 5:00 p.m., ate dinner on the meadow, and then at 7:00 p.m. when all were assembled, "there suddenly flashed into view, as by the touch of an unseen hand, a brilliant electric cross which stood out far above the crowd near the water's edge." Every black man, woman, and child understood the significance of this cross, visible from a "quarter of a mile away" and identifiable as the emblem of the KKK. The Ku Kluxers then began to march and "in regular formation they passed around the electric cross, forming a complete circle. At the command of the officer in charge, the line broke and formed into a hollow square in front of the blazing cross." An initiation of forty-four new members then took place, followed by a speech by a Dr. Johnson of Little Rock, who "made it clear that instead of being a lawless body the Klan stands for obedience to the law." Said to be "a great meeting, conducted by calm, determined men," it boded ill for African Americans in the county.[34] The Ku Klux Klan of the post–World War I period differed from its post–Civil War counterpart, however. African Americans were not the Klan's only and, in some instances, not its main concern. Animated as much by anti-Catholicism, anti-Semitism, and a desire to enforce "order" and "morality" on their communities, they were as likely to harass Catholics and Jews and attack bootleggers and wife beaters as they were African Americans. Still, African Americans found themselves to be targets also, particularly in places where they constituted a large segment of the population.

Given their need for labor in the rapidly expanding plantation system in Mississippi County in the first three decades of the twentieth century, planters like Lee Wilson sought to make certain that such activities did not interrupt their operations. Wilson's attitude toward black labor involved a complicated combination of Old South paternalism and Progressive Era concern for social inferiors. The conservative position on race Wilson and his New South friends subscribed to during the tumultuous 1890s remained a part of this mix and combined with a sense of noblesse oblige that sometimes worked to the advantage of his employees or, on occasion, other African Americans who approached him for assistance. As late as 1932 Wilson extended aid to a black farmer in nearby Jackson County, Arkansas, W. H. Warren, who was a descendant of slaves owned by Lee Wilson's father. Will Warren had accumulated some farmland and earned a reputation as an "honorable colored citizen," but the deepening depression and drop in cotton prices in 1932 made it impossible for him to meet his obligations to the First National Bank of Newport. Although Wilson had no previous contact with Warren and thus no direct relationship with him, he wrote the bank president in Newport, saying, "If he is a good Negro and trying to make a living and pay his debts, I think under those conditions that he should be extended all leniency." Wilson added that given the worthlessness of real estate, it would be pointless to foreclose and "a good business proposition . . . [to] carry him over and give him a chance to work out." Wilson closed his letter to the president pointedly: "Anything you can do for him will be appreciated by me. I would also appreciate your writing me relative to this." The bank president, A. J. Bellinger, complied with the request and "fixed up" the matter. Writing to Warren to express his satisfaction at the turn of events, Wilson took the opportunity to lecture him: "Now what you want to do is cut your expenses and work your crop good and make the bank a substantial payment."[35] Wilson clearly saw himself as a paternalist and certainly exhibited that behavior in a very public way. However, he was also very much a businessman, and his generosity extended only to those who worked hard and made good on their debts.

Given Wilson's reputation for benevolence, many African Americans made the choice to surrender certain ambitions for the safety and security of the Wilson plantation. Henry Lowery was not one of them. Characterized by the *Osceola Times* as "an educated negro, a member of numerous negro lodges and possessed of unusual resourcefulness," Lowery also proved to be a man of strong opinions, the kind of opinions that likely destined him to come into conflict with the powerful white family he worked for.[36] He was a "man on the make," and at forty-three years old probably reckoned that he had little time to waste. He could not have known when he came to work for the Craigs that he was approaching a precipice and would pay the highest price for doing so. On the other side of the precipice was Lee Wilson, the plantation owner with a reputation for better treatment of African Americans recognized even by leading black activists in Arkansas. But Lowery would likely have found the limited opportunities for autonomy and advancement on the Wilson plantation unattractive and Wilson's personal brand of paternalism unacceptably condescending.

Although Wilson purportedly offered a refuge to blacks seeking safety in a violent world, his sanctuary came at a price and held its own drawbacks. Of the approximately 2,500 employees working for him, close to two thousand were farm laborers, most of them African American, and he took them in a variety of arrangements. Tenants, sharecroppers, and day laborers made it possible for Wilson to keep his land in crops and necessitated a complex organizational structure. He employed twenty-nine white farm managers to oversee operations on his fourteen plantations and hired white riding bosses to supervise day laborers. The riding bosses reported directly to Jim Crain, the general farm manager in the company headquarters. The farm managers provided weekly reports to the general manager assigned to each plantation store, who then forwarded them to Crain. Wilson took a keen interest in those reports, frequently followed up with questions and directions, and sometimes visited various farms and fields to check on them personally. Meanwhile, a web of debt bound the tenants and sharecroppers to the company. Food for their families, feed for their livestock, repairs and replacement costs on their equipment, and any doctor bills they incurred were charged back to their account at the company store. The chattel mortgages signed by tenants who owned livestock required them to grow a certain amount of cotton, and when they harvested the crop in the fall, the company tallied their charges, subtracted what they owed from the proceeds of their cotton crop, and gave them the difference. The cotton, of course, was ginned at one of the company gins and marketed through Wilson-Ward and Company in Memphis.[37]

Wilson maintained an intimate knowledge of those who worked for him, particularly the tenants and sharecroppers, and this knowledge served Wilson in numerous ways. Hy Wilson [no relation to Lee] recollected one of his earliest encounters with his boss shortly after coming to work for him. Wilson had hired the Mississippi-born Hy straight out of a Memphis business school, and the young man was eager to demonstrate his skills to the boss. He soon discovered firsthand just how involved Wilson remained in day-to-day operations. One morning Hy was interviewing a tenant who had applied for a chattel mortgage and as he listened, he carefully listed all the man's personal property —mules, cows, implements, etcetera. He issued the tenant a check on the company account and turned to find Wilson standing over him, reviewing the document, a frown on his face. Sheepishly, Hy asked "What did I miss?" "A spotted heifer named Bessie," replied Lee Wilson and then encouraged Hy to do better in the future.[38]

The weekly reports filed by his farm managers reveal the high degree of oversight endured by his tenants and sharecroppers. Pee Wee Morris, one of the farm managers at Armorel, visited no fewer than a dozen farms in one week in the summer of 1932, reporting on the appearance of the cotton and corn crops, the progress on ridding the fields of weeds and grass, and the condition of the livestock on the various farms. He also found it necessary to move "a load of choppers" from one farm to another, and he spent part of his day in the office "checking up trucks and various items." He praised some of his farmers and found fault with others, such as D. K. Morgan, a white man, who had to be notified to better plow and chop his cotton crop and to take better care of his mules. A week later Morris, his patience exhausted, moved Morgan off the farm and into the town of Wilson. Jesse Greer, the local constable, served notice on Morgan, confiscated "all his tools and mules" and moved them to one of the other Wilson farms. Morgan's status in town remains unknown, but he owed too much to the company store and was likely required to work off the debt.[39]

In addition to tenants and sharecroppers, Wilson also used hundreds of day laborers during all phases of the operation: plowing the ground to ready it for planting, running the planter devices to insert the seeds in the ground, and, most important, harvesting the crop. Wilson secured day labor by using labor agents, recruiting locally when he could, bringing some labor from Memphis, and most of the rest from Little Rock or from rural areas elsewhere in Arkansas. Occasionally, particularly during harvest season, the company recruited Mexican families then living in Texas.[40] When necessary, Wilson turned to the U.S. Employment Service. In a series of letters with C. W. Woodman, the assistant director of the farm labor division of the U.S. Employment Service in Fort Worth, Wilson revealed his difficulties securing adequate labor during the harvest of 1932. He needed between 150 and 175 families and indicated that the wages for picking cotton would start at forty cents per hundred pounds and possibly advance to fifty cents per hundred. He indicated he would transport them by truck and promised to house them adequately.[41] Photographs illustrate the degree of supervision endured by the day laborers. One picture shows a supervisor on horseback as black day laborers pulled over sixteen teams of mules, and the caption reads "Boss Lee would often appear on his favorite horse to inspect the work."[42]

Whether it was the appearance of safety on the Wilson plantation or crippling indebtedness, blacks tended to remain despite the fact that they did not succeed in acquiring much property there. The latter was probably the deciding factor for Henry Lowery but, in fact, the prospect of Wilson's closer supervision of labor alone would have driven him to seek opportunities elsewhere, as it happens on a smaller plantation with a less sophisticated organizational apparatus. His troubles began in the fall of 1920, when a precipitous plunge in cotton prices threatened the well-being of many farmers, rich and poor, white and black. Prices had reached unprecedented levels during World War I, encouraging many to assume new indebtedness in order to expand operations. When the postwar recession struck, they were almost uniformly unprepared. Even large planters like Wilson found it necessary to maneuver creatively to address the emergency. Smaller planters and farmers had much less maneuvering room, and many of their tenants and sharecroppers suffered the consequences.

When Lowery received less than he expected from Dick Craig, his direct supervisor, he demanded a written account, exhibiting an unexpected level of assertiveness. The hot-tempered young Craig resented what he viewed as Lowery's defiance of his authority and struck him "and admonished him not to come again for settlement." But Lowery intended to seek a better situation elsewhere and understood that if he left Stonewall Plantation without a reckoning, he risked accusations of absconding owing a debt, and "all his household goods would be 'attached,' and he and his family might be attached, too." Angered by his treatment, he let it be known among the black community that he intended to return and secure his written accounting.[43]

Lowery decided to confront the entire Craig family as they gathered for dinner on Christmas Day, 1920. This may well have been calculated to appeal to them on a day when they might be expected to feel in a more generous mood, but he came armed and apparently ready to do battle if necessary. The Craig family's twenty-five-year-old black cook, Bessie Cornelius, who was "on perfectly friendly terms with 'Mr. Dick,'" had heard that Lowery intended to demand settlement again and warned them of his approach. She had cooked and served the meal and was outside, possibly on the front porch, when she saw Lowery coming up the road. Owen Craig was the first to reach the door, where he found Lowery already on the threshold. They exchanged heated words. Craig "with appropriate language, told him to leave the place, and emphasized his remark with a billet [chunky piece] of wood which he hurled through the door, striking Lowery." The rest of the Craig family, including a daughter, May Belle Craig Williamson (a twenty-seven-year-old married woman), "came pressing through the door" as Lowery backed off the porch. Dick Craig then "rushed out the door and shot Lowery" who, slightly wounded, fired his own gun several times, "unfortunately killing the father and the married daughter and wounding the two sons," Hugh and Dick. Lowery then made his escape, leaving a bloody scene and a decimated white family.[44]

The prominence of the two whites killed in the encounter complicated the response of the white community, and their connection to Lee Wilson pitted certain members of the black community—who came to Lowery's aid—against the most powerful man in the Arkansas delta. Wilson had known Owen Craig since the former's arrival in Mississippi County in 1880, and the men and their families were closely allied. Though older than young Wilson and already married to Mercy Beall in 1880, Craig would later come to rely on Wilson in a variety of ways. He lived in a Wilson house in the early twentieth century while building his place near Nodena Landing, and he relied on loans from the Bank of Wilson, ginned his cotton through a Wilson gin, and brokered it through Wilson-Ward and Company. Sisters Elizabeth Beall Wilson and Mercy Beall Craig had shared a lifetime of both good fortune and grief. They were married to successful men but suffered the disease environment of the swamps. Both lost children, Elizabeth her two-year-old daughter in 1888 and Mercy her young adult son from malaria hematuria in 1897. Elizabeth wrote a heartfelt tribute to twenty-nine-year-old Crate Craig, the oldest of her sister's children, attesting to his Christian principles and promise as a young planter.[45]

Wilson would have viewed the shooting of a neighboring planter and his daughter by a tenant—white or black—as a crime deserving of the most severe punishment, but his outrage was almost certainly greatly heightened because of the familial relationship with the Craig family. The Bank of Wilson posted a $1,000 reward for Lowery, "dead or alive," and soon posses of an estimated one hundred white men went in search of Lowery, and the newspapers carried lurid, if inaccurate, details of the confrontation. Nothing more dramatically demonstrates the unwillingness of whites to discuss the systemic causes of the confrontation than the language used to describe Lowery and the misinformation conveyed in the various reports. None of the newspapers reported the episode as a dispute over settlement of the crop. According to some accounts he was drunk and beating his wife when Owen Craig tried to intervene. These assertions were later retracted by the newspapers that printed them, but still they refrained from a coherent explanation. Instead they described Lowery as a "Negro fiend," a phrase calculated to invoke the image of the black marauder and rapist, an image that preyed on the imagination of white southerners not so far removed from Reconstruction and the alleged primacy of African Americans in that era. The film *Birth of a Nation* had only a few short years earlier (1913) imprinted that idea firmly in the minds of whites, and those whites living in a predominantly black area were particularly susceptible to suggestion.[46]

In the days immediately following the tragedy, Lowery hid in the swamps of Mississippi County while one of his lodge brothers, John T. Williams, "cooked meals" for him and "carried them back and forth."[47] As "scores of armed [white] men scouted thru the bottoms in search of" Lowery, his lodge brothers raised sufficient funds among themselves to enable Lowery to purchase a ticket on the Missouri Pacific train headed south out of Earle, Arkansas, on December 29.[48] He reached El Paso safely, but lacking sufficient funds to cross the border and reluctant to do so anyway without his wife and children, he assumed an alias and went to work as a building janitor in the Texas city. Lowery fully understood the fate that waited him if apprehended but anxious to have his family join him in a planned escape to Mexico, he made a fatal error. He entrusted a letter to an acquaintance, asking him to deliver it by hand to his lodge brother Morris Jenkins in Turrell (Crittenden County) with directions "to go to the home of J. T. Williams . . . and learn the whereabouts of his [Lowery's] wife." They were, in fact, being kept in a house behind the Craig home, allegedly for their "protection," or, more likely, being held as hostages in the hope that Lowery would try to reach them. Jenkins foolishly posted the letter to Williams in the mail instead, and authorities intercepted it, exposing the complicity of his lodge brothers in helping Lowery escape and discovering his alias and his location in El Paso.

Lowery, unaware that authorities now had all the information they needed to secure his arrest, was busy firing a furnace at the bank building in which he worked when two El Paso officials arrested him in mid-January 1921. "Please kill me boss," he said to Captain Claud Smith. "If they take me back to Arkansas, they'll burn me sure."[49] Mississippi County authorities dispatched two sheriff's deputies, Hart Dixon of Osceola and Jesse Greer of Wilson to El Paso to take charge of the prisoner. Greer, a longtime employee of Lee Wilson & Company, was widely regarded as representing Lee Wilson's interests, but the matter took a hopeful turn for Lowery when a prominent black doctor and NAACP activist in El Paso, Lawrence A. Nixon, visited him in jail. Nixon was a major civil rights figure in the state of Texas, mounting challenges to the white primary which led to two important Supreme Court decisions in 1927 and 1933. After visiting Lowery, Nixon was convinced of the likelihood that Lowery's life was in danger if he was returned to Arkansas and secured the services of a notable white attorney in seeking to prevent his extradition to Arkansas. The governor of Texas elicited a pledge from the newly elected governor of Arkansas, Thomas C. McRae, who promised to have Lowery transported to Little Rock where he would receive a fair trial.[50] But Jesse Greer worked for Lee Wilson, not for the two governors. Instead of taking Lowery across Texas through Texarkana and then straight to Little Rock on the Missouri Pacific, they went east across Texas to New Orleans, where they took the Illinois Central train north toward Memphis and straight into the hands of a posse.

Meanwhile, twenty-five armed men in six vehicles closed in on Sardis, Mississippi, arriving half an hour before the train bearing the two deputies and their prisoner. When the train stopped as scheduled, the men boarded it and "caught the deputies by surprise," disarming them and taking Lowery. Governor McRae later said he "could not understand why the Negro should have been taken by the round-about way" and remarked that deputies Greer and Dickson surrendered Lowery "with lamb-like docility." Rumors had circulated earlier that day that Lowery would be paraded through the streets of Memphis and then taken to the scene of the crime and burned at the stake. Memphis officials, alerted that Lowery had been abducted, posted guards on the likely route through the city, but the car carrying Lowery skirted east of Memphis and made Richardson's Landing, to the north in Tipton County, Tennessee, sometime after dark that day. They crossed the river to Nodena Landing where they found a crowd of hundreds—perhaps as high as six hundred—awaiting their arrival, eager for the event that the county sheriff said "every man, woman, and child in Mississippi County" believed should take place.[51]

Descriptions of Lowery's behavior during the fateful trip from Sardis, Mississippi, to Nodena Landing characterize him as a hapless African American, apparently unaware of the gravity of his situation. This hardly credible perspective—especially given Lowery's appeal upon his arrest in El Paso—is another stereotype preferred by the white press. Blacks were "good negroes" or "black fiends" or foolish and even silly. Lowery was said by his captors to have laughed and joked with them in the car carrying him to certain death. A report from a Millington, Tennessee, newspaper offered a more sobering and probably more accurate description of his demeanor. His abductors stopped in Millington to eat and took Lowery into the restaurant to keep him under observation. "A number of Millington citizens were attracted to the restaurant and conversed with him while the white men ate." Lowery "showed the intense strain he was under" and understood that "he was on his way to his death."[52] Black journalist and civil rights advocate William Pickens, who had spent part of his childhood in Little Rock, Arkansas, later published a story in *The Nation* questioning the caricature supplied by Lowery's captors and suggested "it was an evident attempt to lend an air of romance to a bestial crime."[53]

The staging of the event itself was part vengeance, part festival, and, perhaps most important, part warning to the black community. It was announced in the newspapers as planned for 6:00 p.m. on January 26.[54] It took place half an hour late. Describing the scene where the lynching occurred, one journalist called it a natural amphitheater. Virtually in sight of the Craig home, members of the posse chained Lowery to a log and piled dry leaves up around him—according to some sources up to his waist, according to others, up to his head. Two tall men, one of them with a notepad, stood beside him, apparently questioning him. No official transcript of that "testimony" has surfaced although one newspaper account reported "Lowery maintained throughout the trip from Texas, and even while death was creeping over him, that he did not know why he killed Craig and his daughter," but that he was not drunk at the time. Other reports, however, maintained that he was "full of whiskey" at the time of the killings, thus continuing the conspiracy of silence about the actual causes of the dispute between Lowery and Craig.[55] The crowd, said to be made up of Arkansas planters and curious onlookers, included a few women. They waited as Lowery, who asked to see his wife and children, made his farewell. Members of the posse then poured gasoline over the leaves and set him aflame. Lowery "suffered one of the most horrible deaths imaginable," yet "not once did the slayer beg for mercy."[56] In fact, "even after his legs had been reduced to bones he continued to talk with his captors, answering all the questions put to him."[57] Other reports, however, suggest he refused to talk to the men questioning him and remained stoic and silent as his "tearful wife and children" stood nearby. "As the gasoline was poured over his chest and head, the Negro cried out some appeal to one of the many Negro lodges of which he was a member."[58]

The mention of the appeal to Lowery's lodge speaks to the fate of his friends, seven of whom were under arrest for having aided him in his initial escape: John Williams, Mott Orr, Walter Johnson, John Reddick, Frank Capling, Henry Corbin, and Morris Jenkins. Jenkins's wife, Jennie, had also been arrested. Williams and Henry Corbin were in the Mississippi County jail in Blytheville while the remaining prisoners were incarcerated in the Crittenden County jail at Marion. Knowing that the men who had lynched Lowery also intended to do the same to his prisoners, Mississippi County sheriff Dwight Blackwood secured a few dozen men to stand guard around and inside the jail in Blytheville. He later defended his failure to prevent the Lowery lynching, in the face of stiff criticism from Governor McRae, by arguing that nothing could have stopped the crowd of six hundred strong from lynching Lowery, but he could, by staying in Blytheville, keep his two prisoners there safe. McRae, incensed over the savage killing of a man whom he had sworn would be safe if turned over to Arkansas authorities, was doubly determined to prevent further bloodshed. On the governor's orders, Blackwood had Williams and Corbin moved to Little Rock. Those in the Crittenden County jail were moved to Memphis and later to the penitentiary at Little Rock. All but one of them was over fifty and "were considered good Negroes before they were arrested in connection with the escape of Lowery." Jenkins's association with the Wappanocca Outing [hunting] Club in Crittenden County apparently played a role in saving him and his compatriots. He had been employed by the club for over thirty years, and after the Lowery lynching "several prominent Memphians . . . prevailed on Sheriff William Fish [sheriff of Crittenden County] to protect the Negro, who is said to have been a faithful servant." It appears that even the newspapers, which had up to this point conspired to celebrate and advertise the Lowery lynching, now conspired to minimize the threat to the remaining prisoners. The characterizations of them as "good Negroes" and Jenkins as a "faithful servant" were calculated.[59] Although Governor McRae openly criticized Sheriff Blackwood and called for legislation authorizing the removal of law enforcement officials who failed to prevent lynchings, no one was ever called to account for Lowery's death. The legislature, led by the eastern Arkansas delegation, rebuffed the governor in his desire to launch a legislative investigation.

It seems inconceivable that such a well-orchestrated public event as the Lowery lynching could have occurred virtually on Wilson's doorstep without at least his tacit approval. However, a report in the *Chicago Defender* that a $1,000 reward had "been offered for the arrest of Lee Wilson," whom they characterized as an "owner of a sawmill at Wilson, Ark.," seems highly unlikely. The *Defender's* account, in fact, is the only one that mentions Wilson by name as having been implicated in the nefarious affair. According to this account, Wilson sent one of his employees, Jesse Greer, to Texas to bring Lowery back to be lynched. The only other suggestion that Wilson played a role in the plot to place Lowery in harm's way is a postscript in a routine letter, dated January 20, 1921, addressed to the Bank of Wilson and written by a C. H. Dennis requesting the cashier to "advise me when the officers arrive with the Negro Lowery. It might be a favor to me." He obviously had something in mind and thought the cashier would keep him informed.[60]

The lack of direct evidence connecting Wilson to the lynching may reflect Wilson's preference to remain detached, a tendency that kept him personally insulated from the violence perpetrated on his plantation by his riding bosses, farm managers, and labor agents. His silence on the matter of the deaths of his relatives, however, seems almost superhuman. It would have enraged him, and perhaps more important, violated his sense of control over the African Americans in the vicinity of Wilson, Arkansas. Whatever Sheriff Blackwood thought, if anyone could have stopped the lynching, Lee Wilson was that man. As both plantation baron and a member of the bereaved white family, he might have used his authority to orchestrate a very different outcome. The fact that he did not do so suggests that there were very real limits to his ability to control a white mob. Or that, at the very least, he preferred to allow matters to take their course and use the opportunity to remind the black community of the wrath of the white community once aroused. However, in permitting Lowery's *execution*, Wilson risked sacrificing his reputation as a protector of African American labor. On the other hand, the fact that the violence spread no farther suggests an iron will behind the scenes. Seen in that light, Lowery functioned as the sacrifice that satisfied the crowd's thirst for vengeance and retribution. Since Lowery did not work for Wilson his fate illustrates that the security provided by Wilson extended only to those who put themselves directly under his protection. Finally, Lowery's fate may simply illustrate that Wilson's paternalism by no means precluded the use of terror and sanctioning of violence where he deemed it necessary.

The morning after the event exposed the littered grounds of the amphitheater, suggestive of the debris left by a festival or public picnic—except for the aroma of burnt flesh and the remnants of Lowery's corpse. Though reports vary as to whether some in the crowd took pieces of his body as keepsakes, a common enough practice at the time, there was probably little left to reclaim and take back to bury in Magnolia, Mississippi, as Lowery had requested. The mob had repeatedly poured oil over the body even after he was dead, in an apparent effort to obliterate every trace of him.[61] It probably fell to some in the black church community to take on the grim task of gathering what remained of Henry Lowery, for his lodge brothers were either in jail or too reticent to come forward. His wife and children, now free from the "protection" of the Craig family, likely returned to Magnolia with the coffin and the memory of an almost unimaginable horror, one which none of them could easily erase.[62]

The involvement of the black Odd Fellows in Lowery's escape demonstrates that the prevailing system of racism and oppression failed to subjugate all African Americans and that these men, all reported to be responsible and peaceable prior to the incident, understood Lowery's point of view in the matter. That such men would side with Lowery shocked white authorities, and soon they announced plans to "break up Negro lodges" because of the involvement of the Odd Fellows in aiding Lowery. A Memphis newspaper reported that these black organizations "were said to be organized by smart eastern Negroes for the double purpose of inciting the southern Negro and for getting what money they could out of him."[63] Whites in the South simply could not acknowledge the political sophistication of the black population within its midst. Given the semi-secret nature of these organizations, such a crackdown would have been difficult to manage, but the more public activities of a new organization on the horizon in 1921, the Universal Negro Improvement Association, also called Garvey clubs, would have subjected its members to greater scrutiny. The failure of blacks to organize such clubs in south Mississippi County, near the heart of the Wilson empire and, incidentally, near the scene of the Lowery lynching, may or may not be coincidental. In the context of the Lowery lynching, the cluster of Garvey clubs around Armorel, Lee Wilson's valuable northeast Mississippi County plantation, seems inexplicable.[64]

The Garvey clubs, which included as members both male and female African Americans, had branches in rural areas throughout the South. By 1922 thirty-nine such clubs existed in Arkansas, with seven in northeast Mississippi County, including the four near Armorel. Marcus Garvey, a Black Nationalist and immigrant from Jamaica, came to the United States in 1915 and soon became a kind of celebrity. His initial appeal had been made to urban blacks, but neither they nor the NAACP warmed to him or his message. He espoused separatism and a new emphasis on "back-to-Africa." He soon refined his message and found an audience among rural African Americans, many of whom had no desire to leave the South, much less America. Most of those who joined Garvey clubs were more prosperous rural African Americans with at least some small stake in their communities, men and women who had made the decision to remain in the South and make their stand on ground they were familiar with.[65] Garvey's vision, which included a strong measure of the now familiar self-help philosophy espoused by Booker T. Washington, had resonance among this group, particularly coming as it did just after Washington's death. The NAACP's legal strategy was an abstraction to them; it provided them with little solace and no immediate solution to their very real everyday problems. Garvey, on the other hand, promoted self-help, self-defense, and separatism. He celebrated Africa and invited his followers to take pride in their African heritage. He distanced himself from most prominent urban black leaders, particularly those associated with the NAACP, after an infamous meeting with Edward Young Clarke, the imperial wizard of the KKK in Georgia, but Garvey wanted to explain his movement in terms that would make it acceptable to white racists. As it happens, they agreed on one thing only, anti-miscegenation. The self-defense that Garvey promoted was couched in those terms—protecting black women from white men. This was a message that rural African Americans, both male and female, understood.

The absence of Garvey clubs in south Mississippi County suggests that African Americans there dared not engage in public expressions of solidarity, even by an organization that most planters did not view as threatening. But planters in south Mississippi County were on tenterhooks immediately after the Lowery lynching. They grew alarmed at the possibility of a mass migration out of the county. Within weeks after the lynching, the *Osceola Times* ran an editorial advertising the harsh conditions awaiting African Americans in the north: "We have no desire to defend the lynching of Lowery, for it was an outrage, but we do desire to call attention to conditions in the northern part of the country where the Negro is supposed to be among his friends." Reprinting an article about a bombing of a black building owned by a wealthy African American in Chicago in early February 1921, the editor meant to disabuse blacks of any sense that better opportunity awaited them elsewhere.[66]

The strategy employed to keep African Americans from departing Mississippi County in the wake of the lynching involved more than simply advertising the difficulties faced by blacks in the North. Again, planters used prominent African Americans in the community to their advantage, this time employing the services of W. L. Currie, a black farmer and schoolteacher at

Marie, Arkansas, one of Lee Wilson's towns. On February 12, 1921, less than two weeks after the Lowery lynching, prominent whites attended a meeting in Osceola under the pretext of showing their support for the "Negro Country Farmers Organization." Currie served as president of that organization, which had been formed in the summer of 1920 and was connected to the Agricultural Extension Program that Wilson and other planters embraced. Currie opened the meeting but soon turned matters over to prominent whites, including Congressman W. J. Driver and banker Clarence Moore Jr.[67]

White fears of black abandonment went unrealized in Mississippi County. The African American population there continued to increase in the 1920s (+31.3 percent), even as blacks slipped away from Phillips County (-17.1 percent) and most other counties in southeastern Arkansas. The boll weevil may have played a role in the decline in Phillips and surrounding southeastern Arkansas plantation counties, but, whatever the cause, planters in Mississippi County had a much greater need for labor as the land in farms increased by 57,364 acres (+17.1 percent) in that decade while the acreage in farms in Phillips County decreased by 13,869 (-5.8 percent) (table 5.1). Whether the events at Elaine in 1919 explain these different trajectories is difficult to know, but the consequences for plantation agriculture seem apparent. Certainly, planters in Mississippi County discouraged those who attempted to entice African Americans away from their plantations. In April 1923 deputies pulled a black labor agent from his automobile and beat him severely, and the *Osceola Times* used the opportunity to once again remind the county's black community that nothing but trouble awaited them elsewhere, particularly north of Mississippi County.[68] The newspaper also printed a story, meant to serve as a morality tale, about the plight of one Mississippi County African American who had the misfortune to be arrested in southeast Missouri "charged with insulting a white girl." The account celebrated the man's rescue by prominent Mississippi County planter S. E. Simonson—another Wilson confidant—who saved him "from a term in the penitentiary" and brought him home. But there was also a more sinister warning embedded in the story. "If the present agitation continues and the Negroes continue to move north then the only remedy left for the landowners of the South is to import white men to take their places. Once this movement begins nothing will check it and the Negroes themselves will be the losers."[69] In a county with a history of violence against African Americans by whites hungry for their jobs, this had resonance.

The Lowery lynching represented the most dramatic demonstration of white power over black labor in Mississippi County, but it was not the only such demonstration and far from the last. In May 1932 an incident came to light that again questioned the safety of African Americans on the Lee Wilson plantation. Scipio Jones, a prominent African American attorney in Little Rock, wrote Lee Wilson concerning clients he represented. Jones was the best-known black attorney in the state, having represented twelve black men condemned to death in the wake of the Elaine riots of 1919. He was instrumental in securing their release, and he was certainly aware of conditions on the eastern Arkansas plantations. That May he wrote to Wilson that he had "always heard of your eminent fairness to my people," but he was writing in reference to a suit he planned to file in federal court against Wilson "for damages based upon alleged acts of peonage," and his remarks were presumably calculated to elicit concessions from Wilson.

Jones represented two African American men and the estate of a third man, all of whom claimed to have been hired by Wilson's labor agents in Little Rock to pick cotton on one of the Wilson farms in November 1931. They had arrived at the end of the severe drought of 1930–31, a time when planters and those who labored for them endured additional hardships because of the extent of the natural disaster. Jones's clients had been expected to work off their transportation costs to Mississippi County, but "they were interrupted in their picking of cotton by excessive rainfall [a rainfall that was serious enough to interrupt harvest but too late to insure a bountiful one] and being unable to work they were charged for their room and board and began getting deeper into your debt." When they attempted to leave, "they were seized by your agents and employees, brutally beaten and restrained of their liberty until they finally effected their escape. From the beatings administered one of the men died." Jones wrote the letter in the hope that Wilson would voluntarily make restitution, but Wilson's response was curt and dismissive. "Replying to your letter . . . and note contents relative to my men, farm managers or anybody in my employ mistreating any one. This is not true. You can take any steps that you see fit."[70] No record can be found that Jones followed through on filing the suit.

Wilson presided over an enormous enterprise made alive by the labor of 2,500 people, some of whom were unwilling participants. Whether Wilson liked to think so or not, African Americans on his plantation lived in fear of their lives and worked under the threat of violence. What they lived with on a daily basis became manifest to one particular African American who was just passing through on his way from St. Louis to visit family in Mississippi for the 1932 Thanksgiving holiday. Virgil Branch worked for St. Louis financier Thomas Dysart, who had orchestrated a bond issue of $500,000 for Wilson nearly a decade earlier, a connection that influenced Wilson's response to the episode. Dysart's bank had advanced the funds to Wilson and marketed the bonds, a typical procedure in such cases, and Wilson remained indebted to them for nearly $200,000. Two sheriff's deputies, H. B. Carpenter and Jake Thrailkill, stopped Branch as he was driving through the town of Wilson. They accused Branch of speeding and being intoxicated, and as security for a $75 fine they levied against him, they took all the money Branch had ($10), the horn off the car, a pistol, and a shotgun, after which they allowed him to continue on his way. Although the money and the pistol belonged to Branch, the horn and "very fine Remington double barrel shotgun" belonged to Dysart.

Upon learning of what had occurred, Dysart wrote to Lee Wilson, and Wilson, who had yet to pay off his indebtedness to Dysart's company, responded immediately. In fact, Dysart was then an instrumental player in Wilson's latest scheme to recycle his debt and save his company, which, given the agricultural depression and the severe drought of 1931, was near bankruptcy. Dysart described the situation in language that Wilson could understand. "I got this man in Vicksburg many years ago. He is a real southern nigger; was raised by white folks and I know he is honest. He claims that it was raining hard and that he was driving along at a moderate rate of speed. . . . I will certainly appreciate it if you will look into it and see that my gun is returned and also, if possible the nigger's horn and his $10.00." The message Dysart was delivering was clear. He had "gotten" Branch from Vicksburg, thus Branch was "his" man. Furthermore, he was no northern African American who might have behaved in a way unacceptable to southern whites. But Dysart also wanted his property returned as well as that of his "man."

On the day Wilson received Dysart's letter, he responded "I had Carpenter in my office this morning and I told him to get off the premises and stay off and I have also written to the Sheriff of Mississippi County to have his commission as deputy sheriff revoked. I am very sorry this happened." In a revealing admission, Wilson next said, "Of course, I have no control over it, but I am positive that no sons-of-bitches like this can stay on any of my property. I will express to you, this afternoon, your gun and horn." Good at his word, Wilson wrote to county sheriff W. W. Shaver the same day, explaining the circumstances and appealing to him "as a friend and a citizen of Mississippi County to revoke their sheriff deputyship at once." Wilson had reason to believe that the sheriff would comply with his demands as Shaver was indebted to Wilson-Ward and Company and seriously in arrears.[71] To underscore his dissatisfaction with the men, Wilson further announced that "I am today notifying Carpenter and Thrailkill to get out of Wilson and stay out as far as I am concerned. . . . I don't think it is right for people to be robbed in any such manner, and I am sure you don't approve of any such action."[72] Wilson subsequently issued eviction notices to the two deputies and supported the Sheriff when he lodged "hijacking," charges against them, charges which became public.[73]

While Wilson rejected Scipio Jones's assertions abruptly and without ceremony, he responded immediately and decisively to those of Thomas Dysart. The reasons for such different reactions are obvious—Dysart, a powerful white financier with whom Wilson did business, was instrumental in Wilson's refinancing plan; Jones, a black attorney who had had the temerity to represent blacks implicated in the murder of whites during the Elaine Race Riot, had no connection to Wilson. But there were other important differences. Jones's complaint involved day laborers, one of whom allegedly died at the hands of Wilson employees. Day laborers constituted an important part of Lee Wilson's operation, particularly during the harvest, and any publicity about mistreatment had the potential of interfering with Wilson's ability to attract labor. Virgil Branch was just passing through Wilson, had no connection to Wilson's operation, and the offense against him was easily remedied.

Class and racial strife remained a factor in Mississippi County throughout the late nineteenth and early twentieth centuries and sometimes interrupted plantation operations. Wilson and other planters sought, often unsuccessfully, to control the threat to black labor, supporting the 1909 night-riding legislation and using the local authorities to apprehend and punish those who attacked African Americans and, sometimes, the property of plantation owners. Motivated by the desire to maintain an adequate supply of labor in the context of an expanding plantation economy, planters represented themselves as friends to African American laborers. Their friendship, however, came with a price. They used a variety of means to subjugate blacks, including coercion and debt peonage. Planters like Wilson also attempted to attract black labor with better housing, health care, and schools. But even Wilson's construction of a state-of-the art industrial school for the children of his African American employees takes on a different meaning in light of the Lowery lynching. He indicated he owed something to the blacks who worked for him, blacks who remained on the plantation despite the fact that they did not accumulate property at the same rate as local whites. Wilson liked to believe that black labor remained with him because of better treatment and conditions, but much of what they thought and believed remained unknown to him. They created their own institutions outside the view of Wilson and other whites. The willingness of Lowery's lodge brothers to come to his aid suggests an appreciation of the injustices and inequities in the economic arrangement between blacks and white plantation owners and more important, a willingness to defy white authority, to risk everything no matter the consequences.

Josiah Wilson, ca. 1845, Lee Wilson & Company Collection, unprocessed.
University of Arkansas Libraries, Special Collections, Fayetteville, Arkansas

R. E. Lee Wilson, ca. 1920.
Courtesy of the Butler Center, Central Arkansas Library System, Little Rock, Arkansas

R. E. Lee "Roy" Wilson, Jr., in World War I uniform, ca. 1918.
Courtesy of Yale University Manuscripts and Archives

R. E. Lee "Bob" Wilson III, Yale Graduation, 1936.
Courtesy of Yale University Manuscripts and Archives

Wilson mansion, built by Roy Wilson in the mid-1920s, ca. 1939.
R. E. L. Wilson Plantation Photographs (MC1870), Box 2, File 4, Item No. 197,
Special Collections, University of Arkansas Libraries, Fayetteville

Drainage surveyors in Mississippi County swamps, ca. 1913.
Courtesy of Drainage District No. 9, Mississippi County, in Elliott B. Sartain, *It Didn't Just Happen*
(Osceola, Ark.: Grassy Lake and Tyronza Drainage District No. 9, 197?), p. 8.

Black sharecropper family, ca. 1939.
R. E. L. Wilson Plantation Photographs (MC1870), Box 2, File 4, Item No. 144,
Special Collections, University of Arkansas Libraries, Fayetteville

Tom McAfee, Farm manager and his family, ca. 1939.
R. E. L. Wilson Plantation Photographs (MC1870), Box 2, File 4, Item No. 120,
Special Collections, University of Arkansas Libraries, Fayetteville

Negro Industrial School, Wilson, ca. 1939.
R.E. L. Wilson Plantation Photographs (MC1870), Box 2, File 3, Item No. 113,
Special Collections, University of Arkansas Libraries, Fayetteville

Wilson School for whites at Keiser, ca. 1939.
R. E. L. Wilson Plantation Photographs (MC1870), Special Collections,
University of Arkansas Libraries, Fayetteville

Weighing cotton, ca. 1939, Jessie Greer at the scales.
Lee Wilson & Co., *The Story of Lee Wilson & Company* (Pine Bluff, Ark.: Perdue Co., 195?)

Waiting in line at the cotton gin, ca. 1939.
R. E. L. Wilson Plantation Photographs (MC1870), Box 1, File 1, No. 24,
Special Collections, University of Arkansas Libraries, Fayetteville

White workers at the Box Factory, ca. 1939.
R. E. L. Wilson Plantation Photographs (MC1870), Box 2, File 1, No. 70,
Special Collections, University of Arkansas Libraries, Fayetteville

Picking Cotton, ca. 1939.
Lee Wilson & Co., *The Story of Lee Wilson & Company* (Pine Bluff, Ark.: Perdue Co., 195?)

Payday at the company store, ca. 1939.
R. E. L. Wilson Plantation Photographs (MC1870), Box 2, File 4, No. 168,
Special Collections, University of Arkansas Libraries, Fayetteville

Jim Crain behind the desk on the right, Jesse Greer on the left.
Picture Collection, number 4013. Special Collections, University of Arkansas Libraries, Fayetteville.

6

"THE WIND HAVE CHANGED"

The Flood of 1927

Lee Wilson and other planters along the lower Mississippi River Valley expended great effort to control floods and drain the land of back swamps, particularly in the first quarter of the twentieth century. These two goals—flood control and drainage—worked together to create an agricultural bonanza, but drainage also contributed to the increasing severity of floods as drainage ditches dumped greater volumes of water into the Mississippi and its tributaries. Floodwaters reached more threatening and dangerous levels, eventually undermining or overtopping the earliest levees constructed. Years of miscalculation by engineers contributed to the great flood of 1927, and in this context, the class conflict that marked the history of the region once again erupted into violence. Most of the levees built by the Corps of Engineers and St. Francis Levee District (SFLD) remained intact, but some failed miserably. Although Lee Wilson and other planters in Mississippi County posted guards along a vulnerable stretch of levees designed to contain the Little River, "unknown individuals" from Buffalo Island "cut and dynamited repeatedly" the SFLD levee on the plantation side of the river. All across the north end of the county, saboteurs struck other levees protecting plantation lands, and a virtual war broke out between men trying to protect their homes and farms from floodwaters.

Although he had to hire armed men to patrol the levees and to repair those levees damaged by dynamite or floodwaters, Lee Wilson's long campaign to protect his lands from floods and to drain the swamps worked to his advantage during the 1927 flood. He endured minor flooding at his operation at McFerren, but most of his land remained above the flood. Many farmers and planters in the county and elsewhere along the lower Mississippi River Valley were not so lucky. When the Mississippi overflowed its banks in the spring of 1927, water inundated millions of acres of land along the river and its many tributaries, bringing devastation and suffering to tens of thousands of families.

Ironically, Wilson and others like him unintentionally contributed to that devastating event. Historian Pete Daniel argues convincingly that the transformation of the natural environment along the river played an important role in creating the conditions that led to the disaster. As Daniel puts it, "The flood of 1927 happened not only because of unusually heavy rains (though that was the principal cause), but also because of the cumulative tinkering of humans." Just as Wilson and his cohorts in Mississippi County cut, milled, and marketed millions of board feet of lumber there and then turned those cut-over areas to agriculture, lumbermen and farmers played out a similar scenario from one end of the Mississippi River Valley to the other. Again, as Daniel argues, "Loggers for years had cut over the forests along the tributaries and the main river channel, and then farmers had cleared the land, robbing the water of a place to pause before running down to the Gulf." Watersheds like those of Mississippi County allowed the river to spread out when conditions demanded, and their "reclamation" had catastrophic, if then unforeseen, consequences.[1]

John Barry in *Rising Tide* adds another crucial dimension to the story. The Corps of Engineers, acting under the direction of the Mississippi River Commission, did its part to contribute to this catastrophe by adopting a "levees-only" approach to controlling the Mississippi River. One thing that remained in place was their adherence to their original charge: maintain an open channel so that commerce could flow unimpeded along the river. Flood control was a secondary concern. By the time the Mississippi River Commission adopted the levees-only solution, the leading experts, who did not themselves agree on an alternative, "had all violently rejected" it. Instead, they argued, better to preserve (or create) natural outlets and reservoirs. The Commission's ultimate refusal to intrude on property rights by condemning or acquiring land along the river in order to support this alternative approach speaks directly to the role of human agency in the Mississippi Valley region. But, given the propensity of citizens to settle in vulnerable locations along the rich river valley and the tendency of politicians to respond to their demands, the adoption of flood control as a prominent part of the Corps' mission and the ultimate rejection of the "levees-only" approach to controlling the river were almost foreordained. Certainly, it is difficult to conceive of Lee Wilson voluntarily permitting any portion of his property to remain unprotected. It would take the flood of 1927, however, to push the federal government into accepting a major role in the struggle to control floods.[2]

Because of the influence of planters like Lee Wilson, the Corps' levees along the Mississippi River in northeastern Arkansas benefited from the attention of the St. Francis Levee District. Four of the SFLD's eight counties bordered the river: Mississippi, Crittenden, Lee, and Phillips, and few citizens residing there had much confidence that the Corps levees would protect them from floods. After all, the Corps remained hamstrung by a Congress that believed flood control remained the responsibility of landowners or local government. Beginning after the 1897 flood, the SFLD not only improved Corps levees but also constructed some secondary levees where it seemed prudent to do so. Two principal problems preoccupied the SFLD's engineers: the old issue of caving banks and the height and quality of the levees constructed. Harry Pharr, the SFLD's chief engineer, placed the tendency of Golden Lake's banks to collapse into context by estimating that between 1824 and 1945, the riverbank at that point had receded by approximately a mile and a quarter because of banks that simply dissolved into the river. A twelfth-century remedy, the use of willow mattresses affixed to the banks with stones, had proven to be woefully inadequate. A modern answer to the problem of collapsing banks awaited the development of concrete mattresses and the massive machinery necessary to move them onto the banks of the river at vulnerable places. A remedy for the problem of the inadequate height and poor quality of levees was more readily arrived at. After the 1912–13 floods, engineers had raised the standard height of levees to three feet above the local flood stage. However, that solved only half the problem.

The 1927 flood drew the Corps' renewed attention to the composition of soils going into the levees. Both Corps and SFLD engineers began to adhere to a new standard in 1928 that divided soils into three classes—buckshot (clay), loam, and sand—and controlled their use. If buckshot, for example, made up 75 percent of the material going into a levee, the crown width and the slope of the levee was adjusted according to a specific rubric.[3] These measures were not in place in 1922 when yet another serious flood struck the valley and disaster was only narrowly averted. Though landowners and citizens escaped relatively unscathed from that near-miss, some of them demanded greater attention to their situation, but no one in Congress heeded their calls. This was not an era when spending for public works was supported by either the Republican administrations of the 1920s or by Congress. They would have reason to rethink this failure to answer the chorus of voices from the Mississippi River Valley when the 1927 flood occurred and forced a crucial change in government policy. No longer would the Corps of Engineers, for good or ill, be hampered by a public policy that gave only halfhearted support to flood control. The federal government's assumption of responsibility for flood control, together with the important changes in the technique used to construct levees, signaled a new era for those living along the nation's major waterways.

The great calamity that brought about this significant change in policy had begun inauspiciously enough with the spring floods of 1926. Although most levees initially held, the onslaught of unusually heavy rains which ensued in the fall of that year foreshadowed an even more calamitous flood to come. Watersheds along the river valley already stood at capacity, having never fully subsided after record high water earlier that year. Many of the poorly constructed levees were already saturated with water and rendered less reliable. All it would take to bring them down would be another flood, and as if on cue, by January 1, 1927, the Mississippi River "reached flood stage at Cairo [Illinois], the earliest for any year on record."[4] Many of the Mississippi's tributaries began overflowing their banks, placing farms and homes in jeopardy. Thousands of men piled sandbags on top of levees to extend their reach, but as the water pushed up against these saturated structures, it weakened them beyond their capacity to withstand the strain. A relentless series of flood crests that began to roll down the valley early in the year complicated the problem. Flood crests are a rare and menacing phenomenon, and "the most dangerous floods are those that contain several flood crests. The first crest fills the storage capacity of the river, causing later ones to rise higher than they otherwise would."[5] Arkansas, Mississippi, and Louisiana were hit the hardest, but only two breaks actually occurred on the Mississippi River levees along the state of Arkansas, both in southeast Arkansas counties, with most of the serious flooding falling along the Arkansas, St. Francis, White, and Red rivers as levees there failed to stand up against the tide.[6]

The levees along the river's edge in Mississippi County withstood the relentless pressure, thanks in part to the SFLD's investment of tens of thousands of dollars in repairing and enlarging them.[7] The SFLD levees constructed to contain the Little River failed, however, flooding two hundred thousand acres, including ninety thousand acres of cropland. Because of Lee Wilson's aggressive campaign to drain and protect the land he owned, most of the SFLD levees adjacent to his own plantations and towns ultimately stood up against the floodwaters. The farmers in the northwestern corner of the county on both sides of Big Lake endured the greatest flood damage in the county, a reality they bitterly resented. They could not command the labor force necessary to walk their own levees to locate and reinforce weak areas, and so they turned to clandestine activities designed to level the playing field. Needless to say, planters themselves would have resorted to dynamiting levees elsewhere to protect themselves had it been necessary, but their sturdier levees proved safer, and the men they could bring to bear to protect them placed them at a distinct advantage.

Three drainage districts covered the north end of the county, but two of them extended into the southern portion of the county and were designed to drain land there too. The Big Lake levees in the north end of the county, therefore, protected their lands from flooding. Only Drainage District 16, organized in 1914 and covering 56,903 acres, held little significance for the southern end of the county. It covered an area in the northwestern third of the county and encompassed three political townships, including Buffalo Island and the Little River area. This was the district which had required recalculation after the establishment of Big Lake Wildlife Refuge in 1915. Buffalo Island had one of the highest landownership rates in the county in 1930 (28 percent). Some larger farmers acquired enough acreage to hire tenants, but this was not an area of big plantations. The district stretched twenty-seven miles from the

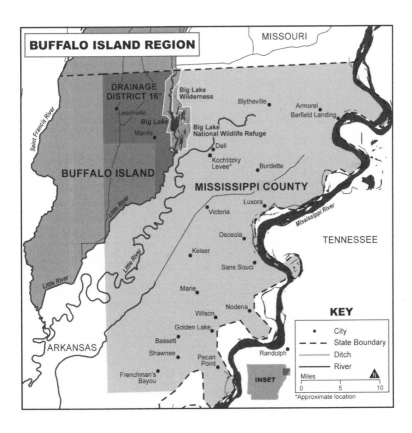

Missouri border southward down to the middle of the county and included a line of ditches meant to drain water off farmlands and into the Little River. That small river passed through Big Lake, and thus the construction of a levee by the SFLD alongside the west side of the river and the lake constituted a crucial component of its efforts.[8]

Drainage District 17, organized in 1918, covered the area just east of Big Lake and the Little River district. It ran eastward to the edge of Blytheville and Dell (the latter was several miles south of the former). It stretched from the Missouri border southward and protected 162,000 acres with 262 miles of ditches and forty-six miles of SFLD-built levees along the eastern side of Big Lake. It consisted of four political townships, two of which were dominated by small farmers. The southernmost township in the district included the large plantation owned by a former governor of Illinois, Frank O. Lowden. Lowden's tract was "three miles wide by five miles long," but he maintained the operation as an absentee landlord, hired a manager, and worked both white tenants and black sharecroppers on the property.[9] Charles Crigger, who managed Wilson's Armorel operation (in District 9), owned a farm within the boundaries of Drainage District 17 and served as one of its directors.[10]

Drainage District 9 was the final of the three districts covering property in the northeastern end of the county but, even more than District 17, it drained lands in the southern plantation area as well. Originally designated the Grassy Lake and Tyronza Drainage District 1 before being revamped and renamed, it was the district that resulted in the courthouse controversy that nearly resulted in violence in 1908. While it overlapped some of the territory covered by District 17, it ran from the far northeastern segment of the county in a southwestward direction and drained 168,156 acres, including the towns of Luxora, Burdette, a portion of Blytheville, and two Wilson-owned towns, Armorel (in the northeast part of the county) and Keiser in the south. Blytheville actually straddled the border of districts 9 and 17. District 9 also drained a plantation Wilson leased and would soon own at McFerren, property he would rename in honor of his daughter Victoria. But Keiser and McFerren/Victoria were particularly vulnerable within the district, as it had become clear twelve years before the 1927 flood that the SFLD levee on the eastern side of Big Lake was weak and could be easily compromised by a serious flood—such as the one which would occur in 1927. Engineer Otto Kochtitzky had designed the district's ditches fifteen years earlier, and he had been commissioned a few years later to construct the Kochtitzky levee as a backup in case the SFLD levee failed.[11] As one observer remarked during the 1927 flood danger, the "district wanted to be doubly safe, so about twelve years ago it had a levee of its own built along the near side of a canal [a District 9 drainage ditch], for just such a case as occurred this spring." The construction of the new levee guaranteed that waters let loose by any breaks in the Big Lake levee would wash up against the new Kochtitzky levee and back onto District 17 lands.[12] Armed confrontations occurred along all three levees—the two levees on either side of Big Lake and the Kochtitzky levee—during the flood of 1927, but it was at the Kochtitzky levee where Wilson's men, who were trying to close a dangerous breach, confronted National Guardsmen in May of 1927.[13]

As the floodwaters approached the critical stage in late March, the Drainage District 16 levees held at first, but the citizens and landowners watched in horror as the water overtopped them and flooded much of the Buffalo Island area. Then breaks in levees in Clay County, designed to keep the St. Francis within its bounds, put them in even greater peril. Water from Clay County, located to the north and west of Mississippi County, swept southward down through Greene and Craighead counties and onto Buffalo Island so that the people there found themselves caught between two disastrous levee failures. But a third blow was soon to follow. As the water flowed over Drainage District 16, it further weakened the levees, and clearly it was only a matter of time before actual breaks would occur. But if the levees on the other side of Little River—Drainage District 17 levees—were dynamited, it might, so some Buffalo Island people believed, save them.[14]

Meanwhile, even without interference, defects in the Drainage District 17 levees became manifest. They were particularly notorious for the many faults of construction that plagued early drainage projects in the county. The woody material embedded in them for nearly a decade eroded further after the heavy rains in January and February 1927 saturated the levees. In addition, too little attention had been paid to the kinds of soils going into the levees. On March 28, 1927, a small rupture opened in a Drainage District 17 levee near the Missouri border and then a much more serious breach developed on April 16, flooding tens of thousands of acres of farmland.[15] The floodwaters quickly inundated or isolated several small towns, including Dell, located on the boundary of District 17 and Drainage District 9, the Grassy Lake and Tyronza district that Lee Wilson had fought so hard to have created nearly two decades earlier. That district remained crucial to the protection of some of his most valuable plantation land, and the people of Dell, who realized that a break in the District 9 levee would drain water off their lands, soon found themselves at the center of a struggle over how to minimize the damages caused by faults in the District 17 levees.[16]

In mid-April, even before the water from the first flood had subsided, breaches of levees at Dorena and New Madrid, Missouri, sent more water rushing toward Arkansas at a rate of about fifteen miles per day. By April 29 "an area about 75 miles long and from 4 to 20 miles wide" in southeast Missouri flooded, and observers predicted the water would reach the Arkansas line by the first of May. Most of the floodwaters ran naturally into the Little River and the St. Francis River. The latter entered Arkansas just west of Mississippi County and compounded the problems faced by the farmers of Buffalo Island. The new flow ran through the breaches in the St. Francis River levees which had already sent water rolling eastward into Mississippi County's District 16 lands. Meanwhile, as the water from the Missouri breaches poured into the Little River, it headed straight into Big Lake, and the additional flow further threatened the people in District 17. The *Osceola Times* raised the alarm, explaining that the Missouri terrain was "covered with huge drainage ditches running east and west about a mile apart which provide veritable locks or reservoirs for the water as it comes. The water rises from five to seven feet in one of these reservoirs and floods on to the next one. This water will ultimately flood over the state line levee from Yarbro west to the head of Big Lake, it is believed, and from there into Kochtitzky Ditch."[17] The moment of truth was at hand. The farmers and planters of District 17, caught between the break in the Big Lake levee and Kochtitzky levee, would find themselves in the middle of a virtual lake. Under these circumstances, some of them began to contemplate sabotage as they looked toward the Kochtitzky levee.

Writer George Moreland traveled southeastern Missouri in early May and provided a valuable eyewitness account of the devastation caused by the Missouri breaks. While still in Missouri, he reported that "the water is coursing through the valley in a wild torrent but is falling." Once the water reached Arkansas, however, the situation in the Big Lake area became much more serious, and Moreland described a county in a state of emergency. He first approached Leachville in the extreme northwestern side of Mississippi County and reported that the water had "submerged much of the country surrounding the town." Leachville, like many other towns in Mississippi County, was located on an old Indian village site and remained above the flood. As he traveled by automobile southeastward toward Manila, Moreland reported water "all along the way" and some water on the roadway "though not too deep to traverse in a car." Like Leachville, Manila was above the flood but had become a virtual island with water "all around." Accompanied by the editor of the *Manila Sentinel*, Moreland continued three miles southeastward to Big Lake, which he described as "a monster now." They crossed over the levee by way of a bridge that had been constructed only four years earlier, a bridge that still stood but was vulnerable. Only the breaks in the spills and levees, which reduced the volume of water running below it, saved the bridge from destruction.

On the eastern side of the lake, he encountered a roadway so flooded that he was forced to take a motorboat through a wind and rain storm. After an uncomfortable and possibly dangerous mile-long journey, he flagged "a truck which was able to navigate the submerged roadway, and in a deluge rain reached the submerged hamlet of Roseland," one town in the area that did not sit above the flood. There he found both the county and circuit judges at the Red Cross station, and together with the judges, he traveled, once again in a motorboat, eastward and deeper into the territory of Drainage District 17 and thence to Dell. "That trip presented one solid expanse of water upon both sides of us. The Frisco tracks are deeply submerged and parts of them washed completely off the right of way." Although Dell was still above the water, Moreland reported that had the flood been "one inch bigger than it is, there would have been water in a part of" the town. As he headed northeast and neared Blytheville, which was positioned above the break in the district levee on Big Lake, he encountered more and more dry land and found that north Mississippi County town above the flood. Although there was no "water within 10 miles of Blytheville," there had been little planting accomplished because of the incessant rain.[18]

The floodwaters covered both sides of Big Lake and brought considerable turmoil and strife to the people of the region. The social and economic ramifications of the disaster exacerbated a long-standing internal feud and complicated the continuing quarrel between the northern and southern ends of the county, but it would be a mistake to think of the citizens of the Big Lake area as anything like united. Farmers west of the lake in Buffalo Island were in a world of their own and at odds not only with the planters in the southern part of the county but also with the planters and farmers east of the lake. "The roots of this ill feeling go back to 1901 when an election was held . . . to decide whether the City of Manila or the City of Blytheville would be the County seat for the northern portion of Mississippi County." Manila appeared to have won the election, but after a celebrated election contest held in the courthouse in Osceola, Blytheville was declared the winner.[19] Whether the victory was legitimate or the election stolen, Blytheville was almost certainly a better choice, for at the time the people of Buffalo Island were practically isolated from the rest of the county. Bounded on the east by Big Lake and on the west by the St. Francis River in Craighead County, Buffalo Island was difficult to reach until the JLC&E built a railroad bridge across Big Lake in 1904, making Manila more accessible. Even then, those on horseback or in an automobile still had to take a ferry across the lake until a single-lane wooden bridge was built across it in 1923. That bridge was undoubtedly the one that Moreland used in his journey through the flood.

The farmers of these five townships on either side of Big Lake, living as they did in the area that had been the subject of the congressional hearings over the sunk lands in 1910, were no strangers to conflict and confrontation. Even as they struggled against lumber barons and plantation owners during the sunk lands controversy, many of them engaged in the battle over the rights to hunt in the Big Lake area, and there they honed their skills in sabotage. The establishment of the Big Lake Refuge in 1915 ended that dispute and an uneasy calm settled over northwestern Mississippi County. Then the 1927 flood struck a nearly fatal blow to many of the small farmers who lived along the northwestern edge of the county. The Red Cross records reported that of the 3,383 families whose homes were flooded in Mississippi County, most of whom would have been in the townships around Big Lake, 3,083 (91.1 percent) of them were owners rather than tenants.[20] This was in sharp contrast to other flooded delta counties, where most of those flooded out were tenants and sharecroppers (see table 6.1). Just as most of the flood refugees in other delta counties were tenants and sharecroppers, they were also mostly African American. Only 935 (6.9 percent) of the 13,562 individuals given aid by the Red Cross in Mississippi County were African American, and none of them found it necessary to seek shelter in camps. Because most of the floodwaters deluged the whitest section of Mississippi County, fully 93 percent of those forced from their homes were white.[21]

Confronted with eminent disaster as the floodwaters threatened to overwhelm them, some of the farmers of northwest Mississippi County took matters into their own hands. All along the Mississippi River Valley, men resorted to the destruction of levees in order to save their own property from imminent flooding. The usual means was to dynamite a section of a levee below the levee fronting their own property. By doing so, they could relieve the stress on the weakened levee adjacent to their acreage as water coursed through the breach and flowed downward and away from them. The most famous incident, chronicled by writer John Barry, resulted from the decision to dynamite levees south of New Orleans in order to save the city. Receiving the sanction of the reluctant Mississippi River Commission, and at the request of the governor of Louisiana, the Corps of Engineers blew holes in levees protecting St. Bernard and Plaquemines parishes. John Barry argues that the decision to do so was ill-founded, that New Orleans was not under the dire threat that the wealthy and powerful New Orleans citizens believed. Nevertheless, reacting to political pressure and against the advice of at least one of their own, the Corps dynamited the levees and flooded tens of thousands of acres of land farmed by "poor crackers." Singer-songwriter Randy Newman memorializes the event in a song which begins, "What has happened down here is the wind have changed," an appropriate metaphor for the controversy that raged all along the lower Mississippi River Valley. The song ends with the refrain, "They're trying to wash us away."[22]

In a scenario that worked in reverse, the Arkansas cousins of the Louisiana crackers took matters into their own hands in Mississippi County and either attempted to or succeeded in dynamiting levees that inundated some acreage owned by wealthy and powerful planters, including Governor Lowden's plantation in Drainage District 17. Some District 17 farmers sought to relieve their own situation by blowing the Kochtitzky levee that protected District 9 lands. S. E. Simonson, the chairman of District 9, who reported the actions of "the vandals who repeatedly cut and dynamited our district levees," revealed an intense struggle which he blamed on "the spirit of jealousy, anarchy and disregard for law smouldering" in two neighborhoods in north Mississippi County: Clear Lake and Bowen townships. He alleged that the deputy sheriff and a constable in Clear Lake Township, on the western side of the Number 9 district, actively participated in these "reprehensible violations of the law." He implicated Oscar Alexander, the justice of the peace of Bowen Township, located north and west of Clear Lake in Drainage District 17. According to Simonson, law officers there arrested nine guards posted to protect the Kochtitzky levee, and Alexander proceeded to convict them in his court on some unspecified charges. Simonson praised the governor, who "upon being fully advised by several citizens in regard to this outrageous decision, promptly pardoned" the men and relieved "them of unjustly imposed fines and costs probably aggregating over $500.00."[23]

Twenty-one-year-old Richard Dawes provided a crucial account of what transpired on the Kochtitzky levee.[24] Morgan Engineering of Memphis, Tennessee, which had secured the contract to maintain and improve the District 9 ditches, hired Dawes in late 1926 and sent the young Wisconsin man to Wilson, Arkansas, in November to work with a surveying party under the direction of H. C. Davidson.[25] That their headquarters were in the town of Wilson was no accident. Lee Wilson intended to make certain that the levees protecting his land were adequately constructed, and he played the critical role in selecting Morgan Engineering and directing their activities. Although young Dawes had no particular surveying skills, he proved to be an able student and a hard worker and was given the opportunity to train for various tasks under Davidson's direction. His training was interrupted in late March 1927, however, when the flood appeared imminent and all hands were required to protect and enhance the levees and to repair breaches. In letters he wrote home to his mother in Wisconsin, he chronicled the experience, including confrontations with would-be saboteurs. After a series of breaks in the levees of the St. Francis River in the next county over, he wrote that "all the engineers and surveyors in the employ of the drainage district were . . . sent out to a levee which protected the district from inundation from the Little River." He wrote that Lee Wilson understood that a failure of the Big Lake levees would threaten some of his own property and the McFerren place he rented. Accordingly, "to protect the estate from flood, Wilson was glad to furnish the district all the laborers needed, and had no trouble in mobilizing a large force immediately in any emergency."[26]

Wilson at first sent men to do what they could to strengthen the District 17 levees along Big Lake, but when it became clear in late April that they were going to fail, he authorized the removal of his men to District 9's Kochtitzky levee. The people of Dell were in a tight spot, and Dawes reported that they were "hostile toward us for keeping the water on their side of the levee. . . . This hostility . . . was something genuine, not a false scare. They would have paid hundreds or thousands of dollars, and killed people, just to open up the levee and let the water off their land and homes. The folks on this side were just as determined to keep them from doing it. The usual method is to blast a hole in the levee. Therefore it was patrolled by armed guards day and night as soon as the water rose." The town of Dell was adjacent to the Lowden plantation, and thus the interests of at least one large (absentee) planter coincided with those of the people of that town. The most critical moment occurred when Dawes and his compatriots attempted to close two fifty-foot crevices that had opened (without the assistance of dynamite) on the Kochtitzky levee despite all their efforts. Those breaks inundated thousands of acres in the middle of the county, including the McFerren plantation, land that was particularly difficult to drain. "At the beginning of the work the Justice of the Peace at Dell tried to arrest us for closing the openings in the levee. Failing in that, the Dellittes persuaded the Captain of the National Guard to try to stop us." They had somehow convinced him that breaching the levee was entirely legitimate, that the levee interfered with a natural watershed. "The Captain sent a detachment down to the levee with some blasting powder to blow a hole in the railroad bank in broad daylight." The St. Louis and San Francisco "Frisco" railroad, with a roadbed raised considerably above the rest of the terrain, served as a convenient northern extension of the Kochtitzky levee in protecting District 9 lands. There the troops "met a group of levee guards. The two groups were wise enough to argue the matter before shooting." The guardsmen withdrew and returned to Dell to report to their captain, who then discovered his mistake.[27]

As this struggle between different factions over the levees took place, over ten thousand individuals in northwest Mississippi County had to be rescued and over two thousand of them cared for in refugee camps. Although the exchanges over the maintenance or destruction of the levees were heated and sometimes violent, there was no hesitancy on anyone's part about the need to save people from drowning. Given that the county was still laced with lakes and swamps, motorboats and rowboats were available in ample abundance. Working through local committees, the Red Cross played a pivotal role in the rescue. Red Cross officials reported that twelve motorboats, three barges, and fifteen "other small boats" came to the aid of the stranded refugees in the county.[28] Anyone who wanted to leave their submerged—or nearly submerged—homes had an opportunity to do so, and only one death, attributed to epilepsy, was reported (in Blytheville).[29] During the first days after floodwaters forced evacuations, most refugees were cared for by citizen-led committees in various towns in the county: Leachville, Manila, Dell, Blytheville, Luxora, and Osceola. Osceola citizens, for example, began receiving refugees on April 22, first housing them in the courthouse, until that building could no longer accommodate them. They next established a campsite on the grounds of the local handle factory in Osceola, and the Red Cross erected tents and constructed a commissary. The Red Cross established two camps nearer to the flooded area: one in the town of Dell and the other at Manila on Buffalo Island.[30] According to Red Cross records, the homes of 3,383 families were flooded, and 13,562 individuals were given aid by that agency. Only 2,276 of them, however, situated themselves in one of the camps, with the rest remaining in their damaged homes or finding safe haven with relatives or friends. Significantly, no African Americans were encamped in the county, and only 935 were listed as given aid by the Red Cross outside of camps. Fully 10,351 whites received such aid. Given the nearly all white demographic of the northwestern townships struck by the flood, these figures ring true. Most of the 935 African Americans receiving aid were likely from the District 17 lands, possibly the Lowden plantation flooded after the breaks in the SFDL's levee on the eastern side of Big Lake. Some of them, however, may have been from Wilson's McFerren operation after the breaks in the Kochtitzky levee.[31]

Historian Pete Daniel has chronicled the complicity of Red Cross officials in keeping black refugees virtually imprisoned in camps until signed out by the planters for whom they worked. Planters were typically fearful of losing their laborers, and because many of their sharecroppers and tenants were in debt to them, they wanted to be sure to be in a position to collect on those debts when the time came. Since no African Americans resided in Red Cross camps in Mississippi County, that controversy never surfaced there, but the camps in Osceola and the other small towns of the area were not without their problems, chief among them being disease. The *Osceola Times* reported on May 20 that of the 175 refugees remaining in their camp "many are suffering from illnesses of various kinds."[32] Red Cross officials held a conference in Memphis in early June to discuss the problem, and physicians representing Arkansas emphasized typhoid, diarrhea, malaria, and pellagra as the most serious threats. Ascertaining the extent of the threat proved difficult, Dr. C. W. Garrison of Arkansas, who attended the conference, admitted, as "the morbidity reports in this state are worthless and tabulations give no accurate idea of prevalence."[33] Fears of an epidemic disease like typhoid preoccupied cautious Red Cross officials. As animal carcasses rotted in floodwaters, the conditions grew ripe for bacteria to grow and endanger an already fragile population. Commonly caused by the bacteria *Salmonella typhi,* untreated cases of typhoid could be fatal, and the common fly rather than the mosquito was the usual vector. Red Cross officials recognized the increased risk of typhoid in disasters like floods and were eager to do what they could to avoid an outbreak. A more endemic disease like pellagra, which could not be unquestionably attributed to the flood, was more problematic and required some discussion and rationalization before a remedy was prescribed.

The immediate problem in Mississippi County, however, concerned "the lack of cooperation from the county health officer," who proved unable to work effectively with the Red Cross workers before a "follow up nursing program began."[34] Subsequently, two public health nurses spent forty-two days assisting the county health officer in tending to the medical needs of the flood sufferers, not only in the camps but throughout northwest Mississippi County.[35] By the end of July, the Red Cross also assigned Dr. Paul Wesson to look after "the population of Mississippi County west of Little River [in Buffalo Island].

He will give two full days a week of his time to visiting in this section besides office calls on other days, for which he will receive $75.00 a month. Similar arrangements will be made if a doctor can be found who will render the service for the territory east of Little River in the same county."[36] No information surfaces in the Red Cross files indicating that a second doctor was located, but at least one of the public health nurses worked in the eastern territory of the county. Whether they found themselves east or west of the Little River and Big Lake, Dr. Wesson and the two public health nurses were overwhelmed. One problem was the rural nature of the county, prompting a health official in Arkansas to urge the establishment of clinics in trading centers since "the population come in droves to these Trading Centers on Saturdays." He also suggested that the existence of the clinics be announced at Sunday Schools.[37] The nurses set up headquarters at clinics in several of the towns but also traveled throughout the flooded area and made house calls in order to carry out their mission. They vaccinated people against smallpox, gave them inoculations to prevent typhoid, and distributed yeast to fight pellagra.

There was a brief delay in dealing with pellagra because Red Cross officials had to be convinced that pellagra was a result of the flood and not a preexisting condition, something they were ultimately able to rationalize on the basis that had the flood not occurred, people would have had fresh vegetables from their gardens. Essentially, rural people all over the South were on the verge of pellagra every spring after a winter of living virtually off fatback and cornbread. Lacking fresh vegetables, they developed a serious niacin deficiency, which could lead to gastrointestinal disorders and possibly death. Once accepting that pellagra was more than simply an endemic disease, the Red Cross promoted the distribution of yeast and advertised the appropriate dosage. Patients were told to take two teaspoons of the powdered yeast, three times a day for six to ten weeks. The Red Cross distributed 950 pounds of yeast to Arkansas counties by August 27, 1927, including fifty pounds sent to Blytheville. It proved difficult, however, to convince people to continue the treatment once the symptoms began to subside and impossible to know for certain whether they were following instructions. The issue with typhoid was similar in that three inoculations had to be given at regular intervals. The nurses found it difficult to follow up effectively with all patients. Some received one or two shots but failed to return to the clinic—or could not be located—for a third injection. Dr. Garrison could confirm that only 4,800 "complete typhoid immunizations" had been administered in St. Francis, Crittenden, and Mississippi counties combined. His figures for other counties in the state were considerably higher. For example, 13,250 complete immunizations had been administered in Drew County (southeast Arkansas).[38]

Although there were differences of opinion among medical professionals about the most appropriate method to pursue to combat malaria, the conferees in Memphis agreed that it was appropriate for the Red Cross to address a disease that was at least as endemic to the region as pellagra. So common was malaria there that young Richard Dawes made a lame attempt to reassure his mother earlier that year, writing "the mosquitoes are bad, just like in Wis., only here they can give you malaria and I might get that, but I'm not unusually susceptible, quinine will take care of that."[39] It is unlikely that his mother was comforted, particularly if she knew anything at all about the disease. Because Dawes was new to the region and thus unprotected by previous encounters with mosquitoes carrying the parasite, he was particularly susceptible and, in any case, quinine merely relieved symptoms. There was no cure, nothing that could "take care" of the tenacious parasite. But whether it was endemic or not, Red Cross officials did not hesitate to accept responsibility for malaria control. They recognized that the additional standing water provided even greater opportunities for mosquitoes to breed. Dr. J. A. LePrince, characterized as "one of the foremost authorities on Malaria Control in the south," warned that the malarial threat was great, citing that during a previous flood, "90% of the families in Alabama were, at one point, down with malaria." Red Cross officials were concerned not only about the local population but also for Red Cross workers in the area. Mostly, they feared a population so devastated by a widespread malarial outbreak that the region would be unable to recover economically from the disaster. Dr. W. R. Redden, a medical officer with the American Red Cross, cited his experience in an earlier disaster, arguing that "quinine kept 100,000 wage earners on their feet and when the Red Cross started to invest money in that game it sought the best advice available from people with years and years of experience."[40] Still, Dr. Redden solicited and received a variety of opinions from the physicians attending the conference. Arkansas's Dr. Garrison insisted that patients should be given ten grains of quinine every night, while other medical professionals from Mississippi suggested smaller dosages. While some remained unconvinced that quinine was sufficiently effective, others understood that its use carried risks, including death if a patient ingested too much. One Arkansas physician who allegedly had experience in the Near East indicated that he could not "justify quinine under any circumstances."[41] In the end, the conferees concluded that fifty thousand people in Arkansas and 120,000 in Mississippi were in need of quinine. They agreed to its distribution but left the question of the appropriate dosage unanswered. They decided to authorize the Red Cross to "furnish sufficient quinine to handle actual cases of malaria and also for use among other people to the extent that it would be administered and handled effectively by plantation owners and on the demand of State Health Departments." Given that many of the flood sufferers in Mississippi County were small farmers who owned their own farms, they probably received their dosages from the county health nurses or Dr. Wesson.[42]

The discussion of the danger of an outbreak of malaria also turned to preventative measures. The conferees briefly debated whether they should encourage the use of mosquito netting and the screening of homes. Dr. Garrison insisted that "these people will not take time to screen their houses. They are busy planting." Neither Garrison nor Redden, in the end, believed that screening houses was practical, and Redden suggested that it was not the responsibility of the Red Cross. "The man should do this himself." Dr. Harrison's motion recommending that "this be left to the discretion of the man in charge of his community" was seconded and passed by the conferees. Instead of screening homes, Garrison advocated the use of netting, and Dr. Redden agreed with him. "If a man is bitten one night and can get mosquito netting the next night he will take it." The good doctors may have been overly optimistic in that assumption, as the use of nets in a hot and humid climate makes it seem hotter and more uncomfortable. The residents of the Arkansas swamps had long lived in the region and had done so without netting for some time. Dr. Garrison seemed aware of that, acknowledging that the value of netting needed to be explained to them. The physicians, finally, debated the treatment of mosquito breeding grounds and the spreading of gamboosia (a type of oil) or automobile oil on the standing water to kill the mosquitoes. Dr. LePrince advocated the application of naphtha in chimneys "to prevent mosquitoes getting in that way." Naphtha, a liquid hydrocarbon manufactured from petroleum, natural gas, or coal tar, was widely used in that area for a variety of purposes (as a solvent, in dry cleaning, as a varnish, or as a fuel) and thus readily available.[43] After a lengthy discussion, however, they left the matter unresolved, and there is no evidence that oil was provided for distribution in Mississippi County.

Connected to the effort to prevent an outbreak of disease was the campaign to destroy the carcasses of the thousands of dead animals left rotting in the standing water or, ultimately, on muddy ground once the flood had receded.[44] Local residents and Red Cross volunteers typically handled the gruesome task, piling the rotting carcasses on top of each other, where possible,

and burning them. The smoke and stench hung like a pall over a blighted landscape. Water remained on much of the flooded ground well into summer—in
some places until late July. Of the two hundred thousand acres flooded in Mississippi County, ninety to ninety-five thousand were normally planted in some
crop. Even those farmers whose land escaped the flood itself found themselves
handicapped by the incessant rain and were unable to plant as normal. With
water remaining so long on the ground, it proved impossible to plant cotton
in some places, and at least some of the land lay fallow for the year. The reduction in cotton acreage is most telling. In 1926, Mississippi County had 49,619
acres in cotton; in 1927, only 19,601 acres in that crop.[45]

Stanley D. Carpenter, the county agent assigned to the south end of the
county, assisted his counterpart in the flooded northern area, J. E. Critz, in
helping farmers there recover from the disaster. Critz reported that he and
Carpenter "made earnest efforts and labored faithfully in our humble way to
show the proper authorities that the flooded territory was going to need an
enormous amount of assistance from some source. We made trips to Little
Rock to conferences, to Memphis to Red Cross headquarters and sat in local
meetings and conferences of local people giving assistance and suggestions
where possible to do so." Still, he reported that "the rehabilitation problem
was a very difficult one due to a second rise and the water remaining on the
land."[46] As Carpenter put it, "On June the first these people had nothing—the
water was just beginning to leave them, and they were advised to plant feed
and food crops first and then some cotton. The most of them did as they were
advised." He attributed much of their recovery to the efforts of the Red Cross,
which "spent approximately $140,000.00" in the county.[47] Indeed, Critz ar-
gued that "farmers will make far more money this year than they did last year,
taking into consideration the entire county." Because the production of crops
was much lower throughout the flooded area, the price for agricultural products was much higher. "The banks have more surplus money at present than
they have had at any other one time in the past two years."[48]

Critz may have been moved to paint such a rosy picture of conditions because of the nature of the report he was writing. The annual narrative reports
of county agents are meant to provide their superiors with the successes of
the programs administered on the county level. Extension agents rarely provided candid appraisals of their failures and, instead, emphasized the obstacles
they had overcome and successes of their programs. What Critz did not report about conditions in Mississippi County is most telling. According to a
document prepared by the Mississippi Valley Flood Control Association, the
devastation to Mississippi County's farmers was significant. In addition to the

loss of livestock and the inability to plant certain cash crops in sufficient abundance, the losses included the following:

1,000 houses damaged	$10,000
1,000 barns damaged	$5,000
1,000 other buildings damaged	$2,500
Damage to farm implements	$10,000
Damage to feed	$50,800
Damage to seed	$21,156
Damage to household goods	$14,000
15 head of horses and mules lost	$2,000
25 head of cattle Lost	$1,000
600 head of hogs lost	$6,000
30,000 poultry lost	$12,000
Damage to land by washing and spreading of obnoxious grasses	$24,000
Loss of rents on lands not cultivated by reason of overflow	$175,000
Damage to fences	$10,000
Damage to growing crops	$145,000
Damage to private roads and bridges	$48,000
Damage to ditches and drains	$24,000
Total property damage	$562,256

These figures are probably much lower than the losses actually sustained in the county. The Flood Control Association's estimates run uniformly lower than those reported by the Red Cross, probably because of the way they were collected. The Red Cross secured their figures from the Red Cross volunteers and from county agricultural agents. The Flood Control Association merely consulted "prominent men" in the flooded communities. In any case, the figures suggest that Critz was probably wildly optimistic about conditions in the county, but there was a grain of truth to his assertion that conditions were, in fact, better than they had been in years. The previous two years had been difficult ones, with the prices received for agricultural products barely keeping pace with the cost of production. Given the aid rendered by the Red Cross and the higher prices for crops, the year 1927 looked pretty good in comparison. Prices rose from twelve cents per pound in 1926 to twenty cents per pound in 1927. Planters like Lee Wilson, who experienced only a modest amount of flooding, planted nearly as much cotton as they usually planted and benefited from the short crop of 1927.[49]

Wilson also benefited from a government program designed to enable farmers and planters to recover from flood damage. He became one of the directors of the Arkansas Farm Credit Company, formed late that summer. Federal authorities like Herbert Hoover, the U.S. commerce secretary who was assigned by President Calvin Coolidge to oversee the relief effort, seemed to be supremely unaware of the boon to cotton growers. As part of the effort to encourage economic recovery from the flood, Hoover advocated the creation of farm credit associations throughout the region. The Arkansas Farm Credit Company, formed late that summer, included some of the most prominent names in Arkansas, including A. B. Banks, a leading banking mogul in Arkansas; Harvey Couch, the president of Arkansas Power and Light; and Joseph T. Robinson, Arkansas's senior senator. Lee Wilson, who was among the original stockholders along with Banks, Couch, and Robinson, also served as an elected director. The directors selected W. A. Hicks, chairman of the American Southern Trust Company, as president. Capitalized with a total of $676,815, with $250,000 coming from the Flood Credits Corporation of New York, the Arkansas Farm Credit Corporation made in 1927 a total of $361,063.65 in loans to 2,566 families in Arkansas. They continued making loans in 1928 to the tune of $217,007.50 to 1,445 families. Of the Mississippi County families, 102 received $23,239 in 1927 to plant 3,802 acres of cotton and 1,846 acres of feed crops. In 1928, 57 Mississippi County farmers received $10,345 to plant 1,724 acres of cotton and 886 acres of feed crops.[50]

The operation of the Arkansas Farm Credit Corporation was not without controversy. The president of the corporation, W. A. Hicks, insisted that all successful applicants for loans from the corporation provide sufficient collateral. It appears that not all of the directors agreed with him. Harvey Couch characterized Hicks's appointment as president as "one of the biggest mistakes made in this disaster." W. Wesselius, the Red Cross reconstruction officer for Arkansas, reported in March 1928 that Hicks's attitude "is entirely that of a cold-blooded banker who looks upon each loan as an instrument which has to be entirely gilt-edged and especially at this time it occurs to me that he is requiring more in the way of collateral than is ordinarily asked for in this country."[51] It is impossible to know for certain what Lee Wilson thought about this matter. On the one hand, he was a close personal friend of Harvey Couch's. On the other hand, he was a hard-nosed businessman and banker himself. In any case, Hicks's point of view prevailed. The Arkansas Farm Credit Corporation functioned like a bank. In the end, the corporation reported very few losses. In Mississippi County, in fact, only one $50 loan—to a farmer with fewer than two hundred acres—went unpaid. Although the record did not provide the farmer's name, he was almost certainly from the heavily flooded northwest part of the county. Given that $10,345 in loans were advanced that year to county residents, the default rate was an enviable .04 percent.[52]

Lee Wilson came through the flood of 1927 relatively unscathed, but the havoc raised by the disaster created hardship for many of the small farmers of the northwestern part of the county and for some planters on the eastern side of Big Lake. Planters like Wilson who were able to put a crop in the ground enjoyed a windfall of high prices in 1927, but there were problems on the horizon arising out of the flood. Even though most of the county's drainage district levees held, some were seriously damaged and fundamentally undermined. Although the Mississippi River Commission and others pleaded for federal funds to repair the levees, very little money was granted, and tax revenues were barely sufficient to meet the bonded indebtedness that the drainage districts had incurred. The Great Depression was still a few years away, but the agricultural sector had been mired in an almost unrelenting slump for almost a decade. Most industries had pulled out of the recession that followed World War I, but agriculture and lumbering, the two chief industries of Mississippi County, had never fully recovered. The flood of 1927 dealt the county—and most other counties in the lower Mississippi Valley—an almost fatal blow. John Barry's *Rising Tide* has focused attention on the lower Louisiana region, and it is certainly true that the work of the Corps of Engineers in constricting the river to a narrower channel has had the greatest impact on that region. Once the federal government fully accepted the responsibility for flood control and once the Corps of Engineers was able to avail itself of more sophisticated technology, it was only a matter of time before the Mississippi River was channeled into a fast-moving monster capable of wrecking havoc at its vulnerable outlet near New Orleans. However, for farmers and planters in Mississippi County and elsewhere in the middle and upper Mississippi region, the policy has paid dividends in the long term. The real losers of flood control and drainage policies—aside from the wildfowl left without a significant wetlands on the north American flyway—were those who could not afford to pay the improvement taxes in the late 1920s and early 1930s and lost their land as a result. Lee Wilson himself found the early 1930s a treacherous period. In an almost biblical turn of events, he survived the flood only to face near ruin as a result of the drought of 1930–31. He managed because he was able to call on the considerable resources at his command: substantial assets and political connections. During the first decades of the twentieth century, Wilson had

created an impressive plantation empire by employing Progressive Era business practices and an ingenious organizational strategy. His business bore the
imprint of the man himself and served to both support and control his family
members. It turns out that the genius of his organizational structure was also
its greatest weakness: his family, something that would become painfully clear
to him in the early 1930s.

7

"GET HARD AND
RAISE HELL"

Despite the disastrous flood of 1927 and a decade-long depression in agriculture beginning after World War I, Lee Wilson appeared to remain impervious to it all. While many other farmers and planters proved unable to weather the almost perfect storm of environmental and economic catastrophes, Wilson consolidated and expanded his operations. In 1930, for example, he purchased the McFerren plantation, which he had been renting for several years, a purchase that, together with some other acquisitions, increased the total acreage he owned from twenty-eight thousand acres in 1920 to over forty-six thousand acres. Yet he too faced difficulties that only grew more serious in the early 1930s. As the Great Depression crept over the country, yet another environmental challenge emerged, one which he had neither anticipated nor prepared for: a catastrophic drought. The unprecedented economic crisis following the stock market crash of 1929 drove many into bankruptcy, and the drought of 1930–31 delivered a devastating blow, pushing some of the survivors over the abyss. Not since the one-hundred-year drought that De Soto witnessed in the mid-sixteenth century had the country seen such a disaster. Crops "burned up" in the field, and Wilson and other agriculturalists harvested less than 50 percent of the acreage they planted in crops. Given the worldwide economic crisis that affected manufacturers and consumers alike, cotton prices, rather than rebounding because so little of the staple reached the market, instead slipped to nearly a nickel a pound by 1931, a historic low.[1]

Illness complicated matters further for Lee Wilson, but he found it impossible to take the time he needed to recover his health. Instead, after a brief respite, he redoubled his attention to the details of running his company, embracing new opportunities arising because of the election of Franklin Roosevelt in 1932. He welcomed New Deal programs without misgivings about the new role the federal government would assume in his affairs. After spending a lifetime making the most of all opportunities and turning them to his advantage, he had supreme confidence in his ability to manipulate federal officials and government programs. Wilson understood that his own wealth and prominence guaranteed him the solicitations of political figures and economic powerbrokers, and as he struggled to save his empire from bankruptcy, he fiercely pursued every connection he enjoyed with such individuals. They included Arkansas's senior senator, Joseph T. Robinson, permanent chairman of the National Democratic Central Committee, who became Senate pro tem in 1932. Wilson also sought assistance from Harvey Couch, the president of Arkansas Power and Light, who became one of the directors of the Reconstruction Finance Corporation (RFC) in Washington, D.C. Couch, Robinson, and other members of the Arkansas congressional delegation aided Wilson in his successful campaign to secure a million-dollar RFC loan, but he also secured the forbearance and aid of his major creditors. Bankers and investment company executives to whom he owed hundreds of thousands of dollars used their influence to facilitate the government bailout, and they signed the waivers necessary to permit Wilson to restructure his company in order to qualify for the RFC loan. Wilson was, in twenty-first-century parlance, too big to fail. His creditors feared his economic collapse almost as much as did he. Wilson simply owed them too much money.

Throughout this latest struggle to save his life's work, Wilson suffered from a prolonged and undiagnosed illness. In fact, his health deteriorated to the point that he retired for a brief period in 1931, in part to recuperate and seek medical treatment but also to keep company with his mistress, the beautiful and youthful Helen May Vaughan. Helen, a twenty-seven-year-old divorcee from Stuttgart, Arkansas, lived in his vacation home in Hot Springs with her two young sons. Neither Wilson's wife, Elizabeth, nor his son, Roy, knew of Helen's existence, but she entertained many of the friends and business associates who visited Wilson in Hot Springs and accompanied him on business trips. As one of his attorneys later wrote, Wilson was an "ultra-modern" man and Helen "went practically everywhere with him."[2]

Regardless of his preoccupation with his young mistress, Wilson's ill health convinced him to surrender "control" of the company to his son, R. E. L. "Roy" Wilson, at a meeting of the holders of certificates of trust on April 3, 1931.[3] His illness had surfaced in December 1930, and it must have been very serious, indeed, because it was just then that the ramifications of the 1930 drought were being fully realized. At that juncture, with a short crop to harvest and cotton prices down to 9.6 cents per pound, he would have had little money to pay his creditors, but he left it to his son to seek solutions to the company's dilemma. He consulted doctors in Memphis, Hot Springs, and at the Mayo Clinic.[4] He grew weary of the remedies offered by his physicians and in May of 1932 "told them all to go to hell."[5] He continued to suffer from ill health, but the situation confronting his company so alarmed him that he returned to active involvement in its affairs and began to reassure nervous creditors. To Benjamin Lang of the First National Bank in St. Louis, with whom he was renegotiating the company's indebtedness, he wrote, "I am hitting the ball every day and busy as I can be, therefore, can't say when I will be in St. Louis, but will see you whenever I come."[6] Late that year, he officially resumed the trusteeship at the annual meeting of the trustees. His creditors probably demanded nothing less than that. Writing R. B. Barton, president of the Memphis Bank of Commerce and Trust, to whom the company owed $120,000, Wilson said in December 1932, "I am glad to report to you that my health has improved to such an extent that I am able to resume business control of Lee Wilson & Company; and my only hope is that God will give me strength to run this business and pull it out of its present condition."[7]

Wilson's mention of God was either rhetorical or a signal of the desperation he felt over his business and his health. He typically saw himself as above or apart from community standards of conventional morality, and it was well known that he had not been "saved" in the Christian sense. Although he sometimes attended the Wilson Baptist Church, he had never been baptized, and it was not uncommon for him to receive communications from people who praised his generosity to good causes but who also prayed for the redemption of his soul. Rev. Paul V. Galloway (a Methodist who was later bishop of Arkansas and Louisiana) of Joiner wrote, "I hope that you will someday soon make your life more consecrated. I know that you have done hundreds of things for people (I have been told of many) and that you have given money to worthy causes." Galloway appreciated all that Wilson had done but believed that if he would "make a forward step religiously it would do more to help this country than all the preaching all of us could do for a year. People look to you financially and empirically and also morally."[8] In the summer of 1932, a Mrs. Walter Wood, who attended a revival in Wilson, wrote to him saying "Bro. Shaffer prayed for you, asking God to restore your health and that you would be saved before the meeting closed. I have thought about you lots, since hearing his prayer. I, too, Mr. Wilson, am praying you will find Christ. . . . And please don't forget that each morning at 9 o'clock I shall be in secret prayer for you for you know Mr. Wilson, prayer will remove mountains."[9]

At the time Wilson received Mrs. Wood's letter, he was sixty-seven years old and had a little over a year left to live. His brush with serious illness gave him a greater sense of his own mortality, however, and to a young friend, John B. McFerren, the heir to the McFerren plantation that Wilson purchased in 1930, he wrote that though he was "considerably better" he felt he had "to conserve my health very closely, for fear that I have a relapse."[10] To another correspondent he wrote, "I am especially anxious to get myself back in shape and live to an old age."[11] Although he seemed to have some premonition that he was not long for this world, he threw himself into the task of salvaging the company he had spent a lifetime building. By this time it was a sprawling enterprise spread over more than fifty thousand acres—including both acreage he owned and acreage he rented—with twenty-five to thirty thousand in cotton cultivation. The company included mercantile establishments, sawmills, planing and box factories, and cotton gins. Wilson employed approximately 2,500 people on his farms and in his other businesses. As a result of the collapse of the banking empire of A. B. Banks and the subsequent bankruptcy of hundreds of smaller banks throughout the state, Wilson stepped in to "rescue" banking operations in the towns of Osceola and Blytheville, essentially assuming control so that by 1932, he controlled four of the five banks in Mississippi County. In addition to the banks he founded in Wilson and, more recently, in Keiser, he now owned the largest shares of the Bank of Osceola and the Bank of Blytheville.

In the town of Wilson, the heart of the company, he employed ninety-four salaried employees, with Frank Gillette acting as the top executive.[12] In response to a query from a man interested in purchasing land in the area, Gillette described Wilson as "a modern little city, the most beautiful between St. Louis and Memphis. The unusual feature is that it is a 'one man town.' In other words, every building at Wilson, except for the depot and ice plant, is owned by the Lee Wilson & Company organization. Wilson is the central point of the plantations of Lee Wilson & Company, reputed to be the largest individual cotton growing plantation in the world."[13] To a creditor concerned about one particular category on the company financial statement—improvements— Gillette responded that they were "all very substantial and do not compare with the average improvements of farming lands as is generally considered by people not familiar with our holdings. The improvements in our towns are especially good. For instance, at Wilson we have a concrete, reinforced two-story administration building, which is approximately 300 x 200; the original cost of which was approximately $120,000. We have a concrete, steel reinforced gin at Wilson, in which is invested approximately $75,000.00; and a saw mill at Wilson, which is steel and concrete, the original cost of which was approximately $250,000.00." Gillette also said that Wilson boasted "an artesian well 1500 feet deep," and that "all of our properties are maintained in excellent condition." Gillette's boast was no exaggeration. Wilson made certain that he had the most modern equipment available and understood how to collateralize every piece of property he owned. His buildings, machinery, and equipment also represented his face to the world, and he fully appreciated the reputation they promoted.[14]

Although the manufacturing of forest products remained a lucrative and important part of the company's profile, by the early 1930s Wilson found it necessary to look far beyond Mississippi County for trees to harvest. The company had no interest in tree farming, something that was soon to become a practice employed by large lumbering concerns elsewhere, as most of the nation's available "virgin" forests were depleted by the mid-twentieth century. Given the fertility of the soil in Mississippi County, Wilson had no incentive to do anything other than turn to more traditional agricultural pursuits, so he looked elsewhere for sources of lumber. In 1932 he purchased a half interest in Peter's Island, located on the eastern side of the Mississippi River in the state of Mississippi. While Wilson hoped to use the island as a hunting preserve, he also intended to harvest the trees. Meanwhile, he employed Frank L. Zander of St. Louis to locate stands of timber that the company could harvest in Missouri, Illinois, Kentucky, and Tennessee, to name but a few states. He authorized Zander to seek rights to harvest timber from property owned by others, to purchase timberlands outright, or to contract with individuals wishing to market the timber they had themselves harvested. Eastern Arkansas lumbering operations, Lee Wilson's among them, were within a decade or so of putting themselves out of business.[15] Far more important to the company's future was the farming side of the business.

Wilson's complex farming operation represented the future of Lee Wilson & Company, and it took all of Lee Wilson's energy and ingenuity to solve the problems confronting agriculture in the era of nickel-a-pound cotton. He expected all his employees to do their bit too, particularly his executives in Wilson and the managers located in his other towns—Evadale, Bassett, Marie, Keiser, Armorel, and including his latest acquisition, McFerren, which he renamed Victoria. They had responsibility for the six subsidiary organizations made up of mercantile establishments, cotton gins, and farm units. The general managers enjoyed considerable autonomy, but they all took orders directly from Lee Wilson and Frank Gillette. James "Jim" Crain served as the executive officially assigned the responsibility for all farming operations for the company, but he must have done most of his supervision in person, for the correspondence files reveal very little written direction from him. Most of the correspondence between Gillette and the general managers involved routine matters, such as those pertaining to payments on merchandise accounts. The managers themselves actually placed orders with hardware and dry goods wholesalers located in Memphis, St. Louis, Chicago, and a variety of other locations, but most of their creditors had come to understand that they were backed by Lee Wilson & Company and turned to the head office when payments were slow. Frank Gillette typically contacted the individual store managers when creditors notified him that a problem existed. When Jesse Clinton, Oscar Hill, and D. D. Waddell got behind on payments to William R. Moore Dry Goods Company in July 1932, Gillette suggested that they "endeavor to give them [Moore Dry Goods] some assistance and satisfy them."[16] Roy Wilson seldom interacted with them but if they fell behind in their payments on the notes they held with the Bank of Wilson, he admonished them. Oscar Hill, for example, had a note due at the bank on October 30, 1932, a convenient date when collections should have been at their best. With cotton prices disappointingly low, Hill was having a difficult time collecting on his accounts. Roy reminded him that he had a $750 note due and wrote: "It is absolutely necessary that Hill & Wilson make a little payment on their obligations at the Bank of Wilson."[17]

The correspondence the store managers had with Frank Gillette and Roy Wilson reflected a crisp and businesslike tone. Lee Wilson's communications with them, alternately familiar and dictatorial, revealed his intimate knowledge of their individual operations. While Gillette and Roy Wilson merely told them to come up with some funds to meet their obligations, Lee Wilson gave specific direction about how to generate funds. For example, he wrote to the store manager at Armorel, William M. Williams, in October 1932, calling his "attention to the matter of collecting accounts due your firm at Armorel. . . . I want you to arrange to put somebody on this collecting that is aggressive and hardboiled. . . . If you cannot get the money take cotton, corn, hogs, or cattle on the account, but see that something is done on collecting these accounts." He made the same point with Jesse Clinton, who managed both the plantation operations and the store at the Idaho Grocery Company at Bassett. Wilson emphasized that their debtors were most likely to have funds during harvest and "if the accounts are not collected now or in the next thirty days, they never will be collected."[18]

Wilson's close supervision of the farming operation barely obscured the complex nature of its structure. Jim Crain was technically in charge of all farming matters, but the general managers exercised oversight over the farm managers under them on their plantations. The farm managers were enjoined to closely supervise the farmers who worked under them. Even in the matter of appointment of farm managers, Lee Wilson exercised direct personal influence. He forbid general manager Charles Crigger from hiring someone as a farm manager that he did not approve of and even lectured him on being "very careful about making trades with managers, and I want every trade made with managers to be made in writing. Take up with me the trades you have made with managers so I can reduce it to writing and have them sign it."[19] The farm managers themselves were reminded of Lee Wilson's connection to them at meetings held at the Community Club House in Wilson. At a May 1932 meeting, the boss lectured them on the importance of keeping their cost of production low while maximizing yields, and he "offered to the man that raises the highest yield of cotton on five acres $100.00 in gold for first prize; $50.00 for second and $25.00 for third. Same thing applying to a five-acre corn yield."[20]

In the final analysis, the crucial players in the management of the company's sprawling farm operations were the general managers, who had a wide variety of responsibilities and a direct personal relationship with the boss. He relied on them to run their operations efficiently and to follow his orders precisely. His management style was to delegate but also to closely supervise his managers. He wanted them to follow his orders, but he also wanted them to have sufficient authority on their own. When it appeared to him that one of his managers was lacking that capacity, he made an issue of it. He wrote to Charles Crigger, "I am informed that some of the Farm Managers and other people at Armorel do not think or consider you their Boss. I am surprised to get this information." Indeed, Wilson's surprise was almost certainly genuine. He had placed Crigger in charge of the Armorel operations precisely because of his maturity and experience. "I wish you would get out there and read the riot act to every God-damn man on all the places, the store crew, office crew and all. Tell them that you are the boss and that I am holding you personally responsible for all the transactions at Armorel, in every line. The thing for you to do is to get hard and raise hell."[21]

Wilson believed that his interests were best served when his managers exercised their own authority, and he depended on them for matters above and beyond the routine concerns associated with running their farms and stores. They were the key figures located on the various plantations, and if he wanted some specific action carried out, it was to them he turned. For example, in March 1932, he sent a directive to them related to the assessment of personal property, something that was actually the province of the county government. To Crigger he wrote, "There is an enormous number of people in this county who do not own any land, but own stock, tools, wagons, cars, etc. and they are not paying any taxes. I want you to get in touch with the tax assessor, J. S. Dillahunty, and when he goes to assess taxes in the Armorel School District, you arrange to have a man go with him and assess every person in that district that has property. I want every share-cropper and everyone in the district assessed. If they only have $50.00 worth of household goods, then assess that."[22] He had two reasons for being concerned about this matter. First, he was one of the organizers of a Taxpayers Association looking into county expenditures and was well aware of the fact that tax collections were woefully inadequate and that the county could not sustain the schools, roads, or drainage ditches.

Wilson's second reason for expressing an interest in assessment of personal property involved its connection to eligibility to vote in the coming Democratic Party primary election. Wilson had a keen interest in seeing not only the election of Junius M. Futrell as governor but also the defeat of Dwight Blackwood. Only those who assessed personal property were eligible to pay the poll tax, and only those with poll-tax receipts could vote. Given that blacks had been excluded from the Democratic Party primary elections since 1906, it was the enrollment of white voters that concerned Wilson, and they were the ones mostly likely to hold personal property. Wilson's reasons for wanting Blackwood defeated are not entirely clear. Blackwood had long been involved in county politics, having held the position of county sheriff from 1919 to 1925. In fact, it was Blackwood whom Governor Thomas McRae criticized in 1921 when Henry Lowery was lynched, but that event had not tarnished his career. He became director of the Arkansas Highway Department in the late 1920s and was serving in that position in 1932. In fact, Wilson professed to be dissatisfied with his handling of that organization, and, indeed, Blackwood's office had been audited for supposed irregularities, and though he had been exonerated, not all of his critics had been satisfied. Although Wilson had never run for public office, he was on a first-name basis with governors, legislators, congressional delegates, and, of course, county officials, for more than two decades. They understood he could deliver votes in key elections.

For whatever reasons, he intended to use his influence to defeat Blackwood. In August 1932, Wilson wrote to F. N. Wilkes, an executive with Arkansas Power and Light Company (AP&L), expressing his concern about the coming election. Wilson had only recently sold Wilson Power and Light to AP&L, earning himself a place on the latter's board of directors. He also enjoyed a close personal friendship with Harvey Couch, the founder and director of AP&L. Wilkes would have found it necessary to listen attentively to Wilson's point of view. Wilson wrote to Wilkes, "Possibly you are not aware that Bill Avera, your manager over in this section, and a lot of your crew are not right in the race for Governor. I think a lot of them are for Dwight Blackwood, and as you already know, if Blackwood is elected it means a continuance for the present rascality and thievery in state office. We are for Futrell over here, heart and soul, doing everything we possibly can to elect him. I think you should get in after the above at once and if there is anything you want me to do on this, in this neck of the woods, just say so." Wilson had prepared a circular titled "Blackwood's Testimony before Audit Commission vs. The Records in Mississippi County and Elsewhere," which he attached to his letter to Wilkes, asking him to distribute it. To W. C. Wall, assistant manager at the Armorel operation, Wilson wrote, "I hope there has been a good many poll tax receipts brought up there and that you will support Futrell for governor." While that statement falls short of an order to see to it that he make certain voters in the Armorel area voted for Futrell, Wall could hardly have been ignorant of Wilson's intentions.[23]

Wilson attempted to make effective use of his relationships with prominent politicians but he was not always successful in securing the favors he hoped for. He supported Futrell for governor because he had known him to be a conservative man of business who had a reputation for honesty. He also knew him to be the chancellor in charge of appointing the "receiver" for the bankrupted Mississippi County drainage districts, and after Futrell's election he sought, unsuccessfully, to convince the new governor to appoint L. L. Hidinger, who was president of Morgan Engineering Company in Memphis, to the position. Wilson had worked closely with Hidinger for more than a decade on the various drainage enterprises in the area—it was Hidinger's men who confronted saboteurs and the National Guard during the flood of 1927—and he wrote letters to several prominent men asking them to encourage Futrell to appoint Hidinger to the receiver's position. When Futrell proved determined to appoint an attorney rather than an engineer and ultimately selected Ernest Ahlfeldt of Stuttgart in July 1933, Wilson took it all in stride. Wilson promptly wrote to the governor's appointee, congratulating him on his appointment and asking to meet with him in Little Rock "to talk matters over."[24] A pragmatic man, Wilson had more pressing matters to address. He meant to use his connection to Futrell and other prominent politicians and businessmen in the effort to save his company.[25]

Lee Wilson faced almost certain ruin in the fall of 1932, and he needed all the help he could muster. He had yet to recover from two bad years in a row, and cotton prices were down to a nickel a pound, well below the cost of production. With debts at unprecedented levels, many of his creditors began to hound him relentlessly, particularly wholesalers, to whom he owed over $100,000. He also owed nearly $1 million to banks, finance companies, and cotton factors. Although his February 1933 financial statement indicated that Lee Wilson & Company was worth $5,347,012.40, most of his assets were tied up in land and buildings, machinery, and livestock. He had only $12,439.80 in cash available, a paltry sum for such a large enterprise. He listed over $1 million in notes and accounts receivable, but much of it was worthless. His assets included another $500,000 in merchandise, including store stocks, lumber, and agricultural products, but there was little prospect of generating much revenue from this source as the depression deepened and few buyers were to be found. The $2.5 million surplus he claimed rested on a shaky foundation of merchant stock, uncollectable debts, and heavily mortgaged real estate and producing lumber and agricultural products for which there was little demand.[26]

Lee Wilson & Company's financial situation had worsened sufficiently by the fall of 1932 that he began to work through a scheme to secure a Reconstruction Finance Corporation (RFC) loan. Some of his largest creditors, in fact, were attempting to assist him. The bankers, to whom Wilson owed nearly $1 million, were eager to find some way to help Wilson save his company from bankruptcy. By the end of 1932, Wilson's debts to six entities alone amounted to $990,902.83:

Bank of Commerce & Trust, Memphis	$150,000
Franklin Life Insurance Co., St. Louis	$100,000
Union Planters Bank, Memphis	$160,000
Mercantile Commerce Bank & Trust Co., St. Louis	$140,000
First National Bank, St. Louis	$195,437.50
Wilson-Ward & Co., Memphis (cotton factor)	$245,465.33

Absent from any of the correspondence he had with them was the stridency he encountered from smaller creditors. Ben Lang, vice president of the First National Bank of St. Louis, to whom Wilson owed $195,437.50, took a solicitous tone. R. B. Barton of the Bank of Commerce in Memphis remained cordial throughout 1932, even as he demanded additional collateral in the form of warehouse receipts for cotton in exchange for continuing to carry Wilson's past-due notes.[27] All of these loans were secured by some form of collateral, but by the end of 1932, most of the value of much of the collateral was doubtful. The First National Bank loan was secured largely by stock in various Wilson entities and in school bonds that were practically worthless.[28]

The cooperation of the bankers in securing the government loan was essential because it required them to subordinate their loans to the RFC. The plan involved a complicated rearrangement of Wilson's finances because the RFC had a narrow interpretation of the kind of entities it could lend funds to, and "southern plantations" did not fit their profile. At issue was that agency's refusal, initially, to accept farmlands as collateral. President Herbert Hoover fashioned the agency in January 1932 to provide loans to banks, trust and insurance companies, building and loan associations, and railroads. He raised its initial capital of $500 million dollars to $3.3 billion by mid-1932. The only agricultural entities it was authorized to make loans to, however, were agricultural credit and livestock credit associations, not farming or plantation operations. Within weeks, perhaps days, of the passage of the legislation creating the RFC, Wilson went to Washington and sought advice from longtime friend and associate Harvey Couch, who was appointed one of seven directors of the RFC. In the end, however, neither Lee Wilson nor his well-placed friend could get beyond the limitations of the legislation creating the RFC.[29]

Although disappointed in his first attempt to secure an RFC loan, Wilson began to formulate a plan for reinventing himself. The revision of the RFC statute in July 1932, which increased the amount it was authorized to lend and broadened its functions, probably served as inspiration. Although this enlargement of its functions did not extend to agricultural enterprises like Wilson's, he was probably encouraged to hope that he could turn it to his advantage and, at the very least, he was inspired by the expansion to $3.3 billion dollars in funds it could lend.[30] In September, he visited his friend Harvey Couch in Washington once again and was told that since the law had been amended Wilson might be able to model himself after the Southtex Mortgage Loan Company of Houston, Texas, which had just secured a $500,000 loan using chattel mortgages and trust deeds on agricultural lands as collateral. Couch suggested he wait until the spring to make application but Wilson thought otherwise.[31] He wrote Senator Hattie Caraway on November 5, 1932, asking her to secure a copy of the Southtex application from Elbert Smith, Couch's assistant. "I suppose that it would not be possible to get a copy out of the files of the RFC, but possibly you could borrow a copy for a day or two and have a copy made for me." He offered to pay all expenses for mailing the report and thanked her in advance for "doing me this favor." Five days later he received the information he requested from Caraway's assistant, with a missive reading "It is always a pleasure for Mrs. Caraway and I to be of assistance to you."[32]

Wilson explained to Couch in December that the fact that his "banker friends" were eager for Wilson to meet his obligations to them influenced his decision to make another attempt to secure RFC funding. Wilson wrote Couch, "I hate to worry you with my affairs, but if I cannot call on a friend under such circumstances, then who in the world can I call on." Couch suggested he create an entity that included "other mortgages as might be easily done." What Couch meant became clear as Wilson began to formulate his plan. With the help of his longtime attorney, Charles T. Coleman of Little Rock, Wilson created the Eastern Arkansas Mortgage Company and formally transferred all of his assets and debts into the new entity. He assigned himself the vice presidency of the company and had Hy Wilson (one of his executives and unrelated to Lee Wilson) occupy the presidency. But bureaucrats at the RFC office handling the paperwork suggested in mid-January that if it could be shown that the "proposed mortgage company had wider field and was ready and able to do regular mortgage business it would be eligible" for an RFC loan. Wilson should have been prepared for this opinion as he had heard it from Couch and also from an Indiana attorney he had consulted, William Bachelder, who had organized finance companies in Indiana, Virginia, Colorado, and other states that had applied for loans.[33] Once the bureaucrats weighed in on the matter, Wilson began to lay the groundwork for compliance and had Hy Wilson round up mortgages held by his many friends in northeastern Arkansas.

Throughout these negotiations, Wilson had been consulting with his largest creditors, seeking both their advice and their cooperation. C. Leroy Sager, James L. Ford, and Ben Lang, all vice presidents at the First National Bank in St. Louis, had been especially involved, and Ford had actually introduced Wilson to Bachelder. But it was Sager who made the key introduction to an attorney in St. Louis in November 1932. The attorney suggested Wilson form a mortgage company along the lines that Wilson eventually settled on.[34] These bankers were motivated by more than the desire to help a friend in need. By this point in time, the Wilson indebtedness to the First National Bank had risen to nearly $200,000, and his default would have had serious consequences for the bank.

The election of Roosevelt in November 1932 had given Wilson reason to hope for some drastic measures on the part of the federal government and prompted him to include in correspondence to friends and family the words, all in capital letters, HOORAY FOR ROOSEVELT, THE COUNTRY IS SAVED.[35] But Wilson's problems demanded more immediate solutions, and he could not wait for Roosevelt to take office. In any case, Roosevelt had been rather vague in precisely what he was going to do in order to rescue the country from the economic morass it was mired in. Meanwhile, the price of cotton had fallen to new lows in the fall of 1932, and Wilson's creditors had become increasingly nervous. They certainly had reason to be concerned. He wrote to the Mercantile Commerce Bank on December 30, 1932, that he was attempting to secure refinancing through the RFC and that, in the meantime, he was not in a position to pay even the monthly interest due on the $140,000 loan he had with them.[36]

The loan he secured from the Mercantile Commerce Bank demonstrated Lee Wilson's lifelong penchant to take chances, even in the worst of circumstances. He negotiated the loan with them in mid-1930 to purchase the "McFerren place," an audacious move considering the condition of his business and the economic situation confronting the country.[37] He had been farming it for the heirs of its previous owner for several years, and after purchasing it, he made a number of improvements and turned it into a showcase plantation. Although this move was audacious, it was also characteristic of Wilson to take advantage of opportunities like the McFerren purchase, which he got for a bargain, even when his own resources were strained to the breaking point. Indeed, one former Wilson employee, M. W. H. Collins, wrote him from Georgia in December 1932 recalling "very distinctly one Sunday afternoon some two or three years ago of you telling me that your most successful enterprises had all had the groundwork laid during periods of 'hard times.'"[38]

Unable to pay even the interest on his outstanding indebtedness, Wilson faced the prospect of lacking sufficient funds to launch the 1933 crop year. With the relaxation of certain rules under which agricultural credit associations operated, rules which had made it impossible for farming enterprises of Wilson's size to obtain loans, he was ultimately able to secure a $161,750 crop loan through the Regional Agricultural Credit Corporation of Pine Bluff, Arkansas. On February 13 cotton factor W. L. "Lucian" Oates wrote a recommendation to support his application for that loan, which reflected the lengths to which his friends were willing to stretch a point: "It is hard to know very much about the financial responsibility of anyone at this time, but I do not hesitate to say that the moral risk is as good as the best, and as my business experience has taught me that a moral risk is much more desirable than a financial risk, I should say that Mr. Wilson is deserving of every consideration."[39] Wilson then had to convince his friends at the First National Bank in St. Louis to sign an "Agreement not to Disturb," meaning they would subordinate the mortgages they had on his property to those of the Agricultural Credit Corporation. Ford was willing to do so only "providing the interest [on Wilson's indebtedness to them] are paid up to date, monthly interest paid regularly thereafter, or, failing your ability to do that, that you assure us you will pay us the interest out of any loan secured and clean up your interest obligation."[40] Once again, his friends had been willing to extend him a special courtesy.

Even as Wilson secured the production loan, he waited impatiently for the RFC application to make its way through what Wilson viewed as a tedious, even tortuous process. By mid-February he had received from Charles Coleman, his Little Rock attorney, the paperwork necessary to hold the first stockholders' meeting of the Eastern Arkansas Mortgage Company,[41] but it was taking longer to finalize the loan than either Wilson or his creditors imagined, and as the weeks turned into months, they grew increasingly uncomfortable. Both James Ford and Ben Lang of the First National Bank of St. Louis wrote him asking for information on his progress.[42] Finally, his papers were approved by the Little Rock office of the RFC on March 30, 1933, and sent to Washington for final approval. Wilson reported to Ford on April 1, "I am leaving for Washington tonight and will sit up with them until they approve the proposition or kick me out."[43] Even Roy Wilson contributed to the effort his father was mounting once the papers reached D.C. Roy wrote a close personal friend and classmate from Yale, attorney George Thompson Jr. of Fort Worth, Texas, asking him to use any influence he might have with Jesse M. Jones. Thompson suggested that he ask his friend Amon Carter, a prominent newspaperman in Houston and a confidant of Jones, to intercede and, in fact, Thompson wrote the letter which Carter signed. Jones, the chairman of the board of directors of the RFC in Washington, was the most crucial figure in the negotiations. The letter that Carter signed attested to Lee Wilson's reputation and asked Jones to meet with Wilson while he was in Washington.[44] Roy also asked his friend Lucian Oates, an executive with the Anderson Clayton and Company branch in Memphis, to use his influence with the president of the head office of that company, located in Houston. Subsequently, W. L. Clayton, who was personally known to Jones, wrote the RFC director, characterizing Wilson as "the biggest planter in the South and is generally looked upon as a kind of king in his community, which comprises a vast tract of the richest land in Mississippi County, which perhaps is the richest land in the United States. He is generally looked upon as a leader in progressive agriculture and seed development, as well as a big-minded citizen."[45]

More hurdles had to be overcome, however, before progress was made. While in Washington in early April 1933, Wilson visited with every member of Arkansas's congressional delegation and spent time with his friend Harvey Couch. He returned from Washington in mid-April and wrote to Couch complaining about questions the RFC legal counsel in Washington was raising about the mortgages. He explained to Couch that "Mr. Reed, the Chief Counsel of the R.F.C., told me to organize a mortgage company with a capital stock of $750,000.00, with $125,000.00 paid in cash, and $625,000.00 in mortgages, which I have done, which takes up the entire capital stock of the Mortgage Company. He told me if I did this, it would be eligible. I have complied with his request to the letter."[46] As the loan application languished in the Washington office, Wilson's creditors grew even more anxious. Ben Lang wrote on May 2 that "I am becoming rather tired of keeping my fingers crossed, so relieve me, if possible." After learning from Frank Gillette a few days later that Wilson was in Washington again, Lang feared the worst. "It would seem to indicate that the loan is not going through as anticipated." Gillette reassured him that "the last information the writer has from Mr. Wilson is most encouraging."[47]

In fact, the application received tentative approval from the RFC legal division in Washington on May 6, 1933. Wilson traveled to Little Rock to meet with Harry Meek, chief legal counsel of the RFC branch office, but Meek wanted to discuss the application with Charles Coleman, the attorney who had drafted the paperwork creating the Eastern Arkansas Mortgage Company. Wilson wrote Coleman to contact Meek, saying, "Now, if you ever did me a favor in your life, then for God's sake do it." Meek was concerned about a law the Arkansas legislature had just passed that provided landowners with four years to redeem property that was foreclosed on. Coleman wrote a brief on the question, which was subsequently sent back to Washington for the legal office there to peruse.[48] Finally, in mid-June, Wilson received official and final approval of his application. On June 19 the Federal Reserve Bank branch in Little Rock notified him that "we have today disbursed proceeds of your loan from the Reconstruction Finance Corporation in the amount of $756,988.90 as follows: Deposited with ourselves $647,938.33 to retire collection and escrow items held by us. Forwarded [is] our Officers check in the amount of $109,050.57 to the Bank of Wilson, Wilson, Arkansas, for your credit and advice."[49] Further disbursements were scheduled to take place later that year to cover improvement taxes (most of them county drainage district taxes) in the amount of $165,000 and the balance on Lee Wilson & Company bonds of $192,400, making the RFC loan total $1,114,398.90. Although Wilson had always been careful to pay his taxes—knowing all too well the possible consequences for not doing so—he had calculatedly refrained from paying them for the years 1932 and 1933, counting upon the RFC loan to come through.[50]

Wilson's banker friends were pleased, of course, as the $647,938.33 "to retire collection and escrow items" covered Wilson's debts to them. James Ford at the First National Bank of St. Louis was almost giddy, writing on June 19, "I am just delighted to death and so are Ben and Roy about your Reconstruction loan. I wish to thoroughly congratulate you." Wilson's friends and business associates who had agreed to place their assets into the Eastern Arkansas Mortgage Company in order to satisfy the demand of the RFC that the entity include other parties received the balance. While they numbered one-third of the mortgages, their total indebtedness was far less than Wilson's, standing at only 14 percent. Wilson had expended considerable energy and used every friend and acquaintance at his disposal in order to secure the loan. As he put it to the president of Mercantile Trust Bank in St. Louis, "Yes, I had a very hard pull with the RFC to get my proposition over. I dare say if I had been a quitter, I would have blown up."[51]

The Reconstruction Finance Corporation gave Wilson the room he needed to reorganize his operation, but even as he celebrated the success of his effort to secure the loan, he welcomed an initiative launched by the newly appointed secretary of agriculture, Henry A. Wallace, to bring relief to farmers and planters in the spring of 1933. Amid the flurry of legislative acts passed in the first hundred days of Franklin Roosevelt's administration was the Agricultural Adjustment Act, a measure which Lee Wilson vigorously endorsed. The act passed Congress on May 12, 1933, and created the Agricultural Adjustment Administration (AAA). It provided for a dramatic remedy for the problems confronting farmers, but many of the details were left to be worked out by the Department of Agriculture. The basic plan was to pay farmers to withdraw certain crops from production, including cotton, with the idea that prices would rise as a result of the imposed scarcity. Farm groups had proposed something like it during the 1920s, but farmers were not united behind it, and the Republican administrations of that period were adamantly opposed to it. Farmers had to be driven to desperation in order to accept mandatory restrictions on crops, but even in 1933 policy makers understood that the program would have to be voluntary. Since crops were already in the ground by May, however, farmers and planters were invited to "plow up" a third of their crops, and it took a public relations campaign to convince many of them to do so. In an era when bread lines and starvation hung like a specter over the country, the destruction of crops in the field and livestock seemed a kind of madness to the public and to many farmers.[52]

Secretary Wallace appointed Cully Cobb to take charge of the "cotton" section of the AAA, an interesting choice given Cobb's background. A graduate of Mississippi A&M College (now Mississippi State University), he worked in agricultural education in that state until he became editor of the *Southern Ruralist* in Atlanta, Georgia, in 1918, where he remained until his appointment to head the cotton division of the AAA in 1933. Wilson's enthusiasm for the AAA first surfaced in a telegram he sent Henry Wallace on June 20, 1933: "We are pleased with your plan for cotton acreage reduction and are willing to cooperate with you to fullest extent to see cotton acreage reduction carried out from twenty to forty per cent STOP Lee Wilson and Company of Wilson, Arkansas have over twenty thousand acres planted to cotton. We are willing to reduce to any extent you suggest STOP Will use our best efforts to get our neighbors to cooperate in reduction STOP You may use this information in any way you see fit." Cully Cobb telegraphed his appreciation on June 22: "Delighted your attitude cotton acreage reduction." Cobb enlisted the advice of prominent planters and farmers across the South in fashioning the plan, including Oscar Johnston of the Delta Pine and Land Company of Mississippi. After a public meeting held in Memphis on September 6, at which Johnston officiated, Wilson wrote, "I am one hundred per cent for your tentative plan for cotton acreage reduction for the years 1934 and 1935." Wilson also publicized his enthusiastic support in both Memphis and Little Rock newspapers and encouraged other farmers and planters to do likewise.[53]

Wilson recognized immediately the advantages of participating in the cotton reduction campaign and actively encouraged his friends and neighbors. The Extension Service's county agricultural agents were charged with the responsibility of administering the program, and they typically utilized the expertise of those who had supported extension work in the past. Wilson had enthusiastically embraced the county agent's program in Mississippi County from its inception, worked with the agents in modernizing his plantation, and sought their assistance in developing the "Wilson Big Boll," a cotton seed that was widely marketed in the region. Unlike some planters, who feared the intrusion of black county agents onto their lands, Wilson welcomed them. He wanted his tenants and sharecroppers to be apprised of the latest agricultural techniques. Wilson's experience working with the county agents gave him no cause for concern about the possible subversive influence of federal programs, perhaps because he had always been able to make them work to his advantage.

Because county governments typically paid half the salaries of county agents, the agents were particularly solicitous of influential individuals. Typically, county agents relied on their County Agricultural Committees to support and promote their programs and turned to them for aid in securing cooperation with the crop-reduction program. The Mississippi County Acreage Reduction Committee included C. D. Ayres of Osceola, R. C. Branch of Pecan Point, and J. F. Tompkins of Burdette. Ayres and Branch frequently did business with Lee Wilson & Company, and Tompkins, who operated his own large and profitable plantation enterprise in north Mississippi County, was a prominent and prosperous operator. In consultation with the county agent, they appointed the important township committeemen, and among those appointed were some of Lee Wilson & Company's general and farm managers. Bob Robinson, for example, who ran the Keiser Supply Company for Wilson, reported in July that he spent weeks "out in the territory inspecting crops" to be plowed up.[54] These men intimately knew the territory they were assigned, and they got to work, talking to each farmer and signing them up to participate in the program. They had the responsibility of determining just what constituted a third of each farmer's cotton crop and thus enjoyed considerable authority. In the end, Stanley Carpenter, the agricultural agent assigned to south Mississippi County, reported that he had 100 percent participation in the plow-up program in that part of the county. Carpenter reported contracts to destroy 38,104 acres and of that total, seven thousand acres, or 18 percent, of the cotton subsequently destroyed belonged to Wilson.

Of the seven thousand acres Wilson plowed up, however, 1,500 were located on his north Mississippi County plantation, Armorel. The county agent assigned to the north end of the county complained that Wilson's Armorel contract was negotiated with Carpenter, but he had other problems to report. He had failed to secure 100 percent participation in the north end of the county. The evidence is not available in sufficient abundance, but it appears that the independent smaller farmers of north Mississippi County distrusted the program, and farmers there destroyed only 24,474 acres. In all, 62,578 acres of cotton were plowed up in Mississippi County, netting a total of $731,747. Wilson's share of that was $89,670. In addition, farmers in Mississippi County opted to hold 26,904 bales of cotton from the previous year off the market and were paid $498,150 in return. Wilson opted to hold 4,200 bales worth $77,742. His support of the AAA program certainly paid off: $167,412.[55] Together with a modest increase in the price of cotton—from six cents in 1932 to nine cents in 1933) in the fall, Wilson's payments for the plow-up and his RFC loan had turned his operation around.[56]

Wilson's long months of focusing on implementing the AAA program and on securing his RFC loan were not at the expense of his typical oversight of the details of his business operations. While Wilson continued this kind of routine oversight, he also introduced another new mechanism of control. In June 1933 he informed his managers in the farm, store, and lumber operations that he believed "it necessary that I keep a considerably closer touch on the operations and business of this company." The memo he sent to J. E. Counts, who ran the retail lumberyard in Wilson, was typical. He said he was "putting into effect a rule, requiring all department managers to give me a weekly report on their business. I want you to keep a memorandum book and pencil with you each day, so that you can jot down where you have been, who you have called on and with what results; the expenses of your department, the way your men are conducting themselves as to their duties, the condition of your stock, what is moving and what is not, information regarding fire hazards, etc. In short I want you to give me a report on everything that occurs in your department. Then on Monday morning of each week I want this report written out and put on my desk without fail. I want this to begin Monday morning, July 3, 1933, and have these reports to me each week thereafter."[57] Soon his managers were complying with his request. He received reports from general managers like Oscar Hill of Bassett, from store managers like George Charlton of Evadale, and from farm managers like Pee Wee Morris.[58]

There is no evidence in the record indicating how Lee Wilson responded to these reports, for he was not destined to live long enough to do much beyond receive them and, perhaps, ponder over the details. The tremendous amount of energy he expended in saving his company and extending his control and oversight over his employees came at a great personal cost. He literally worked himself to death. By late July he was back at the Mayo Clinic, where he was examined and treated for some unrevealed malady. The truly fateful trip was one to Chicago on August 13 to attend the Chicago World's Fair, where he stayed at the Congress Hotel for five days. That city was so overcrowded with fair goers, however, that its infrastructure was seriously compromised, and its water supplies were contaminated as sewage backed up in homes and even luxury hotels like the Congress Hotel. There Wilson apparently contracted amoebic dysentery, as did three to four hundred other hotel patrons. It proved difficult for the family's attorneys to prove that the disease was the result of his sojourn at the Congress Hotel, not only because the source of amoebic dysentery is difficult to trace but also because of other circumstances. He left the Congress on August 18, but the first symptoms did not appear until August 30. His trip to the Mayo, which the attorneys decided to keep to themselves, was another concern. The fact that he had been suffering from ill health for a couple of years was widely known and greatly complicated matters.

After falling ill on August 30, Wilson traveled to Hot Springs and his mistress, Helen May. His physicians there apparently erroneously diagnosed him as suffering from cancer, a diagnosis that was eventually contradicted by doctors in Memphis when he was transferred there on September 20. A subsequent autopsy confirmed that he was suffering from amoebic dysentery. A final complication was the fact that his mistress was with him at the Congress Hotel in mid-August. As Wilson's Mississippi County attorney, Cecil Shane, explained to the Chicago lawyer hired to file suit in Illinois, "I cannot tell you whether the woman with him was registered as his wife, but I rather think she was. She was not a nurse, whether or not she was registered that way." By the time the suit was coming to trial a year later, she had married and was willing to testify that he "took all his meals at the Congress while there and she and we know a number of others who had meals with him at the Congress." In the end, however, the Wilson family accepted a $1,000 settlement. William L. Bourland, the Chicago attorney representing them, advised them against pursuing the suit, saying, "I am sure that the Hotel company was negligent and that they should be held liable, but the disease is one which has a long period of incubation and in order to show the probability that the disease was contracted at the Congress, it would be necessary to take depositions of 300 or 400 people who did contract the disease there, together with an equal number of doctors. . . . The Wilson case is further complicated by . . . the circumstances surrounding his occupancy of the room [the presence of Helen May Vaughan].[59]

Wilson's last weeks of life were marked by suffering described in detail by Dr. J. A. Crisler, who reported that he was placed in his care in Memphis on September 20 suffering from "extreme prostration, fever, great abdominal pain and rapid heart action. In other words the entire picture was that of a man nearing dissolution from an acute dysentery." Dr. Crisler employed appropriate treatment for the time, including "repeated blood transfusion and other intravenous transfusion such as glucose and saline, at frequent intervals, in hoping to sustain him long enough to invite favorable reaction, which, however, in spite of all our efforts, failed to come." He died at 3:50 p.m. on September 27, 1933.[60] He was sixty-eight.

* * *

It was while suffering the agonies described by Dr. Crisler that Wilson wrote a deathbed will that further complicated the lives of his family and threatened the company he had fought so hard to preserve. Shane indicated that Mrs. Vaughan "went everywhere" with Lee Wilson, but his family was apparently unaware of her existence. Not even his son knew of her and angrily confronted Wilson's valet about keeping the secret. Lee Wilson compounded the shattering revelation by altering his will at the last minute. The new will left property and money to Helen, money the company found impossible to deliver, even with the changes in fortune that the federal programs had made possible. All the federal monies received were earmarked, particularly the RFC loans, and the company was characteristically cash poor. But Helen May Vaughan was the least of their problems. Within weeks after his death, white tenants on one of the Wilson plantations filed suit against the company, signaling a new militancy on the part of his plantation labor force. Their dissatisfaction likely always existed beneath the surface, but the uncertainty brought about by Wilson's death finally inspired them to action. These problems were left to Roy Wilson and Jim Crain to work out. While most observers, including company employees like Hy Wilson, expected Roy to be made official trustee of the company upon the boss's death, the certificate holders (Lee Wilson's widow and children) elected Roy and Jim Crain to serve as co-trustees of the company. The circumstances surrounding this decision remain unknown, but it launched a partnership marked by accommodation and strife. The family and the company faced trying times over the next few years as they struggled to deal with the ramifications of Wilson's death.

8

CHANGING OF
THE GUARD

During the last year of his life, Lee Wilson aggressively pursued every avenue open to him to save his company from bankruptcy, confident in his ability to maneuver in a world where the federal government assumed a new and intrusive role in the agricultural economy. Following a lifelong pattern of capitalizing on opportunities whenever they presented themselves, he embraced New Deal programs, expanded operations by purchasing the 7,700-acre McFerren plantation at a bargain, and reduced the company's long-term indebtedness by buying up Lee Wilson & Company bonds for fifty cents on the dollar. In order to pursue the McFerren acquisition, he renegotiated his indebtedness with one of his largest lenders and artfully delayed payments to other creditors. The new plantation enhanced his profile at a time when image mattered most and enabled him to lay the groundwork for the continuing prosperity of his company.[1]

A week before he died, however, he drafted a last will and testament that undermined his son's expectations of taking the helm. Wilson crafted that document in order to establish support for his young mistress, Helen May Vaughan, and to further secure her protection, he bypassed his son and named two of his trusted executives as co-executors. This change positioned one of those executives, Jim Crain, to march into the vacuum left by the founder's death and seize control of the company. Crain, a rough-hewn farm manager from a modest background who recognized that his "main chance" was at hand, undermined Roy Wilson, the effete and pampered son of the founder, who suffered from recurring bouts of malaria and turned to alcohol as a palliative.

Lee Wilson's decision to name Jim Crain and Hy Wilson (no relation) as executors of his estate reflected at least two deeply personal preoccupations that concerned him the week before he died. He intended to provide for Helen in a manner that would prevent his family from interfering with her legacy, and he did not want to place her affairs in the hands of his son, whom he neither fully trusted nor believed sufficiently competent to discharge the responsibility. The carefully crafted will called into question the family's assertion that the "pretty young divorcee" had unduly influenced him on his deathbed. Only one aspect of the document seemed ill considered. The will failed to acknowledge his wife's dower rights, a serious omission. Most of the rest of the provisions of the will seemed entirely in keeping with Lee Wilson's long-felt sentiments. No one questioned Wilson's bequest to his two nieces, Dora Davies Merrill and Eva Davies Elkins, gifts that reflected his deep and abiding attachment to— and sense of responsibility for—the daughters of his sister Victoria. Neither did any party seek to overturn provisions he made for his African American manservant, Rufus Lawrence, to whom he left the use of a one-hundred-acre farm for his lifetime. The fact that he left nothing to his three children, argu- ing in the will "it is my belief that moneys heretofore expended by me to, for and in their behalf amount to fair, just and liberal treatment at my hands," aroused no challenge. They held shares in the trust worth $330,000 each, which provided them with substantial income. Two other matters, however, required negotiation between the legatees: the bequest to Helen and an even larger bequest to the Wilson School District.

Although the award to Helen fascinated the public and certainly caused the family great concern, the more problematic bequest involved the one made to the Wilson School District, a gift that reflected Wilson's enduring interest in education. The will provided the school district with the income from 2,960 shares of stock held in various entities, including two Mississippi County banks, and awarded it the income from a little more than 1,200 acres of farmland located around the original Wilson homestead. More importantly, the will made the district the beneficiary of two thousand shares of the trust. By vesting the district with voting membership in the trust, Wilson not only gave it a stake worth approximately $330,000; he provided it with a role to play in the management of the company, a matter that had previously been strictly a family affair.[2]

Even as the attorneys for the various legatees addressed these questions in intense negotiations behind closed doors over the ensuing months, another drama involving both the trust and Roy Wilson began to unfold. Most observ- ers assumed that Roy would take his father's place as the director of the trust, but, instead, the holders of the certificates of trust (Lee Wilson's wife and chil- dren) elected Jim Crain and Roy Wilson together as co-trustees of the trust. According to one newspaper, the document creating the unusual arrangement by Lee Wilson in 1917 had "failed to disclose machinery set up for the perpetuation of the trust in the event of the death of the self-appointed trustee."[3] In naming Jim Crain and Hy Wilson trustees of the will and thus caretakers of Lee Wilson's shares of the trust, Lee Wilson had set up the circumstances making it possible for Jim Crain to occupy a crucial position within the company trust. Given the ongoing contest over that document, it seemed prudent to have at least one of the trustees of the estate in a signatory position on the board.

Although Jim Crain enjoyed Lee Wilson's confidence, his son did not, and Wilson's disenchantment with his son deepened as he watched Roy's performance as trustee of the company for the fourteen months in 1931 and 1932 during which Wilson had relinquished, temporarily it turns out, control. The company had never witnessed such a disastrous economic situation, and only the most vigorous and calculating response could pull it back from the abyss. Though the elder Wilson suffered from an illness serious enough to convince him of the necessity to step down in late 1931, he reinserted himself in the affairs of the company during this crisis period. To be fair to Roy, even had he mustered the requisite response in ingenuity and energy, he lacked the personal connections with the company's major creditors, relationships that Lee Wilson manipulated in his successful effort to keep them at bay as he negotiated the RFC loan, another arena where long-term associations, this time with politicians, worked to his advantage. But even Roy Wilson's best friend from Yale, George Thompson, seemed to have doubts about the younger Wilson's ability to shoulder the responsibility of running the vast and complicated Wilson company upon the founder's death. Thompson, a prominent attorney in Fort Worth, Texas, served as one of the attorneys representing the family in their negotiations over the will. As he counseled Roy, his college roommate, in working out the compromise concerning the will controversy, he also advised him on his handling of his sisters and mother, the other certificate holders, reflecting a strained relationship between the siblings. In late December 1933, Thompson wrote his old friend about issues Victoria and Marie had raised when he visited earlier that month, advising Roy that he should provide them with "monthly statements of the company's operation" and lecturing him: "Bear in mind, Roy, that there are five stockholders in this business, and each one is as much interested in it as you are, and that you are not only working for yourself, but you are also working for them." Thompson left no doubt as to his meaning. "By keeping Victoria and Marie, as well as your mother, fully informed, you will immeasurably retain their confidence."[4] Rather than resenting his friend's advice, Roy responded by sending him copies of letters he had written to his sisters, letters intent upon keeping them abreast of the company's business.[5] Thompson complimented Roy on the "tactfulness" of his letters to his sisters and urged him to "always be patient with them and . . . not get exasperated at any request, or even criticism they may make."[6]

Thompson further advised Roy on his handling of another delicate matter, one in which he spoke to Jim Crain's primacy. Frank Gillette, a longtime and well-trusted company executive, had suffered a heart attack during the summer of 1933 and remained convalescent. Thompson reminded Roy that "your mother spoke to you and me about the Company tendering back to Mr. Gillette his former position? At the time you will remember your mother almost went to the point of insisting that this be done, and you told her that you were quite willing for Mr. Gillette to resume his position." The advice from Thompson took an unexpected turn at this point. "Now Roy, I believe it would be a fine thing for you to do, at the threshold of this new year, *with the acquiescence and consent of Jim Crain, and with his approval* [emphasis added], to sit down and write a nice letter to Mr. Gillette, expressing to him the appreciation of all members of the Wilson family . . . telling him that his old position is always open to him." Even Thompson, therefore, in the midst of advising his friend about how to gain the confidence of his mother and sisters, acknowledged Jim Crain's unprecedented position of authority in the company. Perhaps he knew his friend Roy too well. At the end of his long letter, Thompson referred to an attached article "that greatly impressed me, entitled 'The Man Who Counts,' by the late Theodore Roosevelt." Thompson probably meant to encourage his friend, but the missive also suggests that he had his doubts about Roy's abilities.

It is not the critic who counts; not the man who points out how the strong man stumbled, or where the doer of deeds could have done better. The credit belongs to the man who is actually in the arena; whose face is marred by dust and sweat and blood; who strives valiantly; who errs and comes short again and again, because there is no effort without error and short-coming; who does actually strive to do the deeds; who knows the great enthusiasm, the great devotions, and spends himself in a worthy cause; who at the best knows in the end the triumph of high achievement; and who at the worst, if he fails, at least fails while daring greatly, so that his place shall never be with those cold and timid souls who know neither victory nor defeat.[7]

In fact, there existed a power vacuum at the top of the company pyramid, and the months following Lee Wilson's death marked a period of transition. Perhaps this was a natural consequence of having a larger-than-life figure like Lee Wilson in control for so long. Standing in his father's shadow, suffering from alcoholism and debilitating bouts of malaria, Roy was in a poor position to assume the kind of mastery over the company that Lee Wilson had practiced. But the company was at a critical juncture with regard to the reorganization of its finances and required decisive leadership. The RFC had disbursed funds to pay off some of their major creditors, but in December 1933, the agency failed to release monies to pay taxes and bondholders. Rather than go to Washington himself to appeal directly to the RFC, Roy decided to send Jim Crain. This was all the opening Crain needed to begin establishing connections to Arkansas's power brokers.

Jim Crain's history with Lee Wilson & Company dated back to late 1912, when he went to work for the Idaho Grocery Company in Bassett. He came from Brandon, Mississippi, located in Rankin County between the delta and the longleaf-pine belt. The second of thirteen children of a farmer and his wife, born in 1888, Crain grew up learning how to work the land and received only a basic education in the county public schools. He briefly attended a business college in Nashville before returning to Brandon and accepting a position as a bookkeeper at a mercantile company. In some ways his background resembled that of Lee Wilson's in that he made his own way in the world despite an inferior education. Both made fortunate marriages, Jim Crain to the daughter of an attorney and judge, Lee Wilson to the daughter of a prosperous farmer and lumberman. However, while Jim Crain was the son of a hardscrabble farmer, Lee Wilson's father was a planter, albeit a frontier planter with few pretensions. Wilson might have claimed more "aristocratic" origins—although it is unlikely he would have done so—but he probably recognized in Jim Crain something of his own drive and ambition.[8]

Crain's arrival on the scene in 1912 coincided with Roy's graduation from Yale and return to Wilson. Although near enough in age—Roy was twenty-two, Crain was twenty-four—the two men could not have been more different. Crain was a tough-minded and vigorous man who understood that he had to shape his own future. Roy Wilson, the over-indulged son of a powerful and dominating father, had received an Ivy League education that hardly prepared him for the future his father had in mind. His class picture at Yale reveals a callow young man, thin, fair-haired, and almost effeminate. He played polo at Yale and began a lifelong association with the sons of other wealthy fathers.

Whether he was already suffering from malaria and a dependence on alcohol is unknown. Nor is it known precisely what work his father put him to upon his return to Wilson. While Jim Crain rose to the position of farm manager over the entire company operation by 1917, Roy Wilson remained a shadowy figure, who surfaced only briefly in the record and then disappeared again. After a trip to Europe in the 1920s, he built a fine Tudor mansion and had Tudor facades placed on the Lee Wilson & Company buildings in Wilson, demonstrating his preoccupation with image. By the time of his father's death, he had assumed certain responsibilities for the company, but he seemed content to play the role of the country gentleman who was more preoccupied with traveling to Kentucky to participate in foxhunts.[9]

From the beginning Roy Wilson seemed eager to allow Jim Crain to assume the responsibilities and obligations of running the operation. One of the first hurdles the company faced upon Lee Wilson's death was the final disbursement of funds held by the RFC, the funds designated to pay Lee Wilson & Company bondholders and to pay off past-due drainage taxes. Roy Wilson promoted Jim Crain and Hy Wilson, as trustees of his father's estate, as the parties responsible for securing the release of the funds. The bond issue dated back to April 1922, when the William R. Compton Company of St. Louis, Missouri, brokered $500,000 in Lee Wilson & Company first-mortgage bonds at 7 percent interest.[10] In the last year of his life, Wilson aggressively purchased thousands of shares of these bonds at forty-five to fifty cents on the dollar, despite his own shaky financial situation.[11] He shrewdly included payment on bonds scheduled to mature on December 1, 1933, in his loan application to the RFC. When the RFC failed to distribute the funds allocated for that purpose, apparently because of some bureaucratic snafu related to Wilson's death, eager bondholders besieged the company, wanting to know when they could expect to receive their funds. As an executive from the National Candy Company indicated in a letter to Roy Wilson on December 8, 1933, "Mr. [Lee] Wilson told me personally just about two weeks prior to his death that funds have been set aside in escrow to meet the full principal and interest requirements up to December 1st. Therefore I am unable to understand why the bonds were not taken up at maturity."[12] Indeed, the RFC had the sum of $178,400 in hand to pay for the bonds scheduled to mature on that date. Roy proposed that they could retire the bonds at sixty to seventy cents on the dollar. Rather than go to Washington himself to discuss the matter with Harvey Couch of the RFC, however, Roy planned to send Jim Crain and Hy Wilson to do the negotiations and wrote a letter of introduction to Couch in order to facilitate the matter.

When Couch paid a visit to Little Rock, presumably on AP&L business (he continued to function as president of that operation), Crain and Hy Wilson met with him there and, subsequently, sufficient funds were dispersed to take care of the bonds.[13] In a sense, then, Roy Wilson contributed to Jim Crain's rise to power and authority.

The most important concession Roy made to Jim Crain remains inexplicable but most significant. He acquiesced to an arrangement making it possible for Crain to transact business for the trust without his co-signature, a favor he did not himself enjoy. As Hy Wilson put it, "when Mr. Roy Wilson signs an instrument for Lee Wilson & Co. it is necessary that Mr. Crain also sign it, but it is not necessary for Mr. Roy Wilson to sign with Mr. Crain." This arrangement dated back to the October 1933 establishment of the co-trusteeship. Although it may have merely reflected Roy Wilson's intention to be absent from company premises for extended vacations—something that did, in fact, occur—it might also indicate a perception that Crain was the most competent of the two men.[14]

Meanwhile, the negotiations involving the compromise agreement over the will reached fruition in March 1934. All parties agreed to settle their differences rather than endure further public scrutiny. The most significant concession came from the school district, which agreed to relinquish the two thousand shares of the Wilson trust it received under the terms of the will. Elizabeth Beall Wilson, as part of her right of widow's dower, received a third of those shares, with the three children dividing the remainder. Marie had already given her son a thousand shares of her own portion.

Elizabeth Beall Wilson	2,666
Roy Wilson	2,444
Victoria Wilson Wesson	2,444
Marie Wilson Harmon Howells	1,444
Joe Wilson Nelson	1,000

This settlement of the shares set up the possibility of an interesting new dynamic to the family business relationship, but, in fact, the shares were not distributed according to the compromise agreement. Instead, Crain and Hy Wilson held the two thousand shares (the school district surrendered) in their positions as executors of the will: "Estate of R. E. Lee Wilson, deceased." They described this arrangement to Guthrie King in July 1935 as follows:[15]

Elizabeth Wilson	2,000 shares
(Wife of R.E. Lee Wilson, deceased)	
Victoria Wilson Wesson	2,000 shares
(Daughter of R.E. Lee Wilson, deceased, Husband is Frank H. Wesson. They live at Springfield, Mass.)	
R.E.L. Wilson, Jr.	2,000 shares
(Wife is Natalie Armstrong Wilson)	
Marie Wilson Howells	1,000 shares
(Daughter of R.E. Lee Wilson, deceased)	
Joe Wilson Nelson	1,000 shares
(Son of Marie Wilson Howells. He is single)	
Estate of R.E. Lee Wilson, deceased	2,000 shares
(J. H. Crain and W. F. Wilson are Executors and Trustees of this estate)	

Although this arrangement rankled Victoria Wilson Wesson, who demanded the distribution of the shares in June 1936, they remained under Jim Crain's control until 1950, when his association with the company ended, a termination accompanied by his relinquishment of authority over the estate. The dispute over the distribution of the shares figured prominently in a growing estrangement between Crain and Lee Wilson's grandsons, both of whom began to work for the company in 1936 only to find that Crain's sons and brother occupied the more important positions while they were shunted aside. But that estrangement did not rise to the level of controversy until after World War II.

The compromise decree concerning Lee Wilson's will had other and more immediate implications. Helen May Vaughan accepted, instead of $100,000 and title to a large bungalow (now on the National Historic Register) next to the golf course in Hot Springs, a $65,000 cash settlement and the retention of the $250 stipend promised to be paid to her monthly during the year following Wilson's death.[16] One other important provision of the will remained intact, at least in principle, that which placed the cash award to Helen in a trust to be administered by Union Planters Bank in Memphis. The trust set up in Helen's name was to provide her with income for a specified length of time—until the youngest of her sons reached the age of twenty-five. At that time, the principal amount was to be awarded to Helen in a lump sum. Some believed that this wording in the will implied that her children had been fathered by Lee Wilson, but there is no evidence of that. She had married at nineteen and bore two children before returning to her parents' home as a divorcee. She began her liaison with Lee Wilson after that time. Wilson's reasons for setting up the trust in this way may have reflected his recognition of that fact that he shared something in common with the two boys, ages seven and eight. Wilson himself was only five, his brother ten, when his own father died, leaving him in the care of a mother who had to fight for the right of her children to inherit so that she could properly support them.

Over the next few years, the two Wilson employees (Crain and Hy Wilson) would have reason to regret their involvement in Helen's affairs, which became increasingly difficult to manage and required advances of cash to permit her to meet her obligations. Helen settled down in St. Louis and married Ernest F. Latta within a year of Lee Wilson's death. Latta was the son of a prominent real estate developer in Hot Springs, a town with a peculiar and thrilling history of its own. As a resort widely known for speakeasies and gambling, it also served as a haven for notorious criminals like Al Capone. It was the kind of town where people looked the other way at aberrant behavior and abided by the notion that "what happens in Hot Spring, stays in Hot Springs." Helen May's quick marriage to Latta suggests—but does not establish—that she was already well acquainted with him. Certainly his prominent position as the co-owner of the Ozark Bathhouse placed him in the first circle of the town's night life. Whatever the circumstances of their marriage, Latta moved to St. Louis with Helen May after they married, and her sons remained in her household. She continued to rely heavily upon the support promised her in Wilson's will, a $250 monthly allotment stipulated to be her due for the first year after Wilson's death. Whether by mutual agreement or for some other reason, Lee Wilson & Company continued sending her $250 per month after the expiration of one year and failed to place the $65,000 in trust in the Union Planters Bank of Memphis in her name.[17]

According to the will, Jim Crain and Hy Wilson were to serve as the trustees charged with the responsibility of administering the will, but the Union Planters Bank in Memphis was to control the $65,000 designated for Helen's benefit, an arrangement in keeping with Wilson's desire to maintain her interests apart from the family and the company. Yet Crain and Hy Wilson were given some authority over Helen's funds, as the bank was required to consult with them concerning the investments made by them on Helen's behalf. Wilson essentially set it up so that Helen's interests were doubly protected. Union Planters Bank had to answer to Jim Crain and Hy Wilson, but neither Crain nor Hy Wilson could act in advance of the bank insofar as Helen's interests were concerned. In fact, Lee Wilson's will had provided that the proceeds from nine insurance policies on his life were to be the basis of the cash settlement to Helen. But most if not all of those insurance policies had been pledged as collateral in that last year or two of his life. The company had secured the RFC payments and paid off its indebtedness to its largest creditors but it was still "cash poor" in the months after Wilson's death. Helen initially accepted the failure of the trustees to place the $65,000 with the bank, but this had the effect of linking her to the company in a way that Lee Wilson had not intended. Although this arrangement worked well enough for the first couple of years, in 1936 she asked Crain and Hy Wilson for a part of the principal sum left to her in order to purchase a home in St. Louis for herself and her sons. Their attorney advised them that the trust would permit them to do so, but the transaction never took place. In 1937 she finally demanded that they place the $65,000 with Union Planters Bank, and by then Jim Crain and Hy Wilson were negotiating with the bank over how to invest the funds on deposit with them. Thereafter, they engaged in a steady stream of correspondence over the investments.[18]

The year 1937 was a particularly difficult one for Helen, as she suffered a serious illness that year, a case of encephalitis.[19] In May Helen sent a telegram to Hy Wilson from the Mayo Clinic in Rochester, Minnesota, announcing that her older son, Alfred, "had a very serious operation" and requesting additional funds.[20] By the end of the year he was enduring his fourth operation, and, since the trust had now been established at the Union Planters Bank, she was writing the trust officer, Thomas Vinton, for an advance on her funds to handle her excessive medical expenses and "to carry me through until such time as the trust provides me with sufficient income."[21] It seems that upon establishing the trust, Crain and Hy Wilson believed their obligation to provide her with a monthly stipend had ended. The bank was now administering her affairs and was unwilling to pay her more than the income generated from the investments. She subsequently complained that they failed to invest the cash in a way that would generate an adequate income. In early 1939 Crain and Hy Wilson, who continued to be obligated under the terms of the original will and the compromise decree to oversee the bank's handling of her affairs, approved a $500 payment to her and secured an agreement from the bank to pay her a monthly stipend of $250. Her medical bills exceeded that amount, however, and Crain and Hy Wilson authorized the disbursement of additional sums.[22] As late as 1940, she was writing Crain, Hy Wilson, and Thomas Vinton desperate for money. "I did not know what the past two years would bring any more than you, yourselves, can know how your cotton crop would yield, namely the decrease in my income, sickness and a lease on an apartment which I could ill afford and from which I have been unable to move."[23] She asked them to authorize the disbursement of part of the capital, but they refused. She took the matter to court and was granted $1,750 of the corpus.[24]

Throughout this period, Jim Crain was tightening his grip on the leadership of the company. After Lee Wilson's death, he became the main signatory, along with Hy Wilson and S. A. Reginold, on letters written concerning a variety of issues. Hy Wilson and Reginold handled routine correspondence with creditors and debtors while Crain handled matters concerning the purchase or sale of land or the securing of agricultural credit.[25] Crain even took over responsibility for maintaining relationships with important political figures like Senator Hattie Caraway, writing her in 1938 concerning federal funds for highway construction in Arkansas, and Senator John E. Miller, with whom he corresponded with on a crucial AAA controversy involving the company.[26]

The volume of correspondence from Roy Wilson, however, decreased as he sank deeper into his alcoholism. To the extent that he surfaced at all in the correspondence, his letters involved plans for participating in foxhunts, the purchase of bird dogs, membership in the Memphis Hunt and Polo Club, arrangements for summer vacations in Colorado, or the purchase of suits from his tailor in New York.[27] In correspondence with some of his old Yale classmates, he made frequent references to his eldest son, Robert E. Lee "Bobby" Wilson, of whom he was particularly proud. He rarely mentioned his younger son, Frank, who proved to be an indifferent student and athlete. From the beginning Bobby Wilson had certain advantages. As the first child born to the class of 1912, he became the recipient of the class's scholarship, and his sponsors had reason to be pleased with his performance at Yale. Referred to as the "class boy," he was the subject of conversation and the object of praise in correspondence Roy had with his old friends in New York.[28] His graduation in the summer of 1936 was a much-anticipated event, and Roy reported to Thelma Engelbrecht, who typically handled his reservations for a summer cottage in Colorado, that their plans were on hold as they were attending the graduation ceremonies. He added, however, that "Bob plans to go to work as soon as he returns home, which leaves Mrs. Wilson and myself on the loose."[29] Indeed, Roy had ceased to find much of interest to employ him at the office and even excused himself from various other responsibilities such as board of trustee meetings scheduled at Arkansas State College in Jonesboro (he had taken his father's place on the board) or a ceremony marking the opening of Arkansas Highway No. 1 in April 1935.[30]

Given Roy's lack of interest in the company and tendency to avoid meetings, his family came to rely upon Jim Crain. Elizabeth Wilson turned to Crain when she wanted representation at a stockholders' meeting of Wilson-Ward Company in July 1935, giving him, rather than her son, power of attorney.[31] The next year she might have chosen another option. By 1936 two of her grandsons, Lee Wesson and Bobby Wilson, were employed by the company, apparently in training for taking over management of its affairs at some time in the future. Lee Wesson went to work for the company in February 1936 and was assigned a variety of tasks, including writing a number of dunning letters to people who owed the company money.[32] He corresponded with vendors over purchases and was handed the responsibility of making reports to the Department of Labor concerning employment statistics for the company.[33] He was identified as the person responsible for matters concerning "turn over" of employees, whether they had resigned or been discharged or laid off. This was a new and, from the point of view of company officials, a burdensome requirement that accompanied New Deal programs related to labor.[34] By 1939 he was involved in negotiations over the purchase of railroad track for a spur the company was building to one of its enterprises.[35]

Bobby Wilson began to work for the company after graduating Yale in the summer of 1936. Like his slightly older cousin, he went to work in the office, but he was also assigned the responsibility of overseeing the construction of a cotton oil mill, an enterprise his grandfather had envisioned in 1932 but abandoned in the wake of the financial disaster then confronting him. Jim Crain revived the idea and even considered the construction of a textile mill. Although the latter never came to fruition, the cotton oil mill became an important part of the company business.[36] It was organized under the Delta Products Company, which filed incorporation papers in May of 1936. Several local prominent planters and businessmen were involved, but Lee Wilson & Company had the largest number of shares:

J. H. Crain, Trustee, Wilson, Ark. (for Lee Wilson & Co.)	750 shares
J. H. Crain & W. F. Wilson, Trustee (for Lee Wilson Estate)	500 shares
Lowrance Bros. & Co., Driver, Ark.	75 shares
D. F. Portis, Lepanto, Ark.	75 shares
C. L. Denton	25 shares
L. P. Nicholsen	25 shares
L. P. Bowden, Joiner, Ark.	25 shares

As Crain explained it to Charles T. Coleman, the Little Rock attorney who had drafted the 1917 trust agreement, "Lee Wilson & Company has taken $100,000 of the stock. . . . Charles Lowrance is president of the corporation, I am vice president and one of the boys here is secretary; the Board of Directors is made up of eleven stockholders." The shareholders were described as Lee Wilson & Company and a number of other large cotton farmers who were "interested in this corporation, each in proportion to the cotton and cotton seed he raises or handles."[37] Crain wrote Coleman for his opinion as to whether the trust could operate in the manner described, and although no answer to the query surfaced in the record, Coleman must have replied in the affirmative because they proceeded to operate as outlined. The mill took more than a year to construct, however, and Bobby Wilson was intimately involved in that part of the process. While Roy Wilson was touring Europe with his wife in August 1937, Bobby was hard at work back in Arkansas overseeing construction.[38]

Neither Bobby nor Lee Wesson was content with the menial tasks assigned to him by Jim Crain, but neither was in a position to challenge Crain. When Victoria Wilson Wesson demanded the distribution of the shares in a letter written to company executive Sumner Reginold in June 1936, she probably had her son's interests in mind. She also shed light on the reason for the delay. "Mr. Crain said that the four hundred forty-four shares of Lee Wilson Co. that came to me from my father's estate were ready to be turned over to us just as soon as the loan to the R.F.C. was paid and the interest for last year credited to our accounts. I do not see that credit on any of my statements? Why haven't I?"[39] Jim Crain negotiated a $500,000 loan with Prudential Life Insurance Company in mid-1935, paying off the RFC loan, and thus, as Victoria suggested, the distribution should have taken place. The company had reduced its indebtedness to the RFC to $471,549.65, and the Prudential loan rendered them free and clear of the tiresome government bureaucracy.[40] Victoria, understandably, expected the distribution of the shares to take place, but correspondence files do not reveal a response to her request from either Crain or Reginold. In fact, the distribution did not take place, and Crain continued to control two thousand shares through his management of Lee Wilson's estate. In 1937, upon the expiration of the trust—it was set up in 1917 so that it expired in twenty years—the renegotiated trust document reflected two additional trustees: R. E. L. "Bobby" Wilson (Roy's son) and Lee Wesson (Victoria's son). A later document confirms, however, that there was no redistribution of the shares either to the existing shareholders or to the two young men, and Crain remained in charge of the shares belonging to the estate.[41]

One of the problems the family endured in their dealings with Crain concerned the control he exercised over Roy. Jim Crain became the classic enabler and often the only one who could, as Victoria put it, "calm brother down more than anyone else."[42] This became painfully clear to all concerned in the summer of 1939. A family drama began to unfold in May as Roy Wilson's alcoholism spiraled out of control, and the family took extraordinary measures. Three documents dated May 31, 1939, provide the first evidence of the family tragedy.[43] Working with the trusted family attorneys, Cecil Shane and Oscar Fendler, Roy's mother, wife, and son filed "Application for Guardianship" with the county clerk, asserting that Roy was "detained and confined in Milwaukee Sanitarium, Wauwatosa, Wisconsin, an institution for the treatment of the insane and feeble minded." Arguing that Roy was "not competent to take care of himself and to look after his personal affairs and personal property" and that "for his well being, it is imperative that he be kept in said institution for some time to come," they asked that Bobby be appointed guardian. A separate guardianship bond, signed by Elizabeth and Natalie, Roy's mother and wife, alleged that they possessed sufficient real estate to allow them to be exempt from actually depositing a $50,000 bond. Judge J. F. Gautney apparently signed the order that day, but a typewritten missive suggests that the entire matter may have been hushed up and the papers merely filed in the circuit clerk's desk drawer. Written by Cecil Shane to Circuit Clerk T. W. Potter, the document reads as follows:

> Please accept and file these papers and keep the matter in strict confidence. You can easily put them in a file with no markings on the back and put them away where others will not be seeing them. It will not even be necessary to record the order or to docket the matter. If you want to make up a docket sheet you can do so and then take it out and put it in the file with the other papers. These people are just trying to render a real service and they do not want to get any publicity about it. Please just don't mention it to anybody. Sign the two certificates attached to the copy and let Mr. [Bobby] Wilson have it. Be sure to attach your seal in both instances.[44]

Despite their best intentions, Roy Wilson proved to be a difficult man to keep incarcerated. Within weeks he was on the loose back in Wilson and creating havoc, his sense of betrayal over the complicity of the "class boy" in his commitment, the son of whom he was so proud, must have cut him to the quick. But his wife, Natalie, and Bobby were in New York City staying at the Barbizon-Plaza Hotel when Roy returned to Wilson. Bobby wrote Cecil Shane from there, saying "Oscar told me you would be at Hotel Pennsylvania [in

New York] Monday, if you are going to be there leave a message at Barbizon-Plaza. Only want you to have a chance to talk to mother before you go home as I feel that you can reassure her." A few days later, Shane apparently made his visit to the Hotel Pennsylvania, only to receive a telegram from Oscar Fendler: "Mr. Crain urged me to wire you to advise Bobby's mother to remain away until you and Bobby returned and arranged matters satisfactorily." Natalie took Crain's advice and went to her sister-in-law's home in Springfield, Massachusetts. Victoria Wilson Wesson wrote to Crain on June 20 that "I hear brother is home and raising hell? I am sick about it not only for his own good but for mother and his family. . . . Natalie is here with me and beside herself with anxiety. I do not think its safe for her to go home until he has cooled off and asks her to come." She spoke of the "awful problem" Bobby had to face but remarked that "I think he has made up his mind to stick to his guns and I truly hope he can but its going to be very difficult. . . . I told Bob that he could tell his father that if he could prove that he could stay off drink for one year he would then consider revoking the papers not otherwise." Victoria asked Crain not to mention her remarks to Roy, saying that she was "assuming the role of a spectator but you know how keenly interested and concerned I am. I am always so apprehensive and fearful of a tragedy."[45]

No terrible tragedy came of the situation, and the stay at the sanatorium must have had some beneficial results. Within a month Roy and Natalie were apparently reunited and planning a vacation to Nova Scotia. He wrote his old friend George Thompson on the third of July that they expected to be there until the first of September. They did go to Nova Scotia, where they visited with a friend of Roy's from New York, E. S. Mills and his family, but they returned to Wilson within a month, and Roy was making plans to attend the Paducah Fox Hunters show scheduled for November, renting three stalls for his horses. In fact, Bobby and his wife accompanied them on that trip, where they enjoyed lavish accommodations at the Irwin Cobb Hotel.

Although the family had achieved a rapprochement, Roy's drinking problem was hardly over. In early 1941 he was again hospitalized for alcoholism, this time in Memphis. Wilson subsequently challenged the $125 medical bill, and the offending physician responded, "I considered you quite an ill man upon admission to the hospital and . . . the skill represented in getting you off of liquor and back on your feet cannot be measured by the few days you were in the hospital." The doctor ultimately settled for $75, but the cure was only temporary.[46] Roy apparently had long periods when he maintained a rather stoic alcoholic haze followed by dangerous drunken rages that left him incapacitated and his family in fear. Eldon Fairley, a young doctor fresh out of medical school and hired by the Wilson company in 1945, recollected his first encounter with Roy Wilson's problem:

> One Sunday afternoon . . . it was in January or February, oh, it was a miserable day. Misting rain and cold. And they called me to come up there. Dr. Ellis was out of town. And I walked in there through that big . . . you feel like you're going in a dungeon or castle . . . that big door clanged behind me. . . . And I looked over to the right, big long dining room, and there was poor Mrs. Wilson sitting there, looked like she was eating a bowl of soup, had one little candle on the table. And I said "if that's not a picture of loneliness, I've never seen one." And I went on upstairs, he was up there drunk. . . . And I got the IV set up."[47]

Fairley characterized Roy Wilson, now fifty-five, as a troublesome and profane man who raged against the indignity of having to endure the doctor's attentions. Nevertheless, Fairley took over administering to his needs, and Roy continued to require periodic infusions of liquids to offset the dehydration that occurred after one of his extended drinking binges. The record does not reveal whether Bobby "revoked" the guardianship or whether it was even necessary to do so. As Shane's missive to Potter suggests, it may never have been made official. In any case, it was only Roy's personal affairs that were at risk as Jim Crain was firmly in place and running the company.

Crain faced a number of challenges apart from having to deal with the Wilson family and the increasingly desperate Helen May Vaughan. The years since the patriarch's death had witnessed a virtual revolution in the way that the operation functioned, as New Deal programs affecting both labor and the running of farming operations had required significant adjustments. Although agricultural labor was exempt from new wage and hours laws, the men working for the sawmills and shook factory were covered, and the company struggled to adjust to new obligations. Of greater concern, however, were the requirements of the Agricultural Adjustment Administration, which had been launched in 1933 with the plow-up campaign, and its successor organization. The massive infusion of cash from participation in that program, together with the RFC loan, had signaled the return of the company to solvency. But Lee Wilson's death had delayed the disbursement of more than the payments on the past-due bonds and overdue taxes that the RFC loan was supposed to take care of. His death interrupted the disbursement of funds owed the company for participation in the cotton option program. Crain had to negotiate with the AAA in order to secure that payment even as he worked to secure the release of the RFC funds. Both were greatly complicated by a suit filed by renters and tenants on the Victoria (McFerren) plantation in the Mississippi County Chancery Court less than two months after Wilson's death.

With a power vacuum at the top, certain employees saw an opportunity to assertively represent their own interests. Indeed, the tears streaming down the faces of African Americans at Wilson's funeral services, regarded by newspaper reporters as a testament of their affection for him, might as likely have been a reflection of their understanding of the uncertainty presented by the patriarch's death. It was clear that a new regime was in the offing and no one could predict the direction it would take or who would be in charge of it. Whether deserved or not, Lee Wilson had a reputation for greater fairness—greater than other planters exhibited, that is—to both his white and his African American employees. It was known that Jim Crain was a hard man, though time would prove that he had absorbed some of Lee Wilson's paternalism. It was understood that Roy was erratic, frequently intoxicated, and often absent. It was in this context, in the weeks and months when the leadership of the company was being sorted out and the will contest was being negotiated, that eighteen white renters and sharecroppers on Lee Wilson's Victoria plantation appeared in court. On November 17, 1933, six renters and twelve sharecroppers lodged a complaint against Roy and Jim Crain as co-trustees of Lee Wilson & Company. Also named in the complaint was the Eastern Arkansas Mortgage Company, the entity Wilson had created in order to secure his RFC loan.[48]

Their complaint was multifaceted and attacked on several levels. First, they were aggrieved at the failure of the company to secure the market price for their cotton or settle with them in a timely manner. Second, they charged that the company had unfairly appropriated all of the proceeds from the plow-up program, some of which should have been paid to renters and sharecroppers who had, at the company's instructions, destroyed cotton that otherwise might have been harvested by them, sold, and the proceeds used to offset their obligations. Third, they alleged the company had moved all its assets into the Eastern Arkansas Mortgage Company in order to defraud its creditors—including its renters and sharecroppers—and was now insolvent.

The first of their grievances revolved around the annual settlement, and the timing of the suit was in keeping with traditional disputes between tenants and planters: harvest. They acknowledged their contractual obligation to gin their cotton at one of the Wilson gins and market it through Wilson-Ward and

Company in Memphis. Their complaint here was twofold. First, that the prices paid for their cotton by Wilson Ward & Co. was from one to two cents less per pound than they could get elsewhere. Apparently, they had exercised a clause in their contracts allowing them to hold back samples of their cotton and secure bids from other brokers. If the price received for the cotton was less than that quoted by the brokers they consulted, the company was obligated to make up the difference. Given that cotton was fetching only approximately eight cents a pound, the one- and two-cent difference was significant enough for concern. Second, they complained that Jim Crain and Roy Wilson had issued a directive stipulating that no renter or sharecropper would receive any proceeds from their cotton until all cotton had been harvested and marketed. The litigants argued that this was "contrary to the express terms of their contract" and cited a provision that the company credit their accounts "with one-half of the net proceeds of the sale" of their cotton and cotton seed at the time it was sold. Then, when a sufficient amount from the sale had "been credited to the account of the share cropper to offset the supplies furnished him," the company was obligated to begin paying him "every two weeks such amount as the books show to be due." Since most of them had delivered "more than enough to pay their accounts" they were entitled to receive payment. Thus, they charged, Roy Wilson and Jim Crain had "breached the contract."

As to the second aspect of their complaint, that having to do with the proceeds of the plow-up campaign, it appears that the failure of the company to share those proceeds contradicted Lee Wilson's intention to do otherwise. In correspondence with an absentee planter during the summer of 1933, Lee Wilson actually proposed that renters and sharecroppers were entitled to receive a portion of those payments. Wilson's intentions were expressed in a series of telegraphed messages in early July with Neal Thomason of the National Life Insurance Company of Dallas, Texas. Lee Wilson & Company handled the administration of several thousand of acres of farmland owned by the insurance company and had asked Thomason to wire instructions concerning the plow-up. Thomason responded with a query on July 6, 1933, "If we should consent [to the plow-up] what do you consider to be a fair division of proceeds as between landlord and tenant?" In his own hand, Wilson scribbled on the bottom of Thomason's telegraph, "We figure 1/3 landlord 1/3 renter, 1/3 sharecropper. It will net each one practically same proportion as when handled the old way." Later that same day, Thomason authorized his company's tenants to sign the plow-up contracts.[49] There is some indication, moreover, that sharecroppers on at least one of the Wilson operations negotiated their portion of the proceeds prior to the plow-up. Charlie Crigger indicated in one of his weekly reports from Armorel in August that in plowing up cotton he was "having some trouble with sharecroppers on amount of settlement."[50] It is impossible to know whether Wilson would have adhered to this proposition had he lived, but it is clear that Jim Crain and Roy Wilson had a different understanding of the company's obligations to its renters and sharecroppers. So did many other planters, and this led to a historic confrontation between planters and their laborers and to the founding of the Southern Tenant Farmers Union (STFU) a year later, in July 1934 in neighboring Poinsett County. The suit filed by the eighteen men on the Victoria plantation in the fall of 1933 anticipated this event and marked the first time that plantation laborers asserted their interests in government payments.

In fact, federal agricultural programs introduced a new player into an old system, a system that had originated with the rearrangement of relations between planters and freedmen after the Civil War. In the sixty-three years since emancipation, freedmen had been joined by poor whites as either tenants (renters) or sharecroppers on plantations, and planters had developed various strategies to maximize both their power over labor and their profit margins. Hamstrung by lien laws that favored planters and inhibited their mobility, black and white plantation laborers were impoverished and made relatively powerless. The introduction of the federal programs offered an opportunity to assert some independence and inspired some to attempt to gain a place at the federal feeding trough. When the company appropriated all of the plow-up payments, the eighteen renters and sharecroppers objected and declared that they were entitled to share in those payments given that they had "suffered the loss of the cotton on which they had worked." They also devised a novel interpretation of the implications of the failure of the company to deliver part of the proceeds to them. They argued that because they had fulfilled their part of their contract, having delivered all or most of their cotton, and paid their indebtedness to the company in full or in part, the company was now in debt to them. This allowed them to argue that they—these renters and sharecroppers—were creditors, and they declared themselves as such. This was a key feature of their complaint and allowed them to assert standing of a different sort to the court.

This assertion of their identity as creditors was a most important part of their complaint and revealed why they had included the Eastern Arkansas Mortgage Company as one of the defendants. They asked the court for permission "to prosecute this complaint as a general creditors complaint on behalf of themselves and also on behalf of all other creditors of Lee Wilson & Company."

Because Lee Wilson & Company had divested itself of all its assets and vested them in the Eastern Arkansas Mortgage Company, it was now insolvent, and they asked "that a receiver be appointed to take charge of, preserve, and conserve the assets of the estate of Lee Wilson & Company and to recover these assets from any person or corporation now holding same."

The lawsuit was doomed from the beginning, however. They initially filed the papers in the federal court at Jonesboro, where Judge Gautney, long a friend to the Wilson family, threw it out on the basis of lack of jurisdiction. They then took it to the Mississippi County Chancery Court, but by the time the matter came up for hearing, the company had undercut their claim by paying the renters and sharecroppers listed in the suit what the company said was due them: altogether $810.76.[51] However, word of the lawsuit reached newspapers, and certain of the company's creditors were made nervous by the assertion that the company was bankrupt. Ely and Walker Dry Goods Company, Continental Gin Company, and Graybar Electric Company had to be reassured that Lee Wilson & Company was not in receivership and was capable of paying their creditors. Officials of RFC also required reassurance, and this must have been the most troublesome aspect of the entire controversy for the company. They were already experiencing difficulties enough with that agency over final disbursement of funds for the payment of its bonds and taxes.[52]

Although Lee Wilson & Company won the opening round in this new struggle between plantation owners and laborers, the battle was far from over. The founding of STFU in neighboring Poinsett County the next summer launched a new phase of an old adversarial relationship. The STFU was thwarted in its efforts to organize on the Wilson plantation in 1934 and 1935, although several chapters appeared in Mississippi County, one at Dyess Colony, the first Resettlement Administration community in the state, which was located just west of Wilson.[53] But Dyess Colony was as close as they got to addressing the problems confronting plantation laborers on the Wilson plantation until the spring of 1936, when the founders of the union took their case to Washington, D.C. On April 12, 1936, the *Washington Post* ran a story that attracted international attention and exposed Lee Wilson & Company to congressional review and, ultimately, AAA sanctions.[54]

Journalist Sidney Olson reported that the R. E. Lee Wilson plantation was the single largest recipient of AAA cotton benefit checks in the country, having received $199,700 for 1933 and 1934 combined. The leadership of the STFU had been in Washington just two weeks earlier attempting to address officials about the problems confronting plantation laborers, and they turned out to be Olson's chief source of information. According to H. L. Mitchell, a founder of the union, the plantation's AAA check for 1935 "had been held up pending an investigation" because the Wilson plantation had "changed from the sharecropping system to a day labor basis in 1934, throwing hundreds of families on relief, and then re-hired workers at 75 cents per day for 12 to 14 hours of work." Mitchell charged that the actions of the Wilson plantation cost the government three times—once when they received the AAA payment, again when the evicted families were forced on to relief roles, and finally when the evictees were placed in resettlement communities. His accusations happened to coincide with a congressional debate over the entire AAA operation. The Supreme Court had recently ruled that the processing tax used to fund the AAA payments was unconstitutional, and Congress had devised a different strategy to continue the funneling of money to farmers. The Soil Conservation Program was characterized as a means of rebuilding the soil, but everyone understood its purpose. Congressional leaders were not of one accord, however, on the implementation of the program. As the *Washington Post* reported, "Both Democratic and Republican political strategists were preparing to capitalize on the steadily-mounting total of names of large benefit recipients. . . . Democrats propose to show that many of those most vigorously opposing the New Deal farm program have meanwhile quietly pocketed fat checks from the AAA. Republicans will strive to show that farm relief under the New Deal has gone to the treasuries of huge Eastern corporations which own agricultural areas." Senator Arthur Vandenberg, a Republican from Michigan, sponsored a resolution calling for the publication of the names of those receiving AAA payments, and it appeared likely to pass.

At least one of the accusations had little basis in fact. The Lee Wilson plantation was on record as in favor of the AAA from the beginning, as were the second- and third-largest recipients of AAA checks, Chapman and Dewey Company in Poinsett County, with whom Wilson had done business, and the Delta Pine and Land Company of Mississippi. The latter, an English-owned corporation, was managed by Oscar Johnston, who had been associated with the AAA from the beginning and served as manager of the "federal Cotton Pool." Two lumbermen from Missouri owned the Chapman and Dewey operation, but one of them had relocated to Memphis three decades earlier.

Whether any of the three entities had shifted from the sharecropping to the day labor system and evicted massive numbers of sharecroppers and tenants is impossible to ascertain given the records available. Lee Wilson & Company defended their treatment of labor in a document entitled "A Review of Lee Wilson & Company operations during the life of the Farm Program," which focused on the period between 1933 and 1938. A self-serving document designed to put the company in the best possible light, it directly challenged the assumption that the company had turned to day labor exclusively, arguing that it understood "the idea and principle of the farm program from its beginning in 1933 was to work to better the condition of all people engaged in making a living on the farm all the way from a day laborer to a farm owner, and as soon as possible raise the status of all tenants." Instead of moving to a day labor system, which would have relieved them of the responsibility of sharing crop subsidy payments, they argued that they went the other direction. "A great many of the cash renters now on our land have been day laborers, share croppers and crop renters, and as soon as any one of our farmers shows he has the ability and is worthy of an advance, we raise his status." Rather than throwing tenants off the plantation and onto the government dole, the company boasted that "there hasn't been a single one of our tenants on the relief or W.P.A. rolls." Finally, the document denied that they employed a commissary system that kept those who labored for them in debt. "Our credit system upon our plantation is far different to the old commissary system. The prices in our stores are competitive with other merchants of the county and when a tenant buys on credit he is charged the same price—cash price—plus interest for the time the account runs." Although the document was clearly written to vindicate the company and must be understood critically, at least some of its assertions seem credible. At the very least, it would have been difficult for the company to have abandoned its long-standing arrangements with over a thousand families who labored on its various plantations. The complex organizational structure put in place decades earlier and refined over the years would have worked against a wholesale abandonment of the renter, tenant, and sharecropping system. On the other hand, Lee Wilson had always intended to maximize his profits, and there is no indication that Jim Crain had any thoughts of doing otherwise. If it had seemed more profitable to him to move in the direction of day labor, something that had always been one feature of the plantation in any case, he would have done so. Only the company's own internal bureaucracy and the watchful eye of the government, made aware by the STFU, would have kept the company from an immediate transformation to day labor.

At least one of the assertions made in the newspaper article about the company was unquestionably accurate. The company's claim for the 1935 parity payment had been held up, but not necessarily for the reasons the STFU suggested. As the comptroller general of the United States later explained to the secretary of agriculture, the controversy began innocently enough when his office reviewed the company's application in late 1935 and noticed that the figures on three of their units were "exactly equal to the Bankhead allotments for these farms" and only slightly in excess of the allotments on four other units. The comptroller doubted "that production on any farm would equal a predetermined allotment" and returned the forms for further clarification. After receiving additional information, his office noticed that there were certain curious omissions of cotton sales figures, omissions that made it impossible for his office to determine the amount owed to the company. Arguing that the correct amount for payment could not be determined by the comptroller's office, his staff passed the matter on "for such administrative action as may be considered proper."[55] As this administrative review was taking place, the STFU made its charges, placing the company under even more intense scrutiny.

As if the trouble over the 1935 parity payments was not enough, questions arose of the 1936 subsidy. Had it not been for the newspaper story, AAA officials might have overlooked certain peculiarities with regard to the company's 1936 application. When those payments were held up too, Jim Crain appealed to Congressman W. J. Driver to intercede on behalf of the company, and Driver did his best. He shared with Crain the information he was able to secure, and the correspondence revealed that the company was caught in another apparent impropriety. As Cully Cobb, director of the Southern Division of AAA, put it to Congressman Driver, "we have been informed that the applications as submitted by Lee Wilson & Company showed an identical production and Bankhead allotment for all but two of the units covered by such applications in 1935." In other words, it appeared unlikely to authorities that all but two of the Wilson plantations would have produced precisely the amount of cotton and withheld the exact number of acres out of cultivation in 1936 as they had in 1935. "The figures of production as entered in the various applications appeared unusual to the state office at Little Rock." The state office discussed the situation with Jim Crain and "refused to approve such applications to the General Accounting Office for payment until Mr. Crane submitted a written statement." Crain ultimately satisfied their objections, but when they passed the paperwork on to the auditor in charge of the Field General Accounting Office, the next intermediary step in the cumbersome bureaucracy, he was also struck by "the similarity of figures," which appeared to "him as being extremely unusual and he therefore sent all of the applications to the office of the Comptroller General [here] in Washington for a decision." There the application languished for four years.[56]

The year 1936 must have been a particularly confusing one for Jim Crain. In the spring the story broke in the *Washington Post*. In the autumn he faced obstacles securing crop reduction payments. In between those two troubling events, in August, in fact, Governor Junius Marion Futrell appointed him to the state Farm Tenancy Commission.[57] The STFU's assertion of massive evictions and charges that planters had failed to share crop subsidy payments with tenants and sharecroppers had made it necessary for Futrell to "investigate" the situation. The sentiments of Futrell, a man who had little sympathy for the poor, believing they had brought their troubles upon themselves, were clear. He hailed from a northeastern Arkansas plantation background and had long enjoyed the support of the Wilson family and other large planters. A look at those appointed to the committee confirms that, with a couple of exceptions, it was loaded with representatives of some of the largest plantations and business interests. Jim Crain might be characterized as the proverbial fox guarding the henhouse, but he was far from the only fox appointed to the commission. Although compelled to include representatives of tenants and sharecroppers like the president of the STFU, J. R. Butler, and the union's attorney, C. T. Carpenter, Futrell did not refrain from packing the committee with those who shared his own point of view and represented the interests of his friends. For example, he appointed W. T. Beall, a longtime employee of the Wilson company (not related to the Beall family with whom the Wilsons were intermarried), and Charles T. Coleman of Little Rock, the firm's attorney. V. C. Kays, president of Arkansas State University, an institution which owed a great debt to the Wilson family, was also appointed. Their report, issued in 1937, was predictable. Although they conceded that some abuses had probably occurred, they essentially produced a document that merely ruminated on the longstanding problems in the tenancy system, placing the blame on the forces of history rather than the actions of particular men.[58]

Meanwhile, Lee Wilson & Company was having no luck securing its parity payment for 1935. Neither had it made any progress in receiving the subsidy or parity payments for 1936. Even as it pursued its claims for those years, it held acreage out of production in 1937, 1938, and 1939 and made applications for payment, but the AAA denied those applications as well. The record does not provide detailed information on precisely why the AAA declined to authorize payments to the company for those later years but does demonstrate Crain's attempts to secure the help of the Arkansas congressional delegation. Hattie Caraway, W. J. Driver, and Senator Joe T. Robinson all attempted to intercede. When John Miller took Robinson's place after the latter's death in 1937, he too wrote letters and made every effort to assist Crain, who, by this point, had established himself as the man in charge of the Wilson plantation, a man to be reckoned with. Crain had clearly assumed the role that Lee Wilson had played, working, just as had the patriarch, to use his influence to secure the company's interests.[59] Still, crop years came and went and the government failed to come through. All this was doubly galling to Crain and costly to the company. Not only were they not receiving the government payments; they had made it a practice to claim the anticipated government payments as income on their tax returns from 1935 through 1939. They then engaged in protracted and complicated correspondence with the Internal Revenue Service when they attempted, without success, to claim those unrealized payments as a loss on the company's 1939 tax return.

By early 1940, however, "the atmosphere had cleared" in Washington, and Crain renewed his efforts to secure the subsidy and parity payments.[60] Once again he called upon prominent Arkansas politicians—Senators John E. Miller and Hattie Caraway—who accompanied him to the office of P. M. Scott, an AAA administrator, and convinced him to launch a new "investigation of all contracts and applications submitted by Lee Wilson & Company and its tenants for the years 1933 through 1939." Those participating in the probe included "a representative from the Southern Division, some representatives from the Arkansas State office, one or more representatives of the Mississippi County office, and two representatives from the General Accounting office." Caraway received assurances from an official that they would expedite the matter and forwarded the correspondence to Jim Crain, saying, "As you will note, they have promised to expedite. . . . If you do not hear further from the matter within a reasonable time, please bring it to my attention."[61] It took longer than Crain would have liked, but in early 1941 the company received the first of two payments totaling $176,310.92 from the AAA, the amount the company argued was owing to them for the years 1937 through 1940. They had apparently accepted the loss of the 1935 and 1936 payments. Whether the STFU can be credited with the six-year lapse in payments is debatable, but the attention they focused on the company was damaging, at least for a time.[62]

Not only did Roy Wilson play no role in that matter; he relinquished responsibility for overseeing the troubling drainage district debacle in Mississippi County. At the time of Lee Wilson's death, the districts were all in receivership or on the verge of bankruptcy. In the last months of his life, Lee Wilson advised the St. Francis Levee District in its efforts to secure an RFC loan to refinance its operations and encouraged Roy to do the same for the drainage districts. It was Jim Crain, however, who took the lead in the latter endeavor and by 1935 had lined up RFC loans for four subdistricts in two Mississippi County drainage districts, three of which were in bankruptcy. The two districts were Carson Lake Drainage District No. 8 and Grassy Lake and Tyronza Drainage District No. 9, both of which covered most of the property owned by Lee Wilson. They had been at the center of the battle between Wilson and those who opposed the establishment of drainage districts in 1908, a battle that Wilson won by 1911 and 1912. Most of the work constructing those districts took place after the floods of 1912 and 1913, and together with levees constructed by the St. Francis Levee District and the Corps of Engineers, protected south Mississippi County from the worst of the 1927 flood. The cost, however, had been high. Quite apart from the destruction of wetlands of Mississippi County—something no one in the early part of the twentieth century raised alarms about—the drainage taxes worked a hardship on farmers after the drought of 1931 drove farm revenues to new lows. By 1935 Crain and Hy Wilson worked out a compromise with the federal receivers so that landowners were "forgiven" their drainage taxes for 1933 and 1934.[63]

Drainage District 11, which was in a category all its own, covered lands "in the vicinity of Wilson and around Joiner, Bassett, Pecan Point and Frenchman's Bayou." Lee Wilson & Company owned much of this acreage, and Hy Wilson served as receiver, an interesting appointment given the fact that the Lee Wilson estate, of which Hy Wilson was an executor, held all of the district's outstanding bonds.[64] A small district on the southeastern edge of the county, it avoided bankruptcy, but, as with districts 8 and 9, as of May 1935, all of the bonds were "past due since 1933" and it had no funds on hand for routine maintenance. Nevertheless, because the bondholders and the receiver were one and the same, Hy Wilson could write "I expect to have this district cleaned up within the next month or two. It will very likely be necessary to make an additional spread of taxes against the lands in this district to take care of its indebtedness. However, I do not think that I will even put this on the County tax books, thereby saving that expense. I will just work it out directly with the land owner." Hy Wilson—and thus Lee Wilson & Company— occupied an enviable position in the case of this small but important little district.[65]

The RFC loans provided a necessary breather for Districts 8, 9, and 11, just as such loans did for districts in other Arkansas counties and in twenty-five other states, but the districts remained substantially in debt. They now had to pay off the bondholders and the RFC, and though tax revenues were up in the second half of that decade, it took entry into World War II and the boon to agriculture that brought about to begin to make a real difference to their economic condition. In 1941, in fact, just months before Pearl Harbor, Jim Crain appealed to the Civilian Conservation Corps about placing a CCC camp in Mississippi County, explaining "Our drainage system is in such bad shape today that we are asking for federal aid. This will be in the nature of rehabilitation and not maintenance. If this could be accomplished with the aid of a CCC camp located in this county, I feel sure afterwards we will be able to carry on."[66] The outbreak of war four months later decimated the ranks of the CCC and turned attention away from the problems confronting drainage districts. Most of them were able to recover over the next several years as tax revenues rose, but by 1947 a new problem emerged, reflecting an unintended consequence of drainage along the St. Francis River. Flooding along the St. Francis River reflected a lack of coordination between various counties in Missouri and Arkansas. Drainage districts in these two states all funneled water into the Little River or the Tyronza River, both of which flowed into the St. Francis and "increased the rate of flow beyond the capacity of the channel southward" causing serious floods in Poinsett and Crittenden counties.[67] The author of this report, the East Arkansas Drainage and Flood Control Association, of which Jim Crain was a committeeman, proposed a renewed effort on the part of the St. Francis Levee District to finish what it had started in 1894.

Jim Crain proved to be an able and resourceful leader. He rose to a position of power and appeared to enjoy unquestioned primacy, but that was an illusion. Lee Wilson's two grandsons, Lee Wesson and Bobby Wilson, had other ideas. Although he would remain in control and continue ably leading the company through the trials and opportunities of World War II, Crain would not be the man to lead the plantation through the next phase of the company's history, the important transition to capital-intensive agriculture. That was the province of Lee Wesson, who came back from government service during the war ready to challenge Jim Crain's leadership, and of Bobby Wilson, the "class boy" who had served with distinction in the military but initially refrained from resuming his position with the company. Only a summons from his father and a sense of responsibility for the empire his grandfather built would bring him back to take over the leadership of Lee Wilson & Company.

9

THE RETURN OF THE "CLASS BOY"

After serving in the army during World War II, Bob Wilson initially chose another life in a different venue rather than return to Wilson, Arkansas, and pursue the destiny laid out for him. Described in 1936 by one of his father's former classmates at Yale as "rather a shy, diffident, lovable, good-looking rascal," he became an able leader of men during the war, distinguishing himself as an infantry officer, winning several medals, including two Bronze Stars, one with cluster. He saw "some rough action," suffering wounds on Utah Beach (D-Day plus 1) but then fighting on through central France, the Rhineland, the Ardennes, and Germany. The experience changed him, enlarged his world, and broadened his perspective. Rather than return to the delta plantation his grandfather had carved out of the swamps, he settled in Loveland, Colorado, and went into the ranching business with a Yale classmate. He loved the outdoor life there on the Anchor Ranch more than he did the executive position he had held with Lee Wilson & Company before the war. Characterized as an expert rider who was "practically born on a horse and rides like a centaur," he yearned for something other than running a plantation operation. He had described his work as an accountant in the family company in the spring of 1941 as "absorbing," but even then he celebrated the fact that "my wife and son have a pernicious habit of luring me away from the pursuit of gold in favor of the pursuit of happiness, and I find life much fuller by not arguing with them too much."[1]

Duty to family and to the legacy created by his grandfather eventually put an end to his dreams of greater happiness in a different venue. The return of the "class boy" to Wilson came as a result of a challenge to Jim Crain's leadership role in Lee Wilson & Company, a challenge launched by Bob's cousin in March of 1946. Lee Wesson served as an executive with American Aviation in Jamestown, New York, during World War II, and his position at the plant deepened his interest in management. He returned to Wilson "loaded for bear" after the war, and with the support of his mother and Aunt Marie, began a campaign to unseat Crain. Roy Wilson, who remained dependent on Jim Crain to run the business, refused to countenance young Wesson's efforts. Although Bob felt no great loyalty to Crain, he understood his father's dependence upon him and knew that joining his cousin in the challenge to Crain amounted to disloyalty to his father. He faced an excruciating dilemma and chose to remain apart from the controversy. During almost three years of family intrigue, Crain's hold over Roy loosened enough so that Roy, apparently on his own, broke free of his control. Roy then summoned Bob, challenging him to return to Wilson and accept the responsibility he had been born to. Otherwise, Roy proposed to sell the property. Out of a sense of responsibility for the legacy his grandfather created, Bob abandoned his own dreams of life in the West and returned to the town of Wilson in 1950 to become president of the company.[2]

In his position of authority Bob steered the enterprise through a fundamental transformation. He adapted to the changes faced by all agricultural operators in the post–World War II era, but the complicated nature of the sprawling Wilson operation necessitated greater adjustments on its part. He presided over a shift from a labor-intensive system employing over two thousand tenants and sharecroppers to capital-intensive scientific agriculture requiring fewer than a hundred wage laborers and marked by the use of machinery and chemicals. By 1964 Bob could confidently and legitimately characterize himself as "not a [feudal] baron but a businessman." Still, the transformation was not complete until his own sons took over the plantation, for Bob Wilson continued to grow some crops designed to provide work for a diminishing labor force. In a sense, the transformation took place in two stages. The first stage occurred between the advent of New Deal programs until approximately 1980, and the second occurred from 1980 to the present. During this latter period his sons leased the lands to local farmers and assumed a management role; these lessees cultivated only cash crops. Although a few of these renters were from outside the region, some were local men with a connection to the area. They owned land nearby and found it cost-effective to lease Wilson property in order to, given the economies of scale, maximize the use of the large machinery they had purchased.[3]

Even before the outbreak of World War II, the Wilson plantation—and plantation agriculture in the South generally—was undergoing a fundamental restructuring, largely because of New Deal programs that funneled cash into the hands of planters, who began a transition to the cultivation of new crops and away from tenancy and sharecropping. World War II accelerated the transformation by deepening an interest in mechanization and the use of chemicals such as pesticides and fertilizers, but some crosscurrents obscured the significant changes underway. A serious labor shortage, resulting from heavy enrollment in the armed services and opportunities in war industries, led Jim Crain to resort to any means necessary to plant, tend, and gather the crops, including the use of German prisoners of war and an unsuccessful attempt to secure the use of Japanese American internees located in southeastern Arkansas. In the absence of Bob Wilson and Lee Wesson, Crain also consolidated his power over the company and by the war's end seemed to have supplanted the Wilson heirs entirely.[4]

Jim Crain ably steered the company through a series of challenges and opportunities, greatly enriching the enterprise but also expanding his own personal holdings. The company benefited from a boon in prices for agricultural and forest products during the war that encouraged an intensification of agricultural development in the county. Cropland harvested in Mississippi County expanded from 393,001 acres in 1939 to 409,490 in 1944, a trend upward that defied a statewide decline by three-quarters of a million acres.[5] Responding to an almost unprecedented rise in agricultural prices, farmers in the county more than doubled their acreage in soybeans, while farmers elsewhere in the state, facing a labor shortage they proved unable to resolve, grew fewer acres in that crop, and, in fact, the total acreage in production in the state declined during the war years. The AAA crop reduction program, which essentially paid farmers to refrain from planting cotton, had encouraged an expansion in the production of soybeans, and the demand for that product rose considerably after the war broke out, as it became a popular substitute for other sources of oil. While cotton acreage rose in the county, also defying a trend in the other direction statewide, the increase was relatively modest, with most farmers responding to a virtual mandate to "grow soybeans and help win the war." Lee Wilson & Company, which operated soybean oil mills in both Wilson and Osceola, reaped the benefits of this trend in soybean production.[6]

The Wilson company had always encouraged their employees to grow kitchen gardens—and even went so far as to inspect them on occasion—but the New Deal crop-reduction programs and the war emergency encouraged the company to experiment with greater production of vegetables and feed crops designed for an outside market. Until the outbreak of the war, however, their efforts in that direction were relatively modest. In early 1943, in response to a rising demand for vegetables, the company expanded its production in that area and opened a dehydration plant. In order to facilitate the marketing of these products to the military, Jim Crain, who had carefully cultivated relationships with key political figures, spent "some weeks in Washington working out the details with government authorities."[7] The company also had to appeal to the County Farm Rationing Committee in 1943 in order to install a modern dehydration plant to process the vegetable crops for the Army, Navy, and Lend-Lease.[8] The firm also benefited from contracts to supply boxes for packaging food products for the Army and Navy and, in fact, packaged much of its own produce for shipment. All of its wire-bound boxes packaged food products, most of which went to the Army and Navy.[9]

In addition to expanding cotton, soybean, and vegetable production, the company intensified its interest in the production of feed crops, and Crain's connections enabled him to secure permission from the War Production Board to circumvent quotas on building materials in order to build alfalfa dehydration plants at Marie, Evadale, and Keiser. Cattle and horses could graze on alfalfa as it grew in the fields, but its uses could be greatly expanded once dried and then made into pellets and packaged for purchase by farmers raising chickens, hogs, cattle, etc., all commodities considered of importance by the military.[10]

The rationing of tires and gasoline threatened to undermine the company's ability to pursue its routine business activities. The increased demand for its agricultural and lumber products just accentuated the problem. Government regulations kept the office staff busy filing "applications for allotment of control materials" and "applications for preference rating" to the War Production Board involving everything from tires to fertilizer. Because of their contracts with the Army and Navy, however, they succeeded in most of their applications for extra allotments of tires, gasoline, and farm machinery.[11] The company sought and received permission to purchase materials to construct a dairy in 1943 but because of a clerical error had to file a bothersome addendum asking for additional materials worth $210. A note of urgency accompanied the application. "It is very essential that we get the dairy into operation as soon as possible as all the dairies in this part of the country have closed down leaving the entire surrounding territory without any dairy products whatsoever."[12]

Although not everyone in the state was united in the effort to secure war industries for Arkansas, top state officials attempted to promote them. One chief obstacle was the dearth of towns and cities with sufficient workers available for hire. Anticipating America's involvement in the growing conflict in early 1941, the War Department issued "general specifications" which stipulated a set of requirements for plant sites, and Governor Homer Adkins traveled to Washington in March that year to meet with officials in the War Department in order to secure additional information on their requirements. Upon his return, he called a meeting of prominent planters and businessmen—including Jim Crain—to discuss what Arkansas might do to fulfill the War Department's expectations but, in the end, the state was unable to paint a portrait sufficiently attractive to secure much more than a token response. However, Mississippi County did secure an Air Force training base near Blytheville, a base that survived until the 1990s and played a role in stimulating the economy of the north end of the county.[13]

Very soon after the war broke out, the serious labor shortage mentioned above loomed and necessitated some extraordinary measures. The number of draftees accelerated in the months after Pearl Harbor, while many others were enlisting in the military voluntarily or were drawn away from the county by the promise of high-paying jobs in one of the war industries. The labor crisis became particularly acute in north Mississippi County, in part because there were too few African American laborers there. For example, whites constituted fully 96.1 percent of the population in the old Buffalo Island area.[14] The farmers of that section of the county had done too good a job maintaining a "white man's country," and it backfired on them. Some plowed under cotton and planted soybeans in June 1943, partly because the price for soybeans rose higher than that for cotton but also because of the labor crisis. Heavy rains kept the few laborers they could secure from the field, but, as the county agent serving north Mississippi County complained, "Where there has been less rain, the crop is in better condition, especially in the southern part of the county, where there also are more negroes [sic] and so not such an acute labor shortage."[15]

With the labor crisis merely getting worse as the war machine drew more and more laborers away from the agricultural sector, Mississippi County's agricultural agents assumed responsibility for aiding farmers and planters in securing labor, but they very soon exhausted the traditional supply of white hill folk. Although they secured more labor in 1943 than ever before, they were "not successful in obtaining large numbers of workers."[16] As early as July that year, planters and farmers began devising plans to secure prisoners of war for labor, initially thinking of Italian prisoners. So desperate were they for labor, they even imagined it might be possible to use American soldiers not yet deployed in harvesting the crop.[17] Ultimately, the crisis was not as great as feared because of a short crop that fall.[18]

The next year the agricultural agents helped farmers and planters secure German prisoners of war camps and thus avail themselves of that source of labor. Lee Wilson & Company secured the first such camp, placing it on their Victoria plantation in February 1944, and plans were afoot to secure at least four others, at Osceola, Luxora, Keiser (all in south Mississippi County), and Blytheville (in north Mississippi County).[19] In April, representatives from Mississippi County farms and plantations attended a meeting sponsored by the Agricultural Extension Service in Brinkley, Arkansas, to discuss the mechanics of applying to the War Department for POW camps, and two Lee Wilson & Company employees attended that meeting: Johnnie Crain, Jim Crain's oldest son, and Eddie Regenold, a company employee.[20] The German POWs worked in drainage and in farm work, harvesting six thousand bales of cotton in 1944. Farmers and planters had managed to survive the crisis in 1944, but they had great fears about what lay ahead.[21] Farmers swamped Agent D. V. Maloch, relatively new to the south end of the county, with requests for labor, but Maloch reported that 1,825 POWs in south Mississippi County helped harvest a bumper crop that fall. He characterized the POWS as having "saved the day."[22] The county agent in north Mississippi County, Keith Bilbrey, had only one prisoner of war camp (in Blytheville) containing only six hundred laborers to draw from but reported that he was able to supply approximately 1,500 laborers altogether. He was not specific as to where he secured the other nine hundred but wrote that Buffalo Island farmers actually resorted to "the lowly goose to furnish the labor hoe help," expecting that "one goose will 'hoe an acre' of cotton."[23]

In the context of this labor crisis, farmers statewide began to abandon the mule in favor of the tractor. A contributing factor involved the transition to soybeans, a crop already subject to mechanical harvesters, unlike cotton, which still necessitated hand choppers and pickers. The number of tractors increased statewide from 12,564 to 26,537 between 1940 and 1945, while the number of mules dropped from 260,424 to 216,174. The rising price of agricultural commodities and the labor shortage combined to favor the tractor over the mule, and Mississippi County mirrored this transition. Farmers and planters in the county doubled their use of tractors, from 1,052 to 2,256, and reduced their use of mules from 17,567 to 12,311. This was the beginning of a trend that only intensified in the postwar period (see table 9.1). Farmers easily secured exemptions to purchase gasoline for their tractors given the military necessity ascribed to the crops they raised. Tractors were particularly important for certain crops, especially soybeans, which were best harvested with combines pulled with tractors.

Lee Wilson & Company faced certain institutional constraints in making any full-scale shift to tractors, although they had been purchasing them since the 1920s. They had thousands of mules and tens of thousands of dollars invested in barns, feeding troughs, and corrals. The handwriting was on the wall, however, and the labor crisis encouraged the company to begin an important transition to mechanized agriculture. The statistics on the number of hired farmworkers tell the tale. In 1940, farmers hired 54,894 such workers across the state. In 1945, the number had dropped to a remarkable 8,856. Mississippi County again mirrored the state statistics. In 1940, 8,051 hired workers labored on the county's farms; in 1945, only 746 did so. In this context, the German prisoners of war, who numbered just over 2,400 in the five camps in north and south Mississippi County combined, hardly made up the difference.[24] A reversal of a trend against the use of tenants and sharecroppers, which had emerged after crop-reduction programs, helped relieve the labor crisis but far from solved it. Between 1935 and 1940, the number of tenants in the county dropped from 8,573 to 6,085, and while their use increased during the war by nearly 1,700 (to 7,791), the number still fell short of the need.[25]

Other forces taking shape during World War II furthered the transformation of southern agriculture initiated by New Deal crop-reduction programs. Historian Donald Holley has labeled the transition from labor-intensive to capital-intensive agriculture the "second great emancipation," freeing planters from the need to maintain sharecropping and tenancy, a repressive institution that had brought much notoriety to the South. Of course, Southern Tenant Farmers' activities suggest that many tenants and sharecroppers objected to evictions, sometimes at great risk to themselves. While historian Jarod Roll finds convincing evidence that many sharecroppers and tenants in nearby southeastern Missouri resisted displacement because they hoped to climb the agricultural ladder to land ownership, it is doubtful that many on the Wilson plantation held such expectations. In accepting a place there, they had embraced— or became resigned to—dependency and landlessness. The reduction of acreage devoted to a crop requiring significant labor inputs enabled planters to transition to crops that necessitated expensive machinery, and the cash funneled into their hands by those same AAA programs enabled them to increase their purchase of machinery. The full mechanization of the cotton crop awaited only a marketable mechanical cotton harvester, something that was to come down International Harvester's assembly line in 1943.[26] The increased reliance on chemicals designed to minimize the irksome weeds that needed chopping was another important component to the transformation that was in the making. No longer would Mississippi County farmers have to resort to using geese as "hoe hands," but there were consequences to the growing reliance on chemicals, consequences that few, if anyone, foresaw.

The first significant experiment with chemicals in Mississippi County during World War II began in 1941 through the application of DDT (Dichloro-Diphenyl-Trichloroethane) by Civilian Conservation Corps workers to fight malaria, but the war demands soon decimated CCC ranks and endangered the program. Their efforts had included work on drainage ditches to eliminate the standing pools of back swamps, still one of the best ways to eradicate malaria. Crain had explained the need for CCC camps in 1941 as a public health measure, arguing that "proper drainage control would also mean proper malaria control for the total population of some 80,000 individuals in this county."[27] Crain's concern about malarial control was genuine. Although the county had been transformed beyond recognition in the thirty years since drainage began in earnest, pockets of standing water remained, providing perfect breeding places for the mosquitoes that carried the parasite that had caused so much havoc and grief to thousands of county residents over the years. It had taken its toll on rich and poor alike, and as Jim Crain well knew, had visited even powerful families, virtually crippling Lee Wilson's son and heir. With the outbreak of World War II, the fight against malaria became a more urgent concern because so many military bases were located in the South, and military authorities aggressively promoted the use of DDT. Billing it as a war measure and arguing that it was necessary to eliminate the mosquitoes that threatened military personnel, the government distributed funds to states to develop malaria control programs featuring the use of the powerful pesticide. Arkansas funneled the money to county programs, and because the Air Force established a training base in Blytheville, a malarial control program operated in Mississippi County.

By late 1945 authorities permitted the expansion of the fight against malaria because of the fear that the disease might be introduced by military personnel deployed in the Pacific, Africa, and even some parts of Europe. The last thing the government wanted was a domestic outbreak that might threaten the return to normalcy after the war. When the federal government ceased funding the program in 1947, Arkansas authorities mandated that county governments would have to raise the necessary funds if they wanted to continue the war against malaria. Roger C. Cooper, the malaria supervisor in Mississippi County, estimated that it would cost $3 for each house sprayed with DDT, and local officials acknowledged the need as "the malaria rate in Mississippi County . . . is still one of the highest in the state."[28] In the end it would take a combination of county, state, and federal funds and a liberal application of DDT to eliminate the mosquito that carried the malaria parasite. By the time that Rachel Carson's *Silent Spring*, published in 1962, exposed the dangers of DDT, malaria was no longer a problem for Mississippi County or any other place in the United States.[29]

Even greater changes followed, not only for Lee Wilson & Company but also for southern agriculturalists generally. The delivery of the first marketable mechanical cotton picker was a harbinger of things to come and further accelerated the trend toward capital-intensive agriculture. However, the adoption of mechanical cotton harvesters occurred only gradually. One of the primary problems involved the inadequacy of existing cotton ginning operations. Planters and farmers continued to use the old gins, replacing them slowly over the next decade with modern operations better able to remove the debris from machine-picked cotton. Therefore, agriculturalists made a relatively slow transition to new modes of production, and they continued to need more labor than was immediately available. With the withdrawal of POWs after the war, planters turned to Mexican laborers hired specifically to chop cotton in the summer months and harvest the crop in the fall. The Bracero program had emerged during World War II to ease labor shortages, but Arkansas farmers had used that source of labor only minimally and no one in Mississippi County employed Mexican laborers through this program. Beginning in 1946, however, Lee Wilson & Company, as well as other planters and farmers in northeastern Arkansas, hired tens of thousands of them to work the land while planters began the transition to capital-intensive agriculture. The erosion of the tenancy and sharecropping system insured a virtual depopulation of the rural countryside. Jim Crain played a pivotal role in securing Mexican laborers for the Wilson company, as well as for his own operation.[30]

A significant demographic shift underway since the mid-1930s intensified in the postwar period. New Deal crop-reduction programs played an important role in this development, pushing many black and white sharecroppers and tenants off the plantations as planters, flush with government subsidies, began a transition to mechanized agriculture. As mechanized and scientific agriculture intensified, the postwar industrial economy in southern, northern, and western cities provided uncertain opportunities for the displaced rural southerners. Mississippi County's population dropped from 80,217 in 1940 to 66,277 in 1960, with the black population falling at a slightly greater rate countywide (see tables 9.2 and 9.3).[31] In Golden Lake, the heart of the Wilson enterprise, the number of blacks dropped more than in any other single place in the county, from 1,072 to 750, representing a decline of from 47 percent to 32 percent (for countywide figures, see table 9.4).[32] African Americans on the

Wilson plantation who held no illusions of climbing the agricultural ladder were likely among the first to leave. In the decades after World War II, the total number of farm operations fell in tandem with the overall drop in the county's population. Most telling, the sharecropper category of farmer completely disappeared from the census by 1960 as planters made the transition to mechanized and scientific agriculture and adopted the wage labor system (see table 9.5).[33]

Census figures and economic data provide valuable information about the transition taking place since the advent of New Deal programs. The experiences of William "Snake" Toney, who worked not on the plantation but in the town of Wilson, provide better insight into the transition's long-term implications for African Americans employed by the company. Toney came to Wilson, Arkansas, in 1931, when he was seventeen, to live with a mother who had left him at a very young age to be raised by her parents. His grandfather was an AME preacher and, as was the custom, moved frequently, but he was usually posted in northeastern Arkansas towns. Snake lived with him for a while in Osceola, just north of Wilson, until he joined his mother and attended the Wilson school for one year. He then went to work at the sawmill and the box factory for a while before being assigned to work with Bob Wilson, who was placed in charge of construction of the cotton oil mill upon his graduation from Yale in 1936.

Snake's experiences with Bob Wilson placed him in a position to become better acquainted with the Wilson family and the company operations than was possible for most African American or even white farm laborers on the plantation. Bob Wilson encouraged Snake to learn all he could about the operation of the cotton mill in order to better his position. He believed that Snake, who stood only five feet six inches tall and weighed under 120 pounds, was too small to survive the grueling work required of ordinary mill hands. In Snake's words, Bob told him that general labor in the cotton oil mill would "kill him" and enjoined him "to get in that mill and learn everything you possibly can." Snake followed this advice and was soon running the mill room, where he had charge of the shakers, which he said would "shake all the impurities . . . the stuff that you didn't want to go into the cotton, it would shake all that stuff out." He later moved over to the lint room, where he took charge of keeping the saws sharpened and worked the midnight to 8:00 a.m. shift.

Snake also used this association with Bob Wilson to best advantage in a way that demonstrates the severe constraints endured by African Americans. Although he remembered Lee Wilson fondly, remarking that the patriarch maintained a close connection to the sawmill operation even in the last years of his life, walking with his workers down the tracks when the morning whistle blew signaling the start of the day, Snake lamented the changes that took place after the patriarch died. In his view, Jim Crain's ascendance to power in the 1930s was accompanied by what he characterized as a "Mississippi" attitude toward the treatment of black labor, a determination to intensify segregation practices by, for example, ending the practice of holding an integrated annual picnic of all company employees. Roy objected to this change, however, and soon the picnics stopped altogether.

Lee Wilson's reputation for better treatment of African Americans may have been undeserved, but Snake Toney's fond recollections of his greater benevolence served as a counterpoint to the story Snake wanted to convey, a story about thwarted ambition and unjust treatment in the succeeding years. One anecdote establishes the benefits Snake derived from his relationship to the Wilson family and another highlights the severe limitations facing company workers, whose interests did not necessarily coincide with those of the company. In one instance, Snake employed his friendship with Bob Wilson to save himself from possible harm, and in the other he was required to take a lesser-paying job in order to benefit the company operation.

By late 1937 Snake enjoyed an enviable position for a young black man when he took over the night shift in the lint room, and he took pride in the fact that he could support his mother, with whom he lived. Although she no longer *had* to work, in the summer of 1938, she took a job chopping cotton in order to secure some extra spending money. One day, however, she decided to stay at home and do her laundry rather than go to the field. Her "boss man," Sonny Lynch, came to her house later in the afternoon to inquire why she had not come to work that day. She told him she simply had some other things to do, which angered Lynch. Snake, who had worked the night shift and was in the other room sleeping, was awakened by the conversation taking place at the front door of the house. Lynch was standing on the front porch; Fanny, Snake's mother, was just inside the house, talking to him through the screen door. In Snake's words, his mother "gave Mr. Lynch some sass," making Lynch angry enough to grab the screen-door handle and attempt to enter the house. Fortunately, the screen door was latched, and at this point, Snake interceded. He had words with Lynch and then picked up his rifle, which was leaning up against the wall by the door. Sonny Lynch swore at Snake and then left the premises, in something of a hurry. Snake knew he was in trouble, however, saying, "You ain't supposed to do that, show a white man a gun." He decided to "hit out and go down to Mr. Bobby's" to seek out his protection. When he reached Bob's house and told him what had happened, Bob said in response, "Are you losing your mind? Are you crazy"? And Snake responded "Mr. Bobby, he made me mad." And Bob replied that he was not supposed "to get that mad." But Bob drove him back to the house where they found three or four white men with guns and ropes waiting for Snake. Their intentions were clear. Bob told them to "clear out," that Snake's mother lived in a cotton oil mill house and did not have to work if she did not want to. He also warned them that nothing was to happen to Snake, that if any harm came to him that they would pay the price. But once the men were gone, Bob turned to Snake and reiterated the foolhardiness of pulling a gun on a white man, essentially telling Snake to abide by the customary observance of white supremacy.

The second anecdote also involved Snake's mother but in a different context. Mrs. C. W. Hoover had asked Fanny to come to work for her as a domestic servant. Fanny told the woman that she did not need to work because her son had a good job at the cotton oil mill. Mrs. Hoover took exception to the refusal and asked her husband, who managed the cotton oil mill, to fire Snake in order to teach Fanny a lesson. When the mill shut down during the summer months, Snake was laid off, when under normal circumstances, because he was a valued worker with a special skill with machinery, he would have been retained to conduct routine maintenance. Snake then went to work constructing a new cotton gin for the company, working for A. J. Landrum. When the summer was over and the gin began operation, Snake was offered a job maintaining the equipment there at the princely sum of fifty cents an hour. He had been making only forty cents an hour at the cotton oil mill. About two weeks into the cotton-ginning season, on a Saturday afternoon, Snake looked up from his machines to see Landrum, Hoover, and Jim Crain standing at the door. Landrum motioned for Snake to join them. He asked Snake, "Didn't you tell me that you were laid off at the cotton oil mill?" Snake answered in the affirmative. Crain turned to Hoover and asked, "If Snake was one of your good men, why did you lay him off?" It only was then that Hoover revealed the reason Snake had been fired. Landrum in response to this revelation said "that's God damned lousy. I'm going to keep Snake on here." But Crain interceded, saying "Snake, the cotton oil mill is very important to the company and we need you there." So Snake had to go back to the cotton oil mill for forty cents an hour. It was what the company needed. In the end, not surprisingly, what the company needed was more important to the company than what was good for one of its valued employees.

It probably goes without saying that a white worker would have been expected to return to the cotton oil mill, too; thus, in this respect, race was not the key factor. Nevertheless, the incident confirmed for Snake Toney that despite the (rather uncertain) safety and security of the Wilson plantation, the decision to remain there had its price. Within a year after he was forced back into the cotton oil mill, Snake married and left Wilson and moved to Louisville, Kentucky, where his new brother-in-law secured him a position repairing machinery in a factory. He entered the military in 1943 and after the war settled in St. Louis. His reminiscences of his years in Wilson have a certain rehearsed quality to them, as though he had gone over and over them in his mind during the fifty-plus years that elapsed from the time they occurred. He had come to terms with his experiences there, still insisted that the Wilson plantation was a good place to work, that Wilson "took care of their own." Despite evidence to the contrary, including his own experiences and those of others, he held to a certain view of the safety and security African Americans enjoyed on the Wilson plantation. At the same time, however, he clearly revealed the dangers and constraints, and he represented the typical response when an alternative presented itself: he left for better opportunities elsewhere. He was not there to witness the transformation that would take place over the next few decades, a transformation that would reduce the labor force to a shadow of its former self and render the operation an entirely agricultural enterprise with little room for talented African American machine workers like Snake.[34]

A metamorphosis of Lee Wilson & Company accompanied the revolution in southern agriculture, not only because plantation operations necessarily changed in the postwar environment but also because of Lee Wesson's reappearance on the scene. Upon his return to Wilson after the war, Wesson, who had gained greater experience in management as an executive with American Aviation, went to work in the company's office, where he chafed under Crain's management. He resented the fact that Crain had placed his own son, John, in the position of assistant manager and employed his brother-in-law, John R. Enochs, in another executive capacity. The Crains were clearly in the process of displacing the Wilsons in the company, and that was something Wesson refused to accept. Although he had important allies in his mother and his Aunt Marie, he did not have Roy Wilson's cooperation or that of his cousin Bob Wilson, the namesake of the founder. It is arguable that Bob's decision to begin a new life in Colorado owed as much to his frustration with his father's alcoholism as to anything else. He had come to understand that he could not rescue or control his father, a father who had grown so dependent on Jim Crain that any attack on Crain would have amounted to an attack on his father. He was in an impossible situation and for a time remained outside the drama that began to unfold in March of 1946.

On March 28, 1946, Lee Wesson, representing the interests of his mother and aunt, filed a petition in federal court in Little Rock requesting the removal of Jim Crain as trustee and general manager of Lee Wilson & Company. The petitioners argued that Crain had "built up vast agricultural and business interests of his own during the past few years" and that it seemed improbable that "he could continue to meet the demands of his own business enterprises and at the same time" manage the Wilson company. Conceding that under Crain's "direction the operations of Lee Wilson & Company have greatly expanded," the petition further stated "that even as capable a man as Mr. Crain cannot satisfactorily serve our best interests and at the same time oversee his own sizeable holdings. Certainly one or the other must suffer and it is only natural, and rightfully so, that a man's first concern be in the interest of his own business." The petitioners explained that they wanted the trust disbanded, a corporation created in its place, and the shares of that new corporation distributed to the existing shareholders. They also wanted the company supervised by someone who is able to devote undivided attention to the undertaking."[35] Lee Wesson was clearly the man they had in mind.

If they entertained any notion that Crain would accept their blandishments about his capabilities, they were to be sadly disappointed. Crain answered their petition with one of his own on May 16, denying that he devoted too much time to his own business enterprises at the expense of Lee Wilson & Company and arguing that the court had no jurisdiction. He also argued that the petitioners did not represent a majority of the trust, a fact that could not be denied. As it stood after Elizabeth Wilson's death (Lee Wilson's widow) in 1943, the shares were divided as follows:

Victoria Wilson Wesson	2,666.66
Marie Wilson	1,333.33
Joe Wilson Nelson	1,333.33
REL Wilson Jr.	2,666.66
Estate of LW	2,000.00

Together, Victoria and Marie owned only 3,999 of the ten thousand shares.[36] Marie had transferred half of her shares to her son, Joe Wilson Nelson. Finally, a telling revelation in Crain's response was the assertion that he had demanded payment of $125,000 he claimed Victoria Wesson owed the company, money she had borrowed but failed to repay. His demand may have played a role in the petition to remove Crain in the first place. Victoria believed that the company had earned greater profits during the war than Crain was revealing and apparently overdrew her account. Although the surprise that Lee Wesson expressed over the fact that Crain was contesting their petition was almost certainly disingenuous, he observed that "litigation in such matters, in my opinion, is always costly and distasteful to all parties, and Mrs. Wesson, Mrs. Howell and I regret that it has become necessary. We are, however, fully prepared to substantiate the claims upon which the petition was predicated and are determined to bring the suit to a conclusion to the best interests of Lee Wilson & Co."[37] Over the next several months they took depositions, demanded and received access to the company's records, and by October were ready to present an amended complaint. This time there was no pretense of cordiality and no reference to Crain's good management of the company's affairs.

The amended complaint alleged a conspiracy between Jim Crain and several of his family members "in the use of Trust funds for their own individual and collective advantages in the building up of large personal estates to the detriment of the Trust" and demanded restitution in an amount exceeding $500,000. The dispute which had begun with a mildly conciliatory tone on the part of the Wilsons had become suddenly vitriolic after Crain's countercharges. The Wilsons, in turn, had responded with the full force of their own long-standing frustration and anger over their loss of control. In the end, however, it was to everyone's best interest that the matter be privately settled. During the next twenty-three months the disputants—now including Joe Wilson Nelson, Marie's son—engaged in intense negotiations. Joe's decision to join his mother and aunt in the controversy likely tipped the balance in favor of the petitioners as now, with his 1,333.33 shares, they represented the majority. Lawyers and family members perused company books and reviewed the various properties, ultimately agreeing to a "partial liquidation," of the trust.[38]

The agreement to partially liquidate the trust, signed in June 1948, marked a significant shift in Wesson's approach. He no longer sought to displace Crain as manager of Lee Wilson & Company and take charge himself. Rather than disband the trust entirely, the petitioners accepted property, leaving Lee Wilson & Company intact and Jim Crain in place as co-trustee with Roy. Victoria Wesson received 7,766.52 acres, including the valuable Victoria plantation in south Mississippi County and dozens of town lots in Osceola. Lee Wesson became the manager of his mother's estate at Victoria, eventually inheriting it after she died in 1952. Marie Howells and her son took an equivalent amount of land, including the valuable Armorel plantation in north Mississippi County and some lots and blocks in Osceola and Bassett. They sold their property and deposited the funds with Roger Babson, an iconoclastic investor and moral crusader who eventually made them millionaires over again. Meanwhile, Roy Wilson continued his alliance with Jim Crain, but that arrangement was soon to come to an end.[39]

The Pandora's box opened by Victoria, Lee Wesson, and Marie loosened Crain's control both over Lee Wilson & Company and over Roy Wilson. Crain announced his resignation as co-trustee of Lee Wilson & Company in early February 1950 "in order to devote more time to his personal business." County court records reveal that some negotiations predated that announcement, and Crain purchased city blocks in Osceola and a few hundred acres from the Trusteeship for $50,000 beginning in January.[40] It is unclear what role the "class boy" played in this final chapter of the controversy, but when Roy Wilson summoned his son from Colorado, declaring that if he did not assume the co-trusteeship he would sell the property, Bob accepted his father's challenge and, like his grandfather before him, actually operated the trust as though he alone were in charge. Though he may have been reluctant to assume the burden and responsibility, Bob soon began his own aggressive management of the company's affairs.[41]

As Bob Wilson took over the administration of a downsized Lee Wilson & Company, Lee Wesson took on a similar role at Wesson Farms. The exact acreage held by the two entities is something of a mystery. The court records suggest that Lee Wesson began with something in excess of 7,700 acres. Newspaper reports indicated that Lee Wilson & Company still operated fifty thousand acres, but a perusal of the county records does not substantiate that number. Rather, they likely retained twenty to twenty-five thousand acres. According to a list produced in 1999 of "Arkansas Business Rankings: Individual/Family Land Owners," Lee Wilson & Company ranked seventh in the state, with twenty-three thousand acres; the Wesson farms ranked seventeenth, with eight thousand acres.[42] However much land they operated, both Bob Wilson and Lee Wesson stood at the helm of institutions that were to successfully negotiate the challenges in the decades to follow, as southern agriculture weathered a series of changes, ranging from the troublesome transition to capital-intensive agriculture to curtailments and reinstatements of federal allotments for agricultural commodities. They managed to survive trade embargoes on certain commodities, particularly on soybeans, which emerged in the 1970s, as agricultural produce became a weapon in the Cold War with Russia. Unfortunately, records are not available to chart Bob Wilson's management of the firm. Indeed, the company's paper trail stops in 1945, just before the dispute that led to the partial liquidation of Lee Wilson & Company.

Bob Wilson took over the stewardship of the company in 1950, the same year the Memphis Cotton Carnival Commission elected him king of the Cotton Carnival, an honor usually reserved for an industry leader. It signaled his arrival as a new force to be reckoned with in the region. Unlike Jim Crain, who had to secure introductions to the political and economic power brokers, Bob Wilson came with credentials in hand, the scion of a distinguished family. Ironically, his tenure as Cotton Carnival king came at just the moment when another crop was about to dethrone cotton, a process that Wilson and others like him presided over. Between 1950 and 1969, the number of acres devoted to cotton steadily declined in Mississippi County—and in the South generally—as the locus of cotton cultivation in the nation shifted to California (see table 9.6).[43]

The decline in cotton cultivation accompanied the maturation of the neoplantation, the system that abandoned the archaic and repressive tenancy and sharecropping system that was itself an echo of the old slave-plantation South. African Americans like Snake Toney played an important role in this transformation as they slipped away in the long Black Migration from the rural South, contributing to a labor shortage that hastened mechanization. By the time that Roy Wilson died in 1958, Bob Wilson had accepted and even embraced the role that he had been born and bred to, and he was the perfect man to oversee its metamorphosis and fully captured the meaning of the phrase when he said "I am no baron but a businessman."

Nevertheless, Bob Wilson was a transition figure. In 1978 he grew a variety of crops designed to "keep a large labor force the year round, instead of scrambling for help at harvest time, as most farmers have to do." While he had eleven thousand acres each in soybeans and cotton, he also grew rice, wheat, alfalfa, pecans, oats, strawberries, and even asparagus and mustard greens. He moved into the Tudor mansion his father built, accompanied by his first wife and four of their five sons. His firstborn, Robert E. Lee Wilson IV, died as a child of leukemia, and he named his fourth boy, born in 1950, Robert E. Lee Wilson V, signifying his acknowledgement of the importance of the Wilson dynasty.[44] In 1978 he wrote to the Yale Alumni Association that he was proud of his children, both those by his first wife and his two daughters with his second wife, Mildred Martin Wilson, whom he married in 1962. In 1978 he said of those two daughters, "They are both complete outdoor girls. They have their own horses which they catch, saddle and bridle and ride by themselves. Both there and at our cattle ranch; they know every facet of a cow's life cycle."[45]

In spite of his aggressive management of Lee Wilson & Company, Bob Wilson always maintained an interest in the cattle business, moving from the Anchor Ranch in Colorado to the Rafter 6 in Wyoming sometime in the early 1950s. In 1968, he began buying land in Nevada, eventually amassing an eleven-thousand-acre ranch there. He reported to the Yale alumni office that year that his thirty-four-year-old son Michael was serving as president of the company. Bob had stepped aside, although he continued to function as chairman of the board. His younger son Steve held the vice-presidency and served as the general farm manager. R. E. L. Wilson V had recently graduated with honors from the University of Virginia, and his youngest son, Frank, was planning on following his own dreams, graduating from Tulane with honors in drama and with hopes for an acting career, moved to California. With Michael and Steve running the company, Bob Wilson spent more time at his ranch in Wyoming. His wife, Millie, accompanied him as they "hunted big game and small all over North and Central America and parts of Free Europe." Bob was proud of her, boasting, "There are over 50 trophies in our hallway and over half of them are hers." In 1986 he wrote that "50 years out of college, no re-grets about anything, would like to start again and do it all over the second time." He died the next year. Michael Wilson, as the oldest son, assumed the mantle of responsibility for the company until his death in 2008, when Steve stepped into the presidency.[46] Steve's half sister Midge and Michael's son, Perry, an attorney in Little Rock, became vice presidents. It was under their leadership that the family decided to sell the estate to banker, landowner, and investor Gaylon Lawrence Jr. in 2010.[47]

By the time that the fourth generation of Wilsons presided over the transi-tion that took place in the decades that followed World War II, a great deal had changed since Josiah Wilson raised a crop in the swamps. By the year 1849, only 8,111 acres produced farm products. In 1880, when fifteen-year-old Lee Wilson crossed the Mississippi River to claim his inheritance, 30,111 acres were in production. By 2007 the swamps had been drained, the land protected from floods by an unbroken chain of levees, the forest cover drasti-cally reduced, and 461,328 acres out of 592,349 were in production. Most of the rest was either reserved land in the Big Lake National Wildlife Refuge, in towns, or along a narrow band between the river and levees. Mississippi County outproduced all other counties in the state in cotton production until the mid-1950s and in soybean production beginning in 1947 until 1964. After 1964, it remained near or at the top. With both soybean and cotton crops highly mechanized and processing facilities for them readily at hand, planters and farmers could finally make some minor adjustments on a year-by-year basis, electing to follow the rise and fall in cotton or soybean prices. Other factors, like foreign competition, weather conditions (in wet years soybeans outperformed cotton), and changes in government subsidies, influenced the selection of crops.[48] In 2007, agriculturalists in the county planted 215,681 acres in cotton and 153,559 in soybeans. They even moved into the rice market, growing 36,412 acres that year, in part because a depletion of the water table in the Arkansas Grand Prairie stressed the rice growers there and in part because the rising price of rice on the international marketplace made it an attractive alternative crop for delta planters. The advent of rice production in the vicinity has brought with it certain new environmental concerns. Because many agriculturalists burn off the rice stubble remaining on the fields after the harvest, they create an air quality hazard. Some farmers, however, are electing to flood their fields, a practice that slowly breaks down the stubble, and, at the same time, provides wetlands for migratory birds. However, rice production lags far behind cotton and soybeans in importance in the Arkansas delta. With cotton prices more than a dollar a pound in 2010—an all time high—farmers and planters made plans to rush back into cotton production in 2011.[49] Whatever crop they elect to grow, the trend to capital-intensive agriculture has contributed to the concentration of land ownership into larger units of production. The average farm size nearly doubled between 1982 and 2007, increasing from 679 acres to 1,250 acres. Even larger harvesters, newly off the market, are further reducing labor needs.[50]

With fewer farmworkers needed, the population of the county began to decline after World War II, falling from 80,286 in 1950 to 70,055 in 1960 to 51,979 in 2000. The Census Bureau's projected figures for 2009 stand at 46,208. While industrial jobs began providing some alternative sources of employment, they were late in arriving and remain insufficiently robust enough to offset the decline in farm labor jobs made obsolete by the advent of scientific agriculture.[51] Only the northern end of the county has exceeded expectations, largely for two reasons. First, given the higher incidence of land ownership and the ability of even landless farmers there to gain at least some personal property, they were the least likely of any population in the county—or elsewhere in the delta—to depart. Second, the placement of the Air Force training base in Blytheville provided an anchor which sustained a population base. When the government closed the base in the early 1990s, things looked bleak for a while but then Nucor [steel] Corporation moved into the area just east of Blytheville in the mid-1990s and turned things around. By 2010, Nucor employed over a thousand people, and a variety of allied industries clustered around the company and employed several hundred more. South Mississippi County, with a Hallmark Cards plant near Osceola hiring less than two hundred employees, together with a few smaller plants, languished by comparison.

The different trajectories of the two areas raise the question of the long-term consequences of plantation agriculture. During its heyday planters relied on a dependent and impoverished work force with little stake in the local economy and even less reason to remain in place once the transition to capital-intensive agriculture began. Whether they were pushed off against their will or willingly joined the southern migration northward or westward, they left little evidence of their previous habitation. Planters burned or bulldozed their tenant houses and planted cotton or soybeans on their foundations in the 1950s, 1960s, and 1970s. Although plantation agriculture had some prominence in north Mississippi County—particularly east of Big Lake with Wilson's Armorel operation as merely the most prominent example—small farmers also operated there and remained in place. As they struggled to stay alive in the capital-intensive economic environment confronting farmers, whether big or small, they served as a ready labor force as Nucor and allied industries moved into the region in the 1990s. Some of the more successful farmers among them, moreover, leased plantation lands like that owned by the Wilson family and stand to serve as managers in the new phase of portfolio agriculture taking shape in the twenty-first century.

Gaylon Lawrence Jr.'s purchase of the Wilson estate stands as one example of the trend toward investors holding agricultural real estate as part of their investment portfolios. Most of these people have little understanding of agricultural production or appreciation for the local communities. Lawrence himself, a banker (with Tennessee Bank and Trust of Nashville) who holds agricultural lands from California to Florida, including part of the old Delta Pine and Land Company in Mississippi, does possess some connection to southeast Missouri and northeastern Arkansas. His family has roots in Sikeston, Missouri, and at some point acquired ownership of Farmers Bank and Trust in Blytheville. This latter connection shadows an earlier association with Lee Wilson, who became a director of Farmers Bank and Trust in 1932. Lawrence's purchase of the Wilson estate mirrors a larger trend among investment firms acquiring agricultural lands, purchases that made sense during the economic crisis which began in 2007. While residential and commercial real estate prices plummeted, agricultural lands rose in value, in tandem with the rise in agricultural prices. Investment firms like the Winchester Group of Champaign, Illinois, and TIAA-CREF, the college pension fund, have increased their holdings in agricultural lands as a way to offset the declining value of other kinds of real estate. The problem is, of course, that agricultural lands themselves can stagnate and decline in value as the price of agricultural commodities sink, as they periodically do, and only prudent investors who intend to hold the land for twenty to thirty years stand to gain. At the same time, "while pension funds are large players in the market, the presence of hedge funds and private equity is growing," and they have a "'buy and sell' strategy, looking to exit the market within ten years." In other words, they may introduce a new level of instability in the value of agricultural lands, and it is hard to predict the outcome of such an economic environment. One thing is clear, however. Invisible investors, whether they come through Winchester, TIAA-CREF, or some other larger concern, or whether they are part of hedge fund operations, are further removed from the communities within which the agricultural lands are located. Most of the people represented by these firms, people who now hold agricultural real estate in their portfolios, have no knowledge of or interest in the agricultural process or the people and communities engaged in them. While Lee Wilson & Company—and others like them—remained committed to the local community, these investment firms and their portfolio holders will regard them distantly and, likely, have little inclination toward supporting them or enhancing the quality of life for local people. Unlike local planters who have established long-standing social relationships with the men who lease their lands, portfolio planters will be interested only in the bottom line. When the enterprise becomes strictly a business transaction, lessees potentially become expendable. Local planters, on the other hand, find it difficult "to take ground back from a farmer when you've known him for years and when you've watched the crop develop during the season." As out-of-state investors looking to maximize their profits, they will have even less interest in the environmental consequences of burning rice stubble or the overuse of certain potentially harmful chemicals.[52]

The ability to maintain agricultural production in Mississippi County is possible only because of the levees and the drainage of back swamps. Soil scientists writing as late as 2002 observed that "no areas in Mississippi County are classified outright as prime farmland," but under certain conditions, fully 91.2 percent of the county achieved that status.[53] As the report suggests, however, "without the levees, the flooding frequency would dramatically change upward," and without careful attention to the maintenance of drainage, the prime farmland category would be greatly reduced.[54] The scientists who provided these details also spoke to potential problems dating back to damage done by the New Madrid earthquakes. The earthquakes created fissures, which filled with sand over time, and some of these fissures descend to the water table and could "transport any water from the surface deep into the soil profile at a very rapid rate . . . and would also allow the rapid movement of dissolved substances in the water such as pesticides and fertilizers. Although many pesticides have biodegradation rates that neutralize the substance over time, rapid movement downward through a sand blow would negate the beneficial surface effects required for biodegradation over time and result in higher concentrations of chemical inputs in sub surface water bodies." In other words, chemicals, potentially harmful to people, could infect the water supply. The potential problem has long been understood—although hotly debated even among scientists—such that in 1925 the Northeast Arkansas Agricultural Experiment Station at Keiser, one of the original Wilson towns, was created. As the authors of the report suggest, the experiment station owes its existence, at least in part, to "public concerns over management of the soils containing sand blows and extreme variability in soil properties within these fields."[55] The current generation of agriculturalists, including Wesson Farms and portfolio planters, must find a way through this dilemma and, at the same time, continue producing important commodity crops for a world market. Only time will tell whether the portfolio planters will be willing to sacrifice a share of their profits for investments in the local environment.

APPENDIX:
TABLES

TABLE 1.1

Land above and below the Floodplain, Mississippi County, 1823–1848

TOWNSHIP RANGE		TOTAL ACRES IN LAND		
		Area Outside of Designated Bodies of Water	Above the Floodplain	Below the Floodplain
10	8	19,840	7,876.5 (39.7)	11,963.5 (60.3)
10	9	13,440	2,042.9 (15.2)	11,397.1 (84.8)
10	10	1,660	189.2 (11.4)	1,470.8 (88.6)
11	8	19,840	2,241.9 (11.3)	17,598.1 (88.7)
11	**9**	**21,120**	**16,938.2 (80.2)**	**4,181.8 (41.8)**
11	10	13,440	-0-	13,440.0 (100)
12	8	19,840	2,737.9 (13.8)	17,102.1 (86.2)
12	9	15,360	276.5 (1.8)	15,083.5 (98.2)
12	10	22,440	5,699.8 (25.4)	16,740.2 (74.6)
12	11	10,880	750.8 (6.9)	10,129.2 (93.1)
13	8	21,760	8,704.0 (40.0)	13,056.0 (60.0)
13	9	19,840	7,241.6 (36.5)	12,598.4 (63.5)
13	10	23,040	5,114.9 (22.2)	17,925.1 (77.8)
13	11	10,240	5,447.7 (53.2)	4,792.3 (46.8)
14	8	23,040	8,478.7 (36.8)	14,561.3 (63.2)
14	9	19,200	9,331.2 (48.6)	9,868.8 (54.4)
14	10	22,720	5,430.1 (23.9)	17,289.9 (76.1)
14	11	22,720	10,451.2 (46.0)	12,268.8 (54.0)
14	12	17,280	8,277.1 (47.9)	9,002.9 (52.1)
15	8	23,040	7,948.8 (34.5)	15,091.2 (65.5)
15	9	8,960	5,456.6 (60.9)	3,503.4 (39.1)
15	10	19,200	1,708.8 (8.9)	17,491.2 (91.1)
15	11	20,480	11,243.5 (54.9)	9,236.5 (45.1)
15	12	19,200	4,185.6 (21.8)	15,014.4 (78.2)
	Total	428,520	137,774.0 (32.1)	290,806.5 (67.9)

Source: Original General Land Office Survey Notes and Plats for the State of Arkansas, 1815–present, Commissioner of State Lands, State of Arkansas, n.d. This table is based upon a close perusal of the GLO field notes of the deputy surveyors. I recorded their characterizations of the land, paying particular attention to comments about its tendency to flood. I calculated the percentage of remarks that classified the land as overflow lands and extrapolated from that to determine the acreage that might have been affected.

Note: Parenthetical numbers in two last columns are percentages. The highlighted line represents the area Josiah Wilson settled in.

TABLE 1.2

Mississippi County Population—1860: Southern vs. Northern Mississippi County, by Race

	WHITE POPULATION	SLAVE POPULATION	WHOLE POPULATION
Southern townships:	1,262 (51.8)	1,208 (77.4)	2,407 (61.2)
Northern townships:	1,174 (48.2)	352 (22.6)	1,526 (38.8)
Total	2,436 (100.0)	1,560 (100.0)	3,933 (100.0)

Source: Manuscript Census of Population, Free and Slave Schedules, 1860, Mississippi County, Arkansas.

Note: Parenthetical numbers are percentages.

TABLE 1.3

Mississippi County Population—1860, by Township and Race

	WHITE POPULATION	SLAVE POPULATION	WHOLE POPULATION
Southern townships delineated:			
Monroe (river)	437 (54.6)	363 (45.4)	800 (100.0)
Scott (inland)	321 (79.7)	81 (20.1)	402 (100.0)
Swain (river)	122 (62.2)	74 (37.8)	196 (100.0)
Pecan (river)	140 (56.5)	108 (43.5)	248 (100.0)
Troy (river)	61 (10.0)	543 (90.0)	604 (100.0)
Carson Lake (inland)	115 (74.7)	39 (25.3)	154 (100.0)
Little River (inland)	66 (100.0)	-0-	66 (100.0)
Total	1,262 (51.0)	1,208 (49.0)	2,470 (100.0)
Northern townships delineated:			
Canadian (river)	378 (67.9)	179 (32.1)	557 (100.0)
Clear Lake (nr river)	162 (79.0)	43 (21.0)	205 (100.0)
Chickasawba	425 (96.2)	17 (3.8)	442 (100.0)
Big Lake	209 (100.0)	-0-	209 (100.0)
Unaccounted for		113	
Total	1,174 (76.9)	352 (23.1)	1,526 (100.0)

Source: Manuscript Census of Population, Free and Slave Schedules, 1860, Mississippi County, Arkansas.

Note: Parenthetical numbers are percentages. Pecan was the only township that lost slave population between 1850 and 1860 and that was due to the collapse of part of the township into the river where the Mississippi turned inward at Pecan Point. The slave population dropped from 167 to 108 in that period. Note also that some pages of the manuscript slave census for Mississippi County are missing, leading to an undercount of 113 in the slave population (see footnote "f" to 1870 Census of Population, p. 86). The missing pages fall between Canadian and Clear Lake. If we apportion those slaves to Canadian Township, which is a likely place for them, then the number of slaves there increases to 292.

TABLE 2.1

Land in Farms

	1880	1900	DIFFERENCE
Improved acres	30,111 (29.5)	76,655 (61.5)	+46,544
Unimproved acres	71,899 (70.5)	48,029 (38.5)	-23,870
Total	102,010 (100.0)	124,684 (100.0)	

Sources: U.S. Bureau of the Census, *Census of Agriculture: 1880,* Table VII, Report on the Production of Agriculture, Tenth Census of the United States, General Statistics (Washington, D.C.: Government Printing Office [GPO], 1883), p. 105; and *Census of Agriculture: 1899,* Census Reports, Twelfth Census of the United States, Agriculture, Table 19, p. 267 (Washington, D.C.: GPO).

Note: Parenthetical numbers are percentages.

TABLE 2.2

Adult Male to Female in South Mississippi County Townships

	ADULT MALE	ADULT FEMALE
1870	526 (49.8)	530 (50.2)
1880	1,706 (58.0)	1,235 (42.0)
1900	2,696 (55.5)	2,160 (44.5)

Source: Manuscript Census of Population, Mississippi County, Arkansas, 1860, 1870, 1880, 1900.

Note: Parenthetical numbers are percentages.

TABLE 3.1

Osceola Times, 1884–1900: Cause of Death and Disease

	MALARIA, FEVERS, CHILLS	COLDS, PNEUMONIA, INFLUENZA	UNSPECIFIED AND MISCELLANEOUS	TOTAL
January	5 (3.4)	20 (16.4)	22 (3.0)	47 (4.7)
February	-0-	28 (23.0)	56 (7.7)	84 (8.4)
March	2 (1.3)	9 (7.4)	51 (7.0)	62 (6.2)
April	2 (1.3)	6 (4.9)	69 (9.5)	77 (7.7)
May	1 (.7)	1 (.8)	26 (3.6)	8 (2.8)
June	4 (2.7)	11 (9.0)	33 (4.6)	48 (4.8)
July	17 (11.2)	3 (2.5)	108 (14.9)	128 (12.9)
August	39 (26.2)	1 (.8)	120 (16.6)	160 (16.1)
September	48 (32.2)	12 (9.8)	78 (10.8)	138 (13.9)
October	16 (10.7)	4 (3.3)	74 (10.2)	94 (9.4)
November	10 (6.7)	4 (3.3)	56 (7.7)	70 (7.0)
December	5 (3.4)	23 (18.9)	32 (4.4)	60 (6.0)
Total	149 (100.0)	122 (100.0)	725 (100.0)	996 (100.0)

Note: Parenthetical numbers are percentages. Some of the diseases and deaths listed as caused by malaria, fevers, and chills, may not have been malaria. Many of the diseases and deaths in the "unspecific and miscellaneous" category were likely connected to malaria.

TABLE 5.1
Acres in Farms from Census of Agriculture, 1920–1930

	1920	1930	DIFF. IN ACRES	PERCENT CHANGE
North of the Arkansas River:				
Crittenden	234,462	245,533	+11,071	+4.5
Lee	206,572	224,432	+17,860	+7.9
St. Francis	190,175	258,824	+68,649	+26.5
Mississippi	277,670	335,034	+57,364	+17.1
South of the Arkansas River:				
Phillips	237,939	224,070	-13,869	-5.8
Desha	124,856	115,772	-9,084	-7.2
Jefferson	290,296	279,720	-10,576	-3.6

Sources: U.S. Bureau of the Census, Census of Agriculture, *Fourteenth Census of the United States: 1920, Agriculture,* Vol. 6, Pt. 2, pp. 560–66 (Washington, D.C., Government Printing Office [GPO], 1922); Census of Agriculture, *Fifteenth Census of the United States: 1930, Agriculture,* Vol. II, Pt. 2 (Washington, D.C.: GPO, 1932), pp. 1136–1141.

TABLE 6.1
Number of Homes Flooded, Arkansas Delta, 1927: Owners vs. Tenants

COUNTIES	TOTAL	OWNERS	TENANTS
Chicot	4,125	625 (15.2)	3,500 (84.8)
Crittenden	1,300	175 (13.5)	1,125 (86.5)
Cross	500	25 (5.0)	475 (95.0)
Desha	5,137	1,027 (20.0)	4,110 (80.0)
Lee	3,650	417 (11.4)	3,233 (88.6)
Mississippi	3,383	3,083 (91.1)	300 (8.9)
Monroe	2,675	535 (20.0)	2,140 (80.0)
Phillips	3,800	380 (10.0)	3,420 (90.0)
St. Francis	3,723	744 (20.0)	2,979 (80.0)

Source: "American National Red Cross, Mississippi River Flood Relief, Statistics Concerning Flood and Losses," Red Cross Records, National Archives, RG 200, Box 735, File 224.08.

Note: Parenthetical numbers are percentages.

TABLE 9.1

Mississippi County Transition to Capital-Intensive Agriculture

	1940	1945	1950	1959	1964
Mules	17,567	12,311	4,675	555	20
Tractors	1,052	2,256	4,712	5,223	4,729
Hired labor	19,446	11,741	5,429	3,753	No record

Sources: U.S. Bureau of the Census, *U.S. Census of Agriculture: 1935*, Vol. II, Pt. 2, Reports for States, Arkansas (Washington, D.C.: Government Printing Office [GPO], 1937) p. 674; *U. S. Census of Agriculture: 1940*, Vol. 1, Part 5, State Reports, Arkansas, pp. 20, 38 (Washington, D.C.,: GPO, 1942); *U.S. Census of Agriculture: 1945*, Vol. I, Pt. 23, Arkansas, Statistics for Counties, pp. 42, 105 (Washington, D.C.: GPO, 1946); *U.S. Census of Agriculture: 1950*, Vol. 1, Pt. 23, Counties and State Economic Areas, Arkansas, pp. 78, 85 (Washington, D.C.: GPO,1952); *U.S. Census of Agriculture: 1959*, Vol. 1, Pt. 34, Counties, Arkansas, pp. 160, 173, 179 (Washington, D.C.: GPO, 1961); *U.S. Census of Agriculture: 1964*, Vol. 1 Pt. 34, County Tables, Arkansas, pp. 265, 313 (Washington, D.C.: GPO, 1967).

TABLE 9.2

Mississippi County Total Population (including Black and White Population)

	1940	1960
South	22,251 (27.7)	17,843 (25.5)
Northeast	32,653 (40.7)	35,513 (50.7)
Northwest	25,382 (31.6)	16,699 (23.8)
Total	80,286 (100.0)	70,055 (100.0)

Sources: Sixteenth Census of the United States: 1940, Population, Vol. II, Pt. 1, p. 486–7; *Eighteenth Census of the United States: 1960, Population*, Vol. I, Pt. 5, p. 15 (Washington, D.C.: Government Printing Office, 1942, 1962).

Note: Parenthetical numbers are percentages.

TABLE 9.3

Mississippi County Population, by Division within County

	1940			1960		
	WHITE	BLACK	TOTAL	WHITE	BLACK	TOTAL
South	10,089 (45.3)	12,162 (54.7)	22,251 (100)	9,533 (53.4)	8,310 (46.6)	17,843 (100)
Northeast	20,711 (63.4)	11,942 (36.6)	32,653 (100)	23,640 (66.6)	11,873 (33.4)	35,513 (100)
Northwest	24,420 (96.2)	962 (3.8)	25,382 (100)	16,170 (96.8)	529 (3.2)	16,699 (100)

Note: Parenthetical numbers are percentages. Table does not include "other nonwhite."

Sources: Sixteenth Census of the United States: 1940, Population, Vol. II, Pt. 1, pp. 486–87; *Eighteenth Census of the United States: 1960, Population,* Vol. I, Pt. 5, p. 15 (Washington, D.C.: Government Printing Office, 1961).

TABLE 9.4

Mississippi County: All Farm Operators, by Race

	1935	1940	1945	1950	1959
White	5,911 (59.2)	5,161 (65.6)	6,215 (64.5)	4,893 (68.5)	2,129 (73.3)
Black	4,078 (40.8)	2,705 (34.4)	3,414 (35.5)	2,247 (31.4)	775 (26.7)
Total	9,989	7,866	9,629	7,140	2,904

Sources: U.S. Bureau of the Census, *U.S. Census of Agriculture: 1935*, Vol. 2, Pt. 2, Reports for States, Arkansas (Washington, D.C.: Government Printing Office [GPO], 1937), p. 674; *U. S. Census of Agriculture: 1940*, Vol. 1, Pt. 5, State Reports, Arkansas, p. 28 (Washington, D.C.: GPO, 1942); *U.S. Census of Agriculture: 1945*, Vol. I, Pt. 23, Arkansas, Statistics for Counties, p. 129 (Washington, D.C.: GPO, 1946); *U.S. Census of Agriculture 1950*, Vol. 1, Pt. 23, Counties and State Economic Areas, Arkansas, p. 73 (Washington, D.C.: GPO, 1952); *U.S. Census of Agriculture: 1959*, Vol. 1, Pt. 34, Counties, Arkansas, p. 153 (Washington, D.C.: GPO, 1961).

Note: Parenthetical numbers are percentages.

TABLE 9.5

Mississippi County Farm Operators, by Land Tenure

	1935	1940	1945	1950	1959
Owners	1,150	1,437	1,605	941	448
Part owners	221	303	174	513	489
Managers	45	41	59	52	49
Tenants	5,253	2,782	4,225	1,991	1,918
Croppers	3,320	3,303	3,566	3,643	No record
All operators	9,989	7,866	9,629	7,140	2,904

Sources: U.S. Bureau of the Census, *U.S. Census of Agriculture: 1935*, Vol. II, Pt. 2, Reports for States, Arkansas, p. 674; *U. S. Census of Agriculture: 1940*, Vol. 1, Pt. 5, State Reports, Arkansas, p. 28 (Washington, D.C.: Government Printing Office [GPO], 1942); *U.S. Census of Agriculture: 1945*, Vol. I, Pt. 23, Arkansas, Statistics for Counties, p. 129 (Washington, D.C.: GPO, 1946); *U.S. Census of Agriculture 1950*, Vol. 1, Pt. 23, Counties and State Economic Areas, Arkansas, p. 73 (Washington, D.C.: GPO, 1952); *U.S. Census of Agriculture: 1959*, Vol. 1, Pt. 34, Counties, Arkansas, p. 153 (Washington, D.C.: GPO, 1961).

Note: For the years 1940 and 1945, cash tenants, share tenants, etc., are delineated. Beginning in 1950, the only distinction made is between tenants (combining all forms of tenants) and croppers. Beginning in 1960, croppers have disappeared as a separate category.

TABLE 9.6
Acres in Selected Crops, Mississippi County, 1949, 1969, 2007

	1950	1969	2007
Cotton acres	284,761	167,453	215,681
Soybean acres	67,479	277,104	153,559
Rice	-o-	1,434	36,412

Sources: U.S. Bureau of the Census, *U.S. Census of Agriculture 1950*, Vol. 1, Pt. 23, Counties and State Economic Areas, Arkansas, pp. 97, 103, 108 (Washington, D.C.: GPO, 1952); *U.S. Census of Agriculture: 1959*, Vol. 1, Pt. 34, Counties, Arkansas, p. 153 (Washington, D.C.: GPO, 1961); *U.S. Census of Agriculture: 1969*, Vol. 1, Pt. 34, County Data, Arkansas, pp. 292, 379 (Washington, D.C.: GPO, 1972); *U.S. Census of Agriculture: 2007*, County Profile, Mississippi County, Arkansas (www .agcensus.usda.gov).

Notes

INTRODUCTION

1. By 1933, Wilson held title to 46,458 acres in Mississippi County, a few hundred acres in neighboring counties, and half of Peter's Island in Tunica County, Mississippi. He also farmed a few thousand acres for out-of-state landowners. This study builds on a rich historiography in southern agricultural, economic, and environmental history. Particularly relevant are books by Pete Daniel, Jack Kirby, Gilbert Fite, James C. Cobb, David Goldfield, and Robert C. McMath. Pete Daniel, in fact, brings the three strands together in his body of work: Daniel, *Shadow of Slavery: Peonage in the South, 1901–1969* (Urbana: University of Illinois Press, 1972), *Deep'n as It Come: The 1927 Mississippi River Flood* (New York: Oxford University Press, 1977), *Breaking the Land: The Transformation of Cotton, Tobacco, and Rice Cultures since 1880* (Urbana: University of Illinois Press, 1985), *Standing at the Crossroads: Southern Life since 1900* (New York: Hill & Wang, 1986), *Race and Class in the American South since 1890* (Oxford and Providence: Berg, 1994), *Lost Revolutions: The South in the 1950s* (Chapel Hill: University of North Carolina Press, 2002), and *Toxic Drift: Pesticides and Health in the Post–World War II South* (Baton Rouge: Louisiana State University Press, 2005). And see Jack Temple Kirby, *Rural Worlds Lost: The American South, 1920–1960* (Baton Rouge: Louisiana State University Press, 1987), *Mockingbird Song: Ecological Landscapes of the South* (Chapel Hill: University of North Carolina Press, 2006), and *Darkness at the Dawning: Race and Reform in the Progressive South* (Philadelphia: Lippincott, 1972); Gilbert C. Fite, *American Agriculture and Farm Policy since 1900 (New York: Macmillan, 1964), and Cotton Fields No More: Southern Agriculture, 1865–1980* (Lexington: University Press of Kentucky, 1984); James C. Cobb, *The New Deal and the South: Essays* (Jackson: University Press of Mississippi, 1984), and *The Most Southern Place on Earth: the Mississippi Delta and the Roots of Regional Identity* (New York: Oxford University Press, 1992); David R. Goldfield, *Black, White and Southern: Race Relations and Southern Culture, 1940 to the Present* (Baton Rouge: Louisiana State University Press, 1900), and *Cotton Fields and Skyscrapers: Southern City and Region* (Baton Rouge: Louisiana State University Press, 1982); Robert C. McMath, *American Populism: A Social History, 1877–1898* (New York: Hill & Wang, 1993), and *Populist Vanguard: A History of the Southern Farmers' Alliance* (Chapel Hill: University of North Carolina Press, 1975). See also Charles Aiken, *The Cotton Plantation South since the Civil* War (Baltimore and London: Johns Hopkins University Press, 1998). Besides the pioneering work done by Daniel and Kirby on the environmental history of the South, several other important studies contributed to my understanding of the issues at work in the transformation of the northeastern Arkansas landscape: Mart A. Stewart, *"What Nature Suffers to Groe": Life, Labor, and Landscape on the Georgia Coast, 1680–1920* (Athens: University of Georgia Press, 2002); Mikko Saikku, *This Delta, This Land: An Environmental History of the Yazoo-Mississippi Floodplain* (Athens: University of Georgia Press, 2005); Lynn A. Nelson, *Pharsalia: An Environmental Biography of a Southern Plantation, 1780–1880* (Athens: University of Georgia Press, 2007); Cynthia Barnett, *Mirage: Florida and the Vanishing Waters of the Eastern U.S.* (Ann Arbor: University of Michigan

Press, 2007); Michael Grunwald, *The Swamp: The Everglades, Florida, and the Politics of Paradise* (New York: Simon & Schuster, 2006); David McCally, *The Everglades: An Environmental History* (Gainesville: University Press of Florida, 1999). For an overview of cotton's importance in the American—not just southern—economy from the antebellum period forward, see Gene Dattel, *Cotton and Race in the Making of America: The Human Costs of Economic Power* (Chicago: Ian R. Dee, 2009). See also an interesting treatment of the worldwide importance of cotton production in the mid-nineteenth century by Sven Beckert, "Emancipation and Empire: Reconstructing the Worldwide Web of Cotton Production in the Age of the American Civil War," *American Historical Review* 109, no. 5 (December 2004): 1405–1538. Beckert's manuscript on the role of cotton in the American and world economy is due out in 2011.

2. For the best study on the Percy family, see Bertram Wyatt-Brown, *The House of Percy: Honor, Melancholy, and Imagination in a Southern Family* (New York: Oxford University Press, 1994). For Chapman and Dewey, see Jeannie M. Whayne, *A New Plantation South: Land, Labor and Federal Favor in Twentieth Century Arkansas* (Charlottesville: University of Virginia Press, 1996). Delta Pine and Land Company ceased operating as a plantation but became a major cotton seed (and soybean) breeder. For Delta Pine and Land Company, see Robert L. Brandfon, *Cotton Kingdom of the New South: A History of the Yazoo Mississippi Delta from Reconstruction to the Twentieth Century* (Cambridge, Mass.: Harvard University Press, 1967), and James C. Giesen, "The Truth about the Boll Weevil: The Nature of Planter Power in the Mississippi Delta," *Environmental History* 14, no. 4 (October 2009): 683–704.

3. Paul M. Gaston, *The New South Creed: A Study in Southern Mythmaking* (New York: Knopf, 1970); and Harold E. Davis, *Henry Grady's New South: Atlanta, A Brave Beautiful City* (Tuscaloosa: University of Alabama Press, 1990).

4. C. Vann Woodward, *Origins of the New South, 1877–1913* (Baton Rouge: Louisiana State University Press, 1951), *The Burden of Southern History* (Baton Rouge: Louisiana State University Press, 1968), and *The Strange Career of Jim Crow* (New York: Oxford University Press, 1955).

5. The literature on the sharecropping and tenancy system is extensive but must begin with Pete Daniel's *Shadow of Slavery,* a crucial book that established the field. Also contributing crucial insights are the following: Harold D. Woodman, *New South, New Law: The Legal Foundations of Credit and Labor Relations in the Postbellum Agricultural South* (Baton Rouge: Louisiana State University Press, 1995); Roger L. Ransom and Richard Sutch, *One Kind of Freedom: The Economic Consequences of Emancipation* (Cambridge [Eng.] and New York: Cambridge University Press, 1977); Leon Litwack, *Been in a Storm so Long: The Aftermath of Slavery* (New York: Knopf, 1979); and Douglas A. Blackmon, *Slavery by Another Name: The Re-enslavement of Black People in America from the Civil War to World War II* (New York: Doubleday, 2008).

6. Steven Hahn, *A Nation under Our Feet: Black Political Struggles in the Rural South from Slavery to the Great Migration* (Cambridge, Mass.: Harvard University Press, 2003); Nan Elizabeth Woodruff, *American Congo: The African American Freedom Struggle in the Delta* (Cambridge, Mass.: Harvard University Press, 2003); Grif Stockley, *Blood in Their Eyes: The Elaine Race Massacre of 1919* (Fayetteville: University of Arkansas Press, 2001); Robert Whitaker, *On the Laps of Gods: The Red Summer of 1919 and the Struggle for Justice That Remade a Nation* (New York: Crown, 2008); Whayne, *A New Plantation South,* p. 52; William F. Holmes, "Moonshiners and Whitecaps in Alabama, 1893," *Alabama Review* 34 (January 1981): 31–49. William Holmes has written several articles that look at a variety of causes connected to night riding in Georgia and Mississippi: "Moonshiners and Collective Violence: Georgia, 1889–1895," *Journal of American History* 67 (1980): 588–611, "Whitecapping in Georgia: Carroll and Houston Counties, 1893," *Georgia Historical Quarterly* 64 (1980): 388–404, "Whitecapping: Agrarian Violence in Mississippi, 1902–1906,"*Journal of Southern History* 35 (1969): 165–85, and "Whitecapping in Mississippi: Agrarian Violence in the Populist Era," *Mid-America* 55 (1973): 134–48. See also Edward Ayres, *Vengeance and Justice (New York: Oxford University Press, 1984)* 260–62.

7. For the flood of 1927, see Pete Daniel, *Deep'n as It Come;* and John Barry, *Rising Tide: The Great Mississippi River Flood of 1927 and How It Changed America* (New York: Simon & Schuster, 1997). For the drought, see Nan Elizabeth Woodruff, *As Rare as Rain: Federal Relief in the Great Southern Drought of 1930–31* (Urbana: University of Illinois Press, 1985). For the South and the New Deal, Daniel, *Breaking the Land;* Kirby, *Rural Worlds Lost;* and Fite, *Cotton Fields No More.* See also Roger Biles, *The South and the New Deal* (Lexington: University Press of Kentucky, 1994); Patricia Sullivan, *Days of Hope: Race and Democracy in the New Deal Era;* Harvard Sitkoff, *A New Deal for Blacks: The Emergence of Civil Rights as a National Issue: The Depression Decade (New York: Oxford University Press, 1978).* For World War II, see Neil R. McMillen, *Remaking Dixie: The Impact of World War II on the American South* (Jackson: University Press of Mississippi, 1997).

8. Kirby, *Rural Worlds Lost.* See also Daniel, *Breaking the Land.*

9. Daniel, *Toxic Drift.*

CHAPTER ONE

1. James David Miller, *South by Southwest: Planter Immigration and Identity in the Slave South* (Charlottesville: University of Virginia Press, 2002).

2. Peter J. Kastor, *The Nation's Crucible: The Louisiana Purchase and the Creation of America* (New Haven, Conn.: Yale University Press, 2004). Patrick Williams, S. Charles Bolton, and Jeannie Whayne, eds., *A Whole Country in Commotion: The Louisiana Purchase and the American Southwest* (Fayetteville, University of Arkansas, 2005). *Randolph Recorder,* June 21, 1834, p. 3. For reference to John Murrell's association with Mississippi County, see James M. Gardner, "Territorial Days in Mississippi County, Arkansas," *Delta Historical Review* (Spring 1995), pp. 46-47.

3. *Randolph Recorder,* June 21, 1834, p. 3.

4. Ibid.

5. Jeannie Whayne, comp., *Cultural Encounters in the Early South: Indians and Europeans in Arkansas* (Fayetteville: University of Arkansas Press, 1995).

6. Marvin D. Jeter, *Edward Palmer's Arkansaw Mounds* (Fayetteville: University of Arkansas Press, 1990), p. 124. Dan F. Morse, ed., "Nodena: An Account of 90 Years of Archeological Investigation in Southeast Mississippi County, Arkansas," Arkansas Archeological Survey Research Series No. 30 (Fayetteville, Ark.: Arkansas Archeological Survey, 1989)), p. 5. Dan F. Morse and Phyllis A. Morse, *Archaeology of the Central Mississippi Valley* (New York and London: Academic Press, 1983), p. 287. Whayne, *Cultural Encounters.*

7. Dan F. Morse, "The Nodena Phase," in *Nodena: An Account of 75 Years of Archeological Investigation in Southeast Mississippi County, Arkansas,* edited by Dan F. Morse, Archeological Survey Research Series No. 4 (Fayetteville: Arkansas Archeological Survey, 1973), p. 84.

8. Ibid.

9. Barbara A. Burnett and Katherine A. Murray, "Death, Drought, and de Soto: The Bioarcheology of Depopulation," in *The Expedition of Hernando de Soto West of the Mississippi, 1541–1543: Proceedings of the De Soto Symposia, 1988 and 1990,* edited by Gloria A. Young and Michael P. Hoffman (Fayetteville: University of Arkansas Press, 1993), pp. 227–37.

10. Dan Morse emphasizes the evidence for disease in Morse and Morse, *Archaeology of the Central Mississippi Valley,* pp. 314–15. See also David W. Stahle, John G. Hehr, and Graham G. Hawks Jr., *The Development of Modern Tree-Ring Chronologies in the MidWest: Arkansas, North Texas, Oklahoma, Kansas, and Missouri* (Fayetteville: Arkansas Archeological Survey and Department of Geography, University of Arkansas, 1982).

11. Margaret Guccione, David W. Stahle, and Charles R. Ewen, "Origin and Age of the 'Sunklands' Using Drainage Patterns, Sedimentology, Dendrochronology, Archeology, and History," Final Report for U.S. Geological Survey Award, p. 9 (Fayetteville: University of Arkansas, 1995?). See also Robert A. Myers, "Cherokee Pioneers in Arkansas: The St. Francis Years, 1785–1813," *Arkansas Historical Quarterly* 56 (Summer 1997): 127–57. See also Willard Rollings, "Living in a Graveyard," in *Cultural Encounters in the Early South,* compiled by Jeannie Whayne (Fayetteville: University of Arkansas, 1995), pp. 38–60.

12. Guccione et al., "Origin and Age," p. 2. Otto W. Nuttli differs with Guccione on the intensity of the quakes, putting them at 7.2, 7.1, and 7.4: Nuttli, "The Mississippi Valley Earthquakes of 1811 and 1812: Intensities, Ground Motion and Magnitudes," *Bulletin of the Seismological Society of America,* February 1973, pp. 227, 230. See also Fuller, *The New Madrid Earthquake (Cape Girardeau, Mo.: Ramfre Press, 1966)* pp. 33–34. Another good source on the New Madrid earthquakes is James Penick Jr., *The New Madrid Earthquakes of 1811-1812* (Columbia, Mo.: University of Missouri Press, 1976).

13. Fuller, *The New Madrid Earthquake,* p. 101.

14. Ibid., pp. 79, 81–82, 83–86. See, especially, Jay Feldman, *When the Mississippi Ran Backwards: Empire, Intrigue, Murder, and the New Madrid Earthquakes* (New York: Free Press, 2005).

15. Ibid., pp. 64–65.

16. Conevery Bolton ValenĐius, *Birth of the Country: How American Settlers Understood Themselves and Their Land* (New York: Basic Books, 2002), pp. 145–52. See also Anthony Wilson, *Shadow and Shelter: The Swamp in Southern Culture* (Jackson: University Press of Mississippi, 2006).

17. Velma Sample, "Mississippi County's First White Settlers," p. 2, WPA County History, Mississippi County, Arkansas History Commission, Little Rock (hereafter WPA County History); Fuller, *The New Madrid Earthquake,* pp. 10, 17.

18. *Randolph Recorder,* June, 21, 1934. Marshall Wingfield, "Town of Randolph," Memphis Information Files, Randolph, Memphis/Shelby County Public Library and Information Center, Memphis, Tennessee, n.d.

19. Ibid. See also *Memphis Commercial Appeal,* November 27, 1949. Between June 15 and June 20, 1834, for example, the *Halcyon, Claibourne, Mohawk, Tippecanoe, Kentuckian, Lancaster,* and *Dover* put into Randolph on their way up the river from New Orleans while the *Walk-in-the-Water* and the *Tennessean* stopped there on their way down. *Randolph Recorder,* June 21,1934.

20. Don Wilson, "Randolph, The Glory Years," *West Tennessee Historical Society Papers* 51 (December 1997):. 98–105. Richard Allin, "Historical Old Randolph Was State's Biggest Port," Newspaper clipping, Memphis Information Files, n.d. See also Gerald M. Capers Jr., *The Biography of a River Town: Memphis; Its Heroic Age* (Chapel Hill: University of North Carolina Press, 1939), pp. 57–59.

21. Wingfield, "Town of Randolph." For information on Sam Wilson, see Harbert Davenport Collection, Texas State Library & Archives Commission, Austin, Texas.

22. William L. Wilson, Last Will and Testament, Will Book A-1, Will No. 42, Tipton County, Tennessee. The will was apparently written and witnessed in October 1836. It was officially filed on November 10, 1837. His exact date of death is unknown. The youngest children were Cassandra, William Henry, Alexander, and Alfred. The oldest were Josiah, Samuel, Parker, and Mary. See Wilson Genealogy, Special Collections, University of Arkansas Library, Fayetteville, Arkansas

23. According to Bruce Vaughan, who examined the photograph, it was probably taken circa 1850 using the ambrotype (or calotype) process, which would have required the subject to remain motionless for about sixty seconds. Interview, Jeannie Whayne with Bruce Vaughan, May 31, 2007, Springdale, Arkansas.

24. Original General Land Office Survey Notes and Plats for the State of Arkansas, 1815–present, Commissioner of State Lands, State of Arkansas (hereafter cited as GLO Records). Roy Minnick, *A Collection of Original Instructions to Surveyors of the Public Lands, 1815–1881* (Rancho Cordova, Calif.:

Landmark Enterprises, 1980). See also Albert C. White, *A History of the Rectangular Survey System* (Washington, D.C.: Department of the Interior, Bureau of Land Management, 1983).

25. Township 11 North, Range 9 East , where Josiah first purchased property, received forty-eight such remarks: thirty at "first rate" and eighteen at "good bottom land." The three closest competitors (T14N/R12E, T13N/R9E, and T15N/R11E) ranked at 40, 35, and 34 such characterizations. GLO Records.

26. Ibid.

27. Wilson "entered" the NW 1/4 of Section 26, Township 11N, Range 9E. The GLO surveyor filed the plats on October 27, 1845. The Archeological Survey list three sites very near that township: Site no. 8 (Bell and Bell plantation) at Section 34, Township 11 N, Range 9 E; site no. 14 (Nettle Ridge, Bell Place) at Section 4, Township 11N, Range 9E; and site no. 15 (Notgrass) at Section 35, Township 11N, Range 9E. Archeologists noted evidence of large villages at two of the three sites (nos. 14 and 15). GLO Land Office Patents, secured from the Web site: Doc. no. 5333, 40 acres, Section 26, Township11N, Range 9E (SWSW); Doc. no. 5641, 80 acres, Section 27, Township 11N, Range 9E (N1/2SE); Doc. no. 5642, 80.68 acres in two parcels, Section 34, Township 11N, Range 9E (NENE) and Section 34, Township 11N, Range 9E (SENE).

28. Wilson sold two hundred acres near Randolph in 1845 for $800, two acres "with improvements" on the edge of the town of Randolph for $300 in1848, and 282.5 acres in 1850 for $600. See Tipton County Deed Records (200 acres), March 5, 1845; Deed Records (two acres with improvements), October 9, 1848; and Deed (282.5 acres), October 9, 1850. His brothers, William H. and Al, were also selling acreage in the 1840s. "Benjamin F. Butler", in *Biographical and Historical Memoirs of Northeast Arkansas* (Chicago: Goodspeed, 1889), p. 479.

29. Sample, "Mississippi County's First White Settlers," p. 2.

30. Margaret Humphreys, *Malaria: Poverty, Race, and Public Health in the United States* (Baltimore and London: Johns Hopkins University Press, 2001).

31. Todd L. Savitt, *Medicine and Slavery: The Diseases and Health Care of Blacks in Antebellum Virginia* (Urbana: University of Illinois Press, 1978), pp. 18–19, 25. David C. Dale and Daniel D. Federman, *Scientific American Medicine,* WebMD, p. 7, http://www.amazon.com/WebMD-Scientific-American -David-Dale/dp/0970390254.

32. Three species of bacteria have been identified: *Haemophilus influenzae, Streptococcus pneumoniae,* and *Neisseria menginitidis,* Trine H. Mogensen, Søren R. Paludan, Mogens Kilian, and Lars Østergaard, *Journal of Leukocyte Biology* 80, no. 2 (August 1, 2006): 267–77. Another source said many different forms can cause it: Streptococcus, Escherichia coli, and Listeria monocytogenes in newborns. kidshealth.org/parent/infections/lung/meningitis.html.

33. Manuscript Census of Population, Mississippi County, 1850. Lonnie Strange, "Civil War and Reconstruction in Mississippi County, Arkansas: The Story of Sans Souci Plantation," Honors Thesis, Spring 2008, University of Arkansas, Fayetteville, p. 8.

34. "Life on Mississippi Colorful When Steam Boats Plied River," WPA County History, no author, p. 1.

35. Donald P. McNeilly, *The Old South Frontier: Cotton Plantations and the Formation of Arkansas Society, 1819–1861* (Fayetteville: University of Arkansas Press, 2000), p. 132.

36. Manuscript Census of Agriculture, Mississippi County, 1850, 1860.

37. Rachel Skoney, "Large Planters and Planter Persistence in Antebellum Arkansas," MA Thesis, University of Arkansas, Fayetteville, 1991, p. 45. The only Wilson slave whose name survives in the record is that of Napoleon Wilson, who provided a detailed and relatively accurate genealogy of the family to a Wilson descendant in 1908. That he possessed intimate knowledge of the family history is not surprising given that he had been a member of the Wilson household since his birth in the mid-1830s and initially belonged to Josiah's father (Manuscript Census of Population, Slave Schedule, Mississippi

County, 1860). Gavin Wright, *The Political Economy of the Cotton South: Households, Markets, and Wealth in the Nineteenth Century* (New York: W. W. Norton, 1978): pp. 139–40.

38. McNeilly, *The Old South Frontier*, pp. 132–33. Orville Taylor, *Negro Slavery in Arkansas* (Durham, N.C.: Duke University Press, 1958). John Blassingame, *The Slave Community: Plantation Life in the Antebellum South* (New York: Oxford University Press, 1972). Charles W. Joyner, *Down by the Riverside: A South Carolina Slave Community* (Urbana: University of Illinois Press, 1984).

39. Manuscript Census, Slave Schedules, Mississippi County, 1860.

40. Manuscript Census of Population, Free and Slave Schedules, Mississippi County, Canadian, Clear Lake, Chickasawba and Big Lake Townships, 1860.

41. U.S. Department of Commerce, *Eighth Census of the United States: 1860*, Population (Washington, D.C.: Government Printing Service, 1862). Linda A. Newston, "A Historical-Ecological Perspective on Epidemic Disease," in William Balée, *Advances in Historical Ecology* (New York: Columbia University Press, 1998), pp. 42–63.

42. Strange, "The Civil War and Reconstruction in Mississippi County," p. 34. Thomas A. DeBlack, *With Fire and Sword: Arkansas, 1861–1874* (Fayetteville: University of Arkansas Press, 2003). Carl Moneyhon, *The Impact of Civil War and Reconstruction on Arkansas: Persistence in the Midst of Ruin* (Baton Rouge: Louisiana State University Press, 1994).

43. Robert Hardin also organized a Confederate regiment from Mississippi County. Sample, "Mississippi County's First White Settlers," p. 2.

44. DeBlack, *With Fire and Sword;* Moneyhon, *The Impact of Civil War and Reconstruction.*

45. Eric Foner, *Reconstruction, 1863–1877* (New York: Harper & Row, 1988). For events on the Mississippi County level, see Charles Bowen's account, written sometime in the early twentieth century, later published in the *Delta Historical Review* (Fall 1996): 8–11. See also "Sartain, Hardin Families Played Important Roles in County History," *Osceola Times*, undated newspaper clipping; "Sans Souci Is a Most Historic Plantation," *Osceola Times*, September 24, 1970, p. 6. *Biographical and Historical Memoirs of Northeast Arkansas* (Chicago: Goodspeed, 1889), p. 457; DeBlack, *With Fire and Sword;* Moneyhon, *Persistence in the Midst of Ruin.* Randy Finley, *From Slavery to Uncertain Freedom: The Freedmen's Bureau in Arkansas, 1865–1869* (Fayetteville: University of Arkansas Press, 1996).

46. Moneyhon, *Persistence in the Midst of Ruin;* DeBlack, *With Fire and Sword.* For the events that transpired in the county, see Bowen, in *Delta Historical Review;* "Sans Souci Is a Most Historic Plantation"; and *Biographical and Historical Memoirs of Northeast Arkansas,* p. 459. Steven Hahn, *A Nation under Our Feet: Black Political Struggles in the Rural South from Slavery to the Great Migration* (Cambridge and New York: Cambridge University Press, 2005).

47. J. B. Murray was county sheriff from 1868 to 1872, *Biographical and Historical Memoirs,* p. 459. H. M. McVeigh, ibid., p. 457. McVeigh was subsequently elected to the state legislature; *Historical Report of the Secretary of State,* edited by Janice Wegener(State of Arkansas, Little Rock, 1978), p. 427.

48. Adah Roussan Blackburn, "Great County," (Memphis) *Commercial Appeal,* June 10, 1925. Adah's first husband, Leon Roussan, was one of the only voices advocating against disfranchisement in 1889.

49. "History of Mississippi County," WPA County History, p. 7; *Biographical and Historical Memoirs,* pp. 457–59, 471.

50. Manuscript Census of Agriculture, Mississippi County, Scott Township, 1870. The entry for the Wilson farm, made on July 12, 1870, just two weeks before Josiah's death, reads: "Napoleon B. Lafont, agent for Josiah L. Wilson."

51. In 1860 eight members of this wealthy family operated sizable plantations in Mississippi County (G. C., C. F., and F. G. McGavock; E. C., W. C., and Jacob Bass; and J. M. and B. B. Harding), with a combined wealth of $428,720. By 1870, only five of them or their heirs remained, holding only $122,600 in real and personal property. See Manuscript Census of Population, Mississippi County, Arkansas, 1860 and 1870.

52. Manuscript Census of Free Population, Mississippi County, Scott and Pecan townships, 1850, 1860, and 1870.

53. Velma Sample, Profile of Dr. F. G. McGavock, , WPA County History, pp. 1–2; "A History of Mississippi County," p. 6.

CHAPTER TWO

1. Shelby County Probate Court Records, Memphis, Tennessee, September Term, 1870. See four documents dated September 1, 7, 14, and 15. The Lafonts also challenged Martha's suitability as guardian of Missouri, whom they characterized as Josiah's "deaf and dumb and idiotic child." Martha responded that Missouri, long domiciled in her household, was competent to make application to remain with her. Shelby County Probate Court Records, September 12, 1870.

2. Young Lee Wilson received $2,749.73. Shelby County Probate Records, Martha M. Wilson, admix vs. Estate of Josiah L. Wilson, dec., "Order of Distribution," March 16, 1871; Guardianship document filed by Martha M. Wilson in the Shelby County Probate Court, Memphis, Tennessee, as guardian of Missouri, William Henry, and Robert Lee Wilson, minor heirs of Josiah L. Wilson, March 16, 1871 (see also Guardians' Certificate filed March 20, 1871). Yet another form, filed on March 16, established that "the only heirs at law" of Josiah Wilson were "Missouri Wilson, Wm Henry Wilson, Robert Lee Wilson, Viola Lafont, wife of N. B. Lafont, and Victoria Davies, wife of J. F. Davies." On March 17, 1871, Napoleon Lafont filed the document providing that Davies represent him in all matters concerning the "distribution of the funds."

3. Mississippi County Probate Court Record, March 20, 1871, Osceola, Arkansas. Josiah's first purchase from the General Land Office was 160 acres the northwest one-quarter of Section 26, Township 11N, Range 9 E. Viola received that piece of property, in addition to another 320 acres (SW 1/4, Sec. 26, T 11, R 9 and SW 1/4, Sec. 23, T11, R 9). She was required to pay the other heirs sufficient sums to equalize the distribution. For land dispersed to other heirs, see Deputy Surveyor, Township 11, Range 9, filed October, 27, 1845 (surveys made in December 1839 and January 1840 and corrected in August 1844), pp. 296–98, 300–302, 306, 496.

4. A report Martha made to the Probate Court in Memphis in March of 1873 suggests a careful stewardship of the funds she controlled in Tennessee. After deducting a modest sum for costs, her three wards were due a total of $660.52 or roughly $220.19 each. William Henry received $220.18. Shelby County Probate Court Record, March, 1873: Gold in Hand, $4,350.00; Interest on Gold (10 percent), $435.00; Currency in Hand, $3,896.19; Interest on Currency (6 percent), $233.77; total, $8,914.96.

5. Lynette Boney Wrenn, *Crisis and Commission Government in Memphis: Elite Rule in a Gilded Age City* (Knoxville: University of Tennessee Press, 1998), pp. 11–12; Gerald M. Capers Jr., *The Biography of a River Town: Memphis, Its Heroic Age* (Chapel Hill: University of North Carolina Press, 1939): pp.107–8.

6. For the statistics on the Irish and African American populations, see Census of Population, Vol. I, Population and Social Statistics, 1870, table 8, p. 380, and table 31, p. 768.

7. Eric Foner, *Reconstruction: America's Unfinished Revolution, 1863–1877* (New York: Harper & Row, 1988), pp. 261–62.

8. Guardian's special bond, Shelby County Probate Court, January Term, 1878, Shelby County Probate Court Loose Records, 1870, File No. 397, Josiah L. Wilson.

9. Capers, *The Biography of a River Town: Memphis*, p. 192.

10. Linton Weeks, *Memphis: A Folk History* (Little Rock; Parkhurst, 1982), p. 56.

11. See Shelby County Probate Court Loose Record, 1870, File No. 397, Josiah L. Wilson, for receipts for payment of taxes.

12. Capers, *The Biography of a River Town: Memphis*, p. 180; Wrenn, *Crisis and Commission Government in Memphis: Elite Rule in a Gilded Age City*, p. 17.

13. John H. Ellis, *Yellow Fever and Public Health in the New South* (Lexington: University Press of Kentucky, 1992), pp. 27–28.

14. Ibid., p. 27.

15. Ibid., p. 28.

16. Ibid., p. 29.

17. Frederick F. Cartwright in collaboration with Michael D. Biddiss, *Disease and History* (New York: Barnes & Noble, 1972). For discussion of waterborne disease, see pp. 154–60. For discussion of malaria and yellow fever, see pp. 141–42, 144–45. See also Ellis, *Yellow Fever and Public Health,*, p. 31.

18. Ellis, *Yellow Fever and Public Health,* p. 31

19. Ibid., p. 18.

20. *Memphis: 1800–1900,* vol. 3: *Years of Courage,* compiled by Carole M. Ornelas-Struve, text by Joan Hassell, editor: a Memphis Pink Palace Museum Book (New York: Nancy Powers, 1982). Ellis, *Yellow Fever and Public Health,* p. 32. J. P. Young, *Standard History of Memphis: From a Study of the Original Sources* (Knoxville: H. W. Crew, 1912), pp. 152, 157; and J. M. Keating, *The Yellow Fever Epidemic of 1878 in Memphis, Tennessee* (Memphis: Printed for the Howard Association, 1879), p. 102.

21. Ellis, *Yellow Fever and Public Health.*

22. Khaled J. Bloom, *The Mississippi Valley's Great Yellow Fever Epidemic of 1878* (Baton Rouge: Louisiana State University Press, 1993), pp. 158–59.

23. Ibid., p. 161.

24. Ibid., p. 160.

25. Ellis, *Yellow Fever and Public Health,* p. 57.

26. Lee Wilson to Mildred Byars, March 11, 1932, Lee Wilson & Co. Correspondence, 1932, B, Folder: Mildred Gregory Byars, Special Collections, University of Arkansas Library, Fayetteville.

27. Gracie Jacobi, "Lee Wilson & Co., at Wilson, Arkansas, $10,000,000 Giant of Agriculture and Industry," as quoted from the *Commercial Appeal,* January 1, 1940, WPA County History, pp. 2, 6; and Mable F. Edrington, ed., *History of Mississippi County,* Arkansas (n.p., 1962), p. 387.

28. Edrington, *History of Mississippi County, Arkansas* pp. 80-83, 383–84.

29. Mississippi County Probate Court Records, July Term, 1881, Book 2, p. 92.

30. On May 29, 1882, he purchased the SW 1/4, of Sec. 8 and the E ½ NW 1/4 and SW 1/4 of NE 1/4 Section 17, all in Township 11 North, Range 10 East. Deed Book 11, p. 169.

31. Mississippi County Circuit Court Records, November 16, 1882, pp. 59–60.

32. Mississippi County Deed Records, March 29, 1883, Deed Book 13, p. 36, Mississippi County Courthouse, Osceola (he purchased the NW 1/4 SE 1/4 Section 33, Township 11 North, Range 9 East).

33. *Osceola Times,* November 1, 1883. See also Mississippi County Probate Court, October 25, 1883, Book 2, pp. 201–4. See also January Term, 1884, Book 2, pp. 212, 220, 222, 229, 230.

34. *Osceola Times,* October 29, 1887. Mississippi County Probate Records, January Term 1885, Book 2, pp. 261, 395, 620.

35. Pearl, who was residing in the household of a Frank and Mary Fisher in 1910, indicated that she had given birth to six children, three of whom were alive. None of them were living with her in the Fisher household, unless the Fishers' four-year-old adopted daughter, Viola M. Fisher, was Pearl's child. Manuscript Census of Population, Memphis, Shelby County, Tennessee, 1910.

36. Mississippi County Probate Records, April Term 1885, Book 2, p. 265.

37. Mississippi County Probate Records; see particularly April Term, Book 2, pp. 297–98, for Lee Wilson's criticisms of Napoleon's handling of her estate. See also pp. 312, 314, 403, 509, 520–21, and 565. He charged the estate $300 per year for boarding her in 1887, 1888, 1889, 1891.

38. Missouri's estate at time of death included 760 acres.

39. Upon Viola's death in the summer of 1885, William K. Harrison took over as administrator of her estate and accepted responsibility for her children, Clarence and Eloise. In 1891 Eloise married Joseph H. Pullen from Memphis, whose brother settled near Lee Wilson at Frenchman's Bayou in 1888 and became a merchant there. The couple subsequently moved to Louisiana, where Pullen became a bank cashier, an executive position in the parlance of the times.

40. Mississippi County Deed Book 15, p. 156, from Clarence Lafont, July 23, 1888, W ½ SE 1/4, Sec. 1; SE 1/4 SE 1/4, Sec 1; SW 1/4 NE 1/4, Sec. 1; N ½ SW 1/4, Sec. 1; E ½ NW 1/4, Sec. 1; SW 1/4 NW 1/4, Sec. 1; N ½ SE 1/4, Sec. 2; SE 1/4 NE 1/4, Sec. 2; S ½ SW 1/4, Sec. 2; N 2/2 NE 1/4, Sec. 10; S ½, Sec. 11; SW 1/4, Sec. 12; N ½ SE 1/4 Sec. 12; N ½ Sec. 13; N ½ SW 1/4 Sec. 13; N ½ Sec. 14; N ½ SE 1/4 Sec. 14; N ½ SW 1/4, Sec. 14; S ½ SE 1/4, Sec. 14. All in Township 11, Range 9.

41. Clarence Lafont to Lee Wilson & Co., December 10, 1933, in Lee Wilson & Co., 1933, K–Mc, Folder: Clarence Lafont; *Osceola Times,* March 15, 1902. "W. H. Pullen," in *Biographical and Historical Memoirs of Northeast Arkansas* (Chicago: Goodspeed, 1889), p. 545; and "Clarence Lafont," in *Biographical and Historical Memoirs,* p. 522.

42. *Osceola Times,* December 7, 1895. According to this newspaper article, Elkins accompanied Lee Wilson on a trip to Cairo and St. Louis to inspect and deliver lumber.

43. Manuscript Census of Population, First District, Cairo, Illinois, 1900, lists the Elkins household. For references to James H Elkins and malaria, see *Osceola Times,* April 2, 1898. For his association with Wilson, see *Osceola Times,* December 7, 1895, and April 1, 1899. Boaz Davies, meanwhile, became a physician like his father before him, practiced medicine in Mississippi County for a while, but eventually sold his property and made a career in the army. Manuscript Census of Population, Fort Bliss, Texas (under the El Paso schedule), 1910, lists Boaz Davies as a soldier working for the hospital corps. For other reference to Boaz Davies, see *Osceola Times,* February 23, 1901 (says "Dr. Boaz Davies is up on a visit to his sister").

44. Quoted in Edward Andrus, "The Pine, the Bluff, and the River: An Environmental History of Jefferson County, Arkansas," MA thesis, University of Arkansas, 2011; *Osceola Times,* January 5, 1889; April 13, 1889; January 11, 1890; May 10, 1890; August 8, 1890; November 2, 1889; April 23, 1892; November 20, 1897; May 14, 1898. "Robert. E. L. Wilson," in *Biographical and Historical Memoirs,* pp. 568–69.

45. *Osceola Times,* August 10, 1895. When Wilson incorporated in 1905, it was reported that the "company owns about 40,000 acres of land." The real property records do not confirm this figure, but it may be that he was including land that he rented and land that he farmed for his nieces and nephew (*Osceola Times,* April 22, 1905).

46. Census of Population, 1900, Table 4, Census Reports, Vol. I, *Twelfth Census of the United States, Taken in the Year 1900,* United States Census Office, Washington, D.C., 1901, p. 10.

47. Some unimproved acres were "old fields," land that had been overused and allowed to grow over. In Mississippi County, however, most of the unimproved acres were in forestland (67,645 acres in forests in 1880 as opposed to 4,254 in "old fields" in that year). Census of Agriculture, 1880, Table VII, Report on the Production of Agriculture, *Tenth Census,* General Statistics Washington, D.C.: Government Printing Office 1883), p. 105; and Census of Agriculture, for year 1899, Census Reports, *Twelfth Census of the United States,* Agriculture, Table 19, p. 267.

48. Only 15.1 percent of the adult men in the southern townships in 1900 had been born in Arkansas. Manuscript Census of Population, 1900, Mississippi County, southern townships.

49. Manuscript Census of Population, 1860, 1870, 1880, and 1900.

50. County court records are unavailable, but a close reading of the *Osceola Times* reveals 105 incidents occurring between 1885 and 1899 and overwhelmingly involving attacks by men against other

men (97.1 percent of the perpetrators and 95.1 percent of the victims were male). More whites than blacks were involved in violent disputes (58.2 percent of the perpetrators were white and 56.9 percent of the victims were white), although the newspaper infrequently reported news of African American matters. And in only 63 of the 105 incidents reported did the newspaper provide enough details to determine the origins of a violent dispute. Liquor and gambling were a component in 20.9 percent of the incidents, robbery was the motive in 8.6 percent of the cases, and seven men (6.7 percent of the cases) fought each other over a woman. The remainder had a variety of causes: an economic dispute between two men over a business transaction, an argument over the ownership of a wool shirt, and a battle between two preachers of different denominations over the meaning and purpose of baptism.

51. Roussan indicated that with the exception of the four years during the Civil War, he had never struck another man in anger. *Osceola Times,* April 5, 1890.

52. Ibid., April 12, 1890.

53. Census of Population, Part I, *Twelfth Census of the United States,* Taken in the Year 1900, Table 19, p. 530.

54. Douglas Blackmon, *Slavery by Another Name: The Re-enslavement of Black People in America from the Civil War to World War II* (New York: Doubleday, 2008).

55. Pete Daniel, *Shadow of Slavery: Peonage in the South, 1901–1969* (Urbana: University of Illinois Press, 1972); Harold D. Woodman, *New South, New Law: The Legal Foundations of Credit and Labor Relations in the Postbellum Agricultural South* (Baton Rouge: Louisiana State University Press, 1995); Roger Ransom and Richard Sutch, *One Kind of Freedom: The Economic Consequences of Emancipation* (Cambridge and New York: Cambridge University Press, 1977).

56. *Osceola Times,* October 13, 20, 1888. The tenants who shot Phillips were arrested and arraigned, but the outcome is unknown. For incidents elsewhere, see Jeannie M. Whayne, *A New Plantation South: Land, Labor and Federal Favor in Twentieth Century Arkansas* (Charlottesville: University of Virginia Press, 1996), pp. 56–59; see also Vincent Vinikas,"Specters in the Past: The Saint Charles, Arkansas, Lynching of 1904 and the Limits of Historical Inquiry," *Journal of Southern History* 65, no. 3, (August 1999), pp. 535–64

57. Historian Harold Woodman traced the conflict that developed between planters and merchants over whose lien was superior, the planters or the merchants. Southern courts (including those in Arkansas) held that the planters' lien was superior. As time passed, merchants acquired land and became planters, and planters acquired or opened stores and became furnishing merchants as well as planters. Woodman, *New South, New Law.*

58. Ransom and Sutch, *One Kind of Freedom.*

59. Mississippi County Deed records, Book 12, February 29, 1884, pp. 503–5; Mississippi County Courthouse, Osceola, Arkansas.

60. Ransom and Sutch, *One Kind of Freedom.*

61. *Osceola Times,* January 14, 1916; January 5, 1917.

62. A rich historiography on populism exists. For southern populism, see Steven Hahn, *The Roots of Southern Populism: Yeomen Farmers and the Transformation of the Georgia Upcountry, 1850–1890* (New York: Oxford University Press, 1983). For an excellent general treatment of populism, Robert McMath's two books: *Populist Vanguard: A History of the Southern Farmers' Alliance* (Chapel Hill: University of North Carolina Press, 1975) and *American Populism: A Social History, 1877–1898* (New York: Hill & Wang, 1993). For Arkansas populism, see Raymond Arsenault, *Wild Ass of the Ozarks: Jeff Davis and the Social Bases of Southern Politics* (Philadelphia: Temple University Press, 1984).

63. For published accounts of the Agricultural Wheel in Arkansas, see three articles by Francis Clark Elkins: "The Agricultural Wheel in Arkansas, 1887," *Arkansas Historical Quarterly* 40 (Autumn 1982): 249–60, "The Agricultural Wheel: County Politics and Consolidation," *Arkansas Historical Quarterly* 29 (Summer 1970): 152–75, and "State Politics and the Agricultural Wheel," *Arkansas Historical*

Quarterly 38 (Autumn 1979): 248–58. See also Clifton Paisley, "The Political Wheelers and Arkansas' Election of 1888," *Arkansas Historical Quarterly* 25 (Spring 1966): 3–21; and Theodore Saloutos, "The Agricultural Wheel in Arkansas," *Arkansas Historical Quarterly* 2 (Summer 1943): 127–40. For more recent work on the Wheel, see Jason McCollom, "The Agricultural Wheel, the Union Labor Party, the 1889 Arkansas Legislature," *Arkansas Historical Quarterly* 68 (Summer 2009): 157–75.

64. *Osceola Times,* July 14, 1888. Mississippi County Wheel officials in June of 1888 included J. F. Ruddle, president; C. W. Griffin, vice president; Harry Carmack, secretary; John B. Driver, treasurer; Reginald Archillion, lecturer; L. E. Martell, chaplain; F. B. Hale, trading agent; L. D. Rozzell, steward; J. W. Price, steward, Van Dorn Brett, sentinel; George Morrow, sentinel; and Joe Hayes, conductor. Several of these men were prominent citizens frequently mentioned in the newspapers, with Rozzell and Driver most prominent among them. All of them were farmers with the exception of pastor F. B. Hale and surveyor and engineer Reginald Archillion. They ranged in age from the mid-thirties to the mid-sixties, and they were from different parts of the county. Ibid., June 16, 1888.

65. Ibid., September 22, 1888.

66. Ibid., September 10, 1898; see also ibid., July 14 and 28, 1888; August 11, 1888; September 8 and 15, 1888; August 16 and 30, 1890; December 6, 1890; April 9 and 23, 1892; September 3, 1892; June 2, 1894; September 17, 1898; October 1, 1898.

67. John Giggie, *After Redemption: Jim Crow and the Transformation of African American Religion in the Delta, 1875–1915* (New York: Oxford University Press, 2008).

68. Ibid.

69. *Osceola Times,* January 5, 1889. Roussan enlisted in the First Confederate Battalion in 1861, was captured and paroled, and then enlisted in the Forty-second Tennessee Regiment, where he served as lieutenant, *Biographical and Historical Memoirs,* pp. 549–50.

70. John William Graves, *A Question of Honor: Election Reborn and Black Disfranchisement in Arkansas* (Charlottesville: University of Virginia Press, 1971), and *Race Relations and Urban Development in Arkansas, 1865–1905* (Fayetteville: University of Arkansas Press, 1990).

71. *Osceola Times,* November 13 and 20, 1897; December 4 and 18, 1897; January 15 and 22, 1898; June 18, 1898.

72. Ibid., July 14 and 28, 1900; August 4 and 19, 1900; September 15 and 22, 1900; October 13, 20, and 27, 1900; November 24, 1900; December 8, 1900; January 1, 1901; March 23, 1901; April 20, 1901; June 8, 22, and 29, 1901; July 27, 1901. On December 8, 1900, the newspaper reported that the DCC had been reorganized. For radical racism, see C. Vann Woodward, *The Strange Career of Jim Crow* (rev. edition, Oxford University Press, 1974).

73. Edrington, *History of Mississippi County,* p. 86.

74. *Osceola Times,* March 17, 1916.

75. Ibid., September 26, 1890; January 23, 1892; November 4, 1893; November 3, 1894. For rumor of a "race war" see ibid., April 5, 1890.

76. For secondary literature on whitecapping and night riding, see Jeannie M. Whayne, *A New Plantation South: Land, Labor and Federal Favor in Twentieth Century Arkansas* (Charlottesville: University of Virginia Press, 1996); William F. Holmes, "Moonshiners and Whitecaps in Alabama, 1893," *Alabama Review* 34 (January 1981): 31–49, "Moonshiners and Collective Violence: Georgia, 1889–1895," *Journal of American History* 67 (1980): 588–611, "Whitecapping in Georgia: Carroll and Houston Counties, 1893," *Georgia Historical Quarterly* 64 (1980): 388–404, "Whitecapping: Agrarian Violence in Mississippi, 1902–1906,"*Journal of Southern History* 35 (1969): 165–85, and "Whitecapping in Mississippi: Agrarian Violence in the Populist Era," *Mid America* 55 (1973): 134–48. See also Edward Ayres, *Vengeance and Justice: Crime and Punishment in the NineteenthCentury American South (New York: Oxford University Press, 1984),* 260–62.

77. Whayne, *New Plantation South,* p. 52; and Act 12, *Acts of Arkansas,* 1909, p. 315.

CHAPTER THREE

1. Cynthia Barnett, *Mirage: Florida and the Vanishing Waters of the Eastern U.S.* (Ann Arbor: University of Michigan Press, 2007); Michael Grunwald, *The Swamp: The Everglades, Florida, and the Politics of Paradise* (New York: Simon & Schuster, 2006); David McCally, *The Everglades: An Environmental History.* (Gainesville: University Press of Florida, 1999). Two special issues of *BioScience* contain several articles on the ecological impact of mankind upon rivers and lakes. Particularly relative are Richard E. Sparks, John C. Nelson, and Yao Yin, "Naturalization of the Flood Regime in Regulated Rivers: The Case of the Upper Mississippi River," *BioScience* 48 (September 1998): 706–20; Barry L. Johnson, William B. Richardson, and Teresa J. Naimo, "Past, Present, and Future Concepts in Large River Ecology: How Rivers Function and How Human Activities Influence River Processes," *BioScience* 45 (March 1995): 134–41; and James A. Gore and E. Douglas Shields Jr., "Can Large Rivers Be Restored?" *BioScience* 45 (March 1995): 142–52. See also Robert W. Harrison and Walter M. Kollmorgen, "Land Reclamation in Arkansas under the Swamp Land Grant of 1850," *Arkansas Historical Quarterly* 6 (Winter 1947): 369–418; and Robert W. Harrison, "The Formative Years of the Yazoo-Mississippi Delta Levee District," *Journal of Mississippi History* 13, no. 3 (1951): 236–48, "Early State Flood-Control Legislation in the Mississippi Alluvial Valley," *Journal of Mississippi History* 23, no. 2 (1961): 104–26, and "Clearing the Land in the Mississippi Alluvial Valley," *Arkansas Historical Quarterly* 13 (Winter 1954): 352–71.

2. A variety of records provide death dates or approximate death dates for each member of his family. For Josiah's death date, see Mississippi County Probate Records and Memphis/Shelby County Archives. For Martha's death, see Yellow Fever record book kept at the Memphis/Shelby County Archives; for Dr. Davies, see Mississippi County Probate Records; for Victoria's death, see both the Mississippi County Probate Records and the *Osceola Times, September 1, 1883.* For William Henry's death, see Mississippi County Probate Records. For Napoleon, see the Mississippi County Probate Records and the *Osceola Times*, February 21, 1885. The newspaper listed Viola as "dangerously ill" and Probate Court Records demonstrate that she had died by July that year, *Osceola Times*, February 21, 1885.For Tiny Wilson, see *Osceola Times,* August 18, 1888, and her headstone in the Bassett Cemetery.

3. Department of Commerce and Labor, Bureau of the Census, Special Reports, *Mortality Statistics, 1900 to 1904* (Washington, D.C.: Government Printing Office, 1906). American Association for the Advance of Science, Bulletin 15, *A Symposium on Human Malaria with Special Reference to North America and the Caribbean Region* (Washington, D.C.: Smithsonian Institution, 1941), p. 10. The local newspaper routinely reported diseases and deaths in Osceola and the various hamlets about the county. These notices appeared in sections devoted to news from Golden Lake, Bardstown, Frenchmen's Bayou, Pecan Point, and the like, and also in the newspaper's community section named, ironically, Raindrops, which was mostly devoted to news about Osceolans.

4. The incidences of illnesses in March and April were generally unspecified. Added to the problem of quantifying the number of people suffering from disease were remarks like this:"Considerable sickness reported by the physicians in the surrounding country—mostly of a malarial character." Quantifying "considerable" proved baffling. Equally problematic were notices that entire families were suffering: "Mr. J. W. Quinn . . . reports all of his family suffering from the effects of malaria," and "Capt. E. M. Ayres and his entire family suffering from malaria," and, also "Mr. James Feezor and family have been suffering from chills and malaria fever " (*Osceola Times,* September 11, 1897, August 13, 1898, August 26, 1899, and September 2, 1899). *Osceola Times,* July 6, 1889. Such references were repeated often: the "doctors have been kept pretty busy this week," February 1, 1890; the "doctors kept busy," September 13, 1890; the "physicians are constantly on the go," September 27, 1890; the "physicians are kept busy night and day," October 1, 1892. Ibid., August 27, 1898.

5. Ibid., August 3, 1898; April 14, 21, and 28, 1894. For Mrs. Wilson's illness, see January 19 and March 27, 1895. For mention of Victoria's illness, see March 23, 1895; April 27, 1895; October 23, 1897;

September 8, November 19, and December 17, 1898. For mention of Hot Springs, see July 9, 1898; for mention of Glenn Springs, see August 26, 1899, and September 16, 1899. A subsequent story, published on September 30, 1899, indicated that the family, in fact, made the move. Dora Merrill joined that household within months.

6. Ann Vileisis, *Discovering the Unknown Landscape: A History of America's Wetlands* (Washington, D.C.: Island Press, 1997), pp. 71–73. Andrew A. Humphreys, Joseph B. Eads, and Charles Ellet Jr. were the three most prominent and influential engineers operating in the nineteenth century; see John M. Barry, *Rising Tide: The Great Mississippi Flood of 1927 and How It Changed America* (New York: Simon & Schuster, 1997).

7. Although Crittenden, Lee, and Phillips counties were the hardest hit in Arkansas, Mississippi County endured the worst flood of its history. Floyd M. Clay, *A Century on the Mississippi: A History of the Memphis District, U.S. Army Corps of Engineers, 1876–1981* (Memphis District: U.S. Army Corps of Engineers, 1986), p. 28; *Osceola Times*, February 25, 1882; Corps of Engineers, *Annual Highest and Lowest Stages of the Mississippi River and Its Outlets and Tributaries to 1960* (Vicksburg, Miss.: U.S. Army Engineers, 1960), p. 53; ,*Twelfth Census of the United States, 1900*, Population, Pt. 1, General Tables, Table 4, Population of States and Territories by Counties at each Census, 1790 to 1900, Arkansas (Washington, Government Printing Office, 1901), p. 10. According to this census, Mississippi County had 7,332 residents in 1880. Since the population more than doubled between 1880 and 1890, an incremental increase of roughly 1,700 between 1880 and 1882 is reasonable. *Osceola Times*, February 25, 1882; and Clay, *A Century on the Mississippi*, p. 29.

8. Corps of Engineers, *Annual Highest and Lowest Stages*, p. 53; *Osceola Times*, March 11, 1882; Clay, *A Century on the Mississippi*, pp. 14–17.

9. Clay, *A Century on the Mississippi*, p. 31.

10. Mark Twain, *Life on the Mississippi* (Oxford and New York: Oxford University Press, 1990; originally published in 1883), p. 192.

11. *Osceola Times*, January 20, 1894.

12. Ibid., March 30, 1895.

13. Ibid., April 5, 1895.

14. Corps of Engineers, *Annual Highest and Lowest Stages*, pp. 53–54. *Osceola Times*, March 12, 1898; November 19, 1898; January 29, 1898.

15. For Wilson purchases, see Senate Report 746, "Title to Certain Lands in Mississippi County, Ark.," February 20, 1919, 65th Congress, 3d. Session, pp. 1,2; and *U.S. v. Lee Wilson & Co.*, F. 630, 651 (E.D. Ark. 1914), pp. 633–34.

16. St. Francis Levee Board, *History of the St. Francis Levee District* (West Memphis, Ark.: St. Francis Levee Board, 1946), p. 1.

17. Elliott B. Sartain, *It Didn't Just Happen* (Osceola, Ark.: Grassy Lake and Tyronza Drainage District No. 9, 197?), p. 18. See J. M. McKimmey, B. Dixon, H. D. Scott, and C. M. Scarlat, *Soils of Mississippi County, Arkansas*, Arkansas Agricultural Experiment Station, Research Report 970, Division of Agriculture, University of Arkansas, p. 25, for soil types, and page 27 for soil permeability. See A. F. Barham, "As I Saw It: The Story of the Development of the Drainage and Flood Control in the St. Francis Basin of Arkansas" (n.p., 1964), pp. 15–16, for detail on weaknesses in the levees in Mississippi County.

18. Clay, *A Century on the Mississippi*, p. 56; Trudy E. Bell, *Images of America; The Great Dayton Flood of 1913* (Chicago: Arcadia, 2008). Barham, "As I Saw It," pp. 7, 14, 54. Flood stage at Memphis, Tennessee is thirty-four feet, and the floods of 1912–13 ran forty-five and forty-six feet, respectively. See Corps of Engineers, *Annual Highest and Lowest Stages*, p. 54; *Osceola Times*, April 11, 1912, April 11, 1913.

19. Robert Robinson to J. H. Robinson, April 24, 1912, letter in possession of author. Thanks to Michael Wilson for providing me with a copy of this letter, which he received from Randall and Charlotte Gerdes. *Osceola Times*, April 25, 1912, p. 2.

20. *Osceola Times,* April 8, 1913.

21. *Osceola Times,* April 8, 1913; Barham, "As I Saw It," p. 17; Mabel F. Edrington, ed., *History of Mississippi County, Arkansas* (n.p., 1962), pp. 385–86.

22. Barnett, *Mirage*; Grunwald, *The Swamp*; McCally, *The Everglades*; Sparks et al., "Naturalization of the Flood Regime in Regulated Rivers"; Johnson et al., "How Rivers Function" ; and Gore et al., "Can Large Rivers Be Restored."

23. McCally, *The Everglades.*

24. Charles H. Barnard and John Jones, *Farm Real Estate Values in the United States by Counties, 1850–1982,* Statistical Bulletin Number 751, Economic Research Service (Washington, D.C.: U.S. Department of Agriculture, March 1987), pp. 8, 10. In 1930, only one other delta county, Crittenden, located immediately south, came close, at $74. Prices in Pulaski County, home to the state capital, stood at $65 per acre, but from there it dropped to $61 in Craighead, $58 in Chicot, $55 in Cross, and $50 in Desha (all delta counties).

25. For land prices, see ibid. Before the collapse in prices during the 1920s, the price of land there had risen to $145 per acre. According to the agricultural censuses, the number of farms increased from 1,720 to 10,583 between 1900 and 1930, with acres in cotton increasing from 34,380 to 218,701 and bales harvested increasing from 22,577 to 140,862. Note particularly that while the number of all tenants increased from 1,204 to 9,561, the number of white tenants increased from 525 to 4,583 and the number of black tenants increased from 679 to 4,978. For number of farms in 1900, *Twelfth Census of the United States,* 1900, Agriculture, Vol. 5, Pt. 1, Table 25, p. 418; for acres in cotton in 1900, see Vol. 6, Pt. 2, Table 10, p. 430 (Washington: Government Printing Office, 1902). For 1930 figures, see *Fifteenth Census of the United States:* 1930, Agriculture, Vol. II, Pt. 2, Table V (Washington: Government Printing Office, 1932), p. 1171.

26. *Osceola Times,* November 30, 1901; Whayne, *A New Plantation South,* p. 104.

27. Ibid., September 27, 1902, p. 4; April 9, 1904, p. 2; Sartain, *It Didn't Just Happen,* p. 22.

28. The Grassy Lake and Tyronza began at Barfield in the north. The engineers constructed a main channel southwest, intersecting with the Tyronza River, essentially turning that small tributary of the St. Francis River into a drainage ditch. Well over a dozen smaller ditches fed into the main channel and took the excess water out of the county and, eventually, into the St. Francis River at a point in Poinsett County. Sartain, *It Didn't Just Happen,* p. 40.

29. Ibid., p. 7 (also quoted in Edrington, *History of Mississippi County,* p. 76, 81).

30. Edrington, *History of Mississippi County,*, p. 82.

31. Mississippi County Deed Book 11, p. 169, from Ben N. Bacchus, County Clerk, May 29, 1882, South Part SW 1/4, Sec. 8; E ½ NW 1/4 and SW 1/4 of NE 1/4, Sec. 17, Township 11N, Range 10E. Barham, "As I Saw It," p. 12; and *Osceola Times,* July 30, 1908, p. 1.

32. Edrington, *History of Mississippi County,* p. 75; Barham, "As I Saw It," p. 12; St. Francis Levee District, *History of the Organization,* pp. 316–17.

33. Barham, "As I Saw It,", p. 12; *Biographical and Historical Memoirs of Northeast Arkansas* (Chicago: Goodspeed, 1889), p. 550; *Osceola Times,* March 12, 1908, p. 1, March 26, 1908, p. 1. The newspaper editor, a strong supporter of drainage, lauded Gladish's appointment. See also Edrington, *History of Mississippi County,* pp. 84–86, which gives an account of Gladish's appointment and the role of drainage supporters.

34. *Osceola Times,* June 3, 1909, p. 2, and June 17, 1909, p. 1. Edrington, *History of Mississippi County,* p. 86. For reference to the convict farm on Wilson's property, see *Osceola Times,* October 26, 1911.

35. Hearings Held before the Committee on Public Lands of the House of Representatives on H.R. 19637 (Washington, D.C.: Government Printing Office, 1910; hereafter *Hearings*), p. 235.

36. Edward A. Bowers, Acting Commissioner, to Secretary of the Interior, August 25, 1894. Exhibit 2, *Hearings,* pp. 36–37, 259, 261.

37. Gore et al., "Can Large Rivers Be Restored?" pp. 143.

38. Senate Report 746, "Title to Certain Lands," p. 2; *U.S. v. Lee Wilson & Co.*, F. 630, 651 (E.D. Ark. 1914).

39. *Hearings*, pp. 8, 259. See also in *Hearings*, "Chronological Statement of Proceedings Affecting Arkansas 'Sunk Lands,'" pp. 3–4; and "Public Lands—Arkansas Sunk Lands," December 12, 1908, vol. 37, pp. 345–50.

40. *Hearings*, pp. 36–37, 259, 261. For quote, see Exhibit 16, p. 265.

41. Ibid., pp. 33, 200–203.

42. Ibid., pp. 146–47.

43. Ibid., pp. 279–395.

44. Manuscript Census of Population, Mississippi County, 1910. Of the thirty who indicated age, 26.7 percent were in their twenties, 33.3 percent in their thirties, and 33 percent in their forties.

45. Ibid. Only seventy-three indicated how many months they had been on their homesteads, and of those, fifty-seven (78.1 percent) were there twelve months or less. Fifty-nine (54.1 percent) had wives and/or children or extended family with them.

46. *Hearings,* pp. 279–395. There were eight hog pens, eight chicken houses, and six corncribs. Two men had smokehouses, four had stables, and one had a corral.

47. Ibid.

48. Ibid., p. 343.

49. Ibid., pp. 279–395.

50. Manuscript Census of Population, Mississippi County, 1910. Of the other two homesteaders, one was born in England and the other in Indiana.

51. H. C. Hall to Secretary of the Interior, February 17, 1911, General Land Office Records, Arkansas Sunk Lands, Interior Department, Appellate Case Files, National Archives (hereafter NA), RG 67.

52. Jeannie M. Whayne, *A New Plantation South: Land, Labor and Federal Favor in Twentieth Century Arkansas* (Charlottesville: University of Virginia Press, 1996), p. 92.

53. Alexander Voglesby, first assistant secretary of the interior, to the Secretary of the Interior, January 27, 1916, p. 16, General Land Office Records, Arkansas Sunk Lands, Interior Department, Appellate Case Files, NA, RG 67. For Wilson's deposition testimony, see *U.S. v. Lee Wilson & Co.*, F. 651 (E.D. Ark. 1914).

54. Clay Fallman, Commissioner, General Land Office, to Register and Receiver, Little Rock, Arkansas, January 29, 1920, p. 3, General Land Office Records, Arkansas Sunk Lands, Interior Department, Appellate Case Files, NA, RG 67. *Lee Wilson & Co. v. United States,* 245 U.S. 24 (1917). Justice White, who had been appointed to the court by President William Howard Taft in 1910, was a Louisiana native who almost certainly knew all about swamplands.

55. "Preference Right to Purchase Certain Lands in Mississippi County, Ark.," Report 1007, House of Representatives, 65th Congress, 3d Session, Jan. 29, 1919, p. 2; and "Title to Certain Lands in Mississippi County, Ark.," Report No. 746, Senate, 65th Congress, 3d Session, February 20, 1919. *Journal of the House,* p. 212.

56. *Congressional Record,* House Bills, 67th Congress, 1st Session, p. 514.

57. For reference to Wilson's clubhouse, see "Buffalo Island: Drainage District No. 16 of Mississippi County, Arkansas . . . Then and Now!" (n.d., n.p.), p. 3. For information on other clubs at Big Lake, see Lynn Morrow, "A 'Duck and Goose Shambles': Sportsmen and Market Hunters at Big Lake, Arkansas," *Big Muddy: A Journal of the Mississippi River Valley* 4, no. 1 (2004): 76; and Steve Bowman and Steve Wright, *Arkansas Duck Hunter's Almanac* (Fayetteville: Ozark Delta Publishers, 1998), p. 62. For Wilson and young Wesson, see Jeannie Whayne interview with Lee Wesson, July 25, 2008, Memphis, Tennessee.

58. "Buffalo Island: Drainage District No. 16 of Mississippi County, Arkansas . . . Then and Now!" p. 3; Frank Graham Jr., *Man's Dominion: The Story of Conservation inAmerica* (New York: M. Evans, 1971), p. 210; Bowman and Wright, *Arkansas Duck Hunter's Almanac,* pp. 57, 61.

59. Bowman and Wright, *Arkansas Duck Hunter's Almanac*, p. 62.

60. Ibid., p. 62; Morrow, "A 'Duck and Goose Shambles,'" pp. 71–72, 76; William T. Hornaday, *Our Vanishing Wild Life: Its Extermination and Preservation* (New York: New York Zoological Society, 1913; reprinted under same title, New York: Arno Press and the New York Times, 1970), pp. 270–71.

61. Bowman and Wright, *Arkansas Duck Hunter's Almanac*, p. 62. See also Hornaday, *Our Vanishing Wildlife*, p. 252.

62. *Osceola Times*, October 18, 1915.

63. For Wilson's hunt on Peter's Island, see, for example, R. E. Lee Wilson to Charles T. Coleman, October 10, 1932, Wilson Correspondence, Special Collections Division, University of Arkansas Libraries. For the legacy to the Boy Scouts, see James E. West (Chief Scout Executive, Boy Scouts of America) to D. Fred Taylor (attorney in Osceola, Arkansas), January 25, 1934, Wilson Correspondence, Special Collections Division, University of Arkansas Libraries. "Last Will of R. E. Lee Wilson," Section VII, describes the property as "the preserve" and indicates it is located in Sections 3 and 4, Township 11 North, Range 9 East, near the property his father originally homesteaded (Mississippi County Probate Records, Osceola District).

64. Owners increased from five hundred to 1,088; tenants from 1,204 to 5,440. According to Census of Population published in 1920, which included detail on 1910 as well as 1920, the population rose from 16,384 to 47,320. *Fourteenth Census of Population*, Vol. 1, Pt. 2, Agriculture, County Table I, Arkansas (Government Printing Office, 1922), p. 565.

65. Robert H. Moulton, "Wilson—That's All: A Whole Town Is the Result of One Man's Idea," *Illustrated World* 38 (February 1923): 845–47, 942.

CHAPTER FOUR

1. Charles T. Coleman to J. H. Crain, Trustee, N.D., Box Lee Wilson & Company, 1941, C–D, Folder Charles C. Coleman, Special Collections, University of Arkansas Library, Fayetteville. This document is actually misfiled in the 1941 box. In a letter written in September 1936, Coleman mentions that Jim Crain, who was running the company after Wilson's death, had asked him to write an evaluation of the trust agreement. See Charles T. Coleman to James H. Crain, September 5, 1936, Box Lee Wilson & Company, 1936, C–Cor, Folder Charles T. Coleman.

2. For "farmer prince," see C. E. Collins, "Arkansas, 'Wonder State,'" *National Reclamation Magazine*, May 1922, pp. 64; for "feudal baron," see Robert H. Moulton, "Wilson—That's All: A Whole Town Is the Result of One Man's Ideas," *Illustrated World*, February 1923, v. 38, p. 847; and "Robert E. Lee Wilson Rules Small Village Like Baron of Feudal Times," *Akron Weekly Pioneer Press* (Akron, Washington County, Colorado), December 22, 1922, p. 6; for "benevolent dictator," see "Reality and Myth in Wilson, Arkansas," *Arkansas Times*, March 1983, p. 23.

3. These references to his acreage are reported in numerous publications but three are as follows: For forty-thousand-acre references see "Robert E. Lee Wilson Rules Small Village Like Baron of Feudal Times," p. 6, and Moulton, "Wilson—That's All." For reference to sixty-five thousand acres, see Seymour Freedgood, "The Man Who Has Everything—in Wilson, Ark.," *Fortune*, August 1964, p. 144.

4. Robert L. Brandfon, *Cotton Kingdom of the New South: A History of the Yazoo Mississippi Delta from Reconstruction to the Twentieth Century* (Cambridge, Mass.: Harvard University Press, 1967), pp. 61–62, 129–31; Lawrence J. Nelson, *King Cotton's Advocate: Oscar G. Johnston and the New Deal* (Knoxville: University of Tennessee Press, 1999).

5. Moulton, "Wilson—That's All," pp. 847–48.

6. *Osceola Times.*, April 9, 1904, p. 3.

7. Ibid., November 8, 1902.

8. Ibid., April 22, 1905.

9. Coleman to Crain, September 5, 1936. See Charles T. Coleman to James H. Crain, September 5, 1936, Box Lee Wilson & Company, 1936, C–Cor, Folder Charles T. Coleman

10. Moulton, "Wilson—That's All," pp. 847–48.

11. Arthur F. Sheldon, *The Science of Efficient Service; or, The Philosophy of Profit-making* (Chicago: Sheldon School, 1915). Sheldon published a number of books, including *The Science of Business Building: A Series of Lessons Correlating the Fundamental Principles and Basic Laws Which Govern the Sale of Goods and Services for Profit* (Chicago: Sheldon School, 1911). This may have been the course that Lee Wilson was having his employees take. Other books include *Elements in Success* (Chicago: Sheldon School, 1909), *The Measure of Value* (Chicago: Sheldon School, 1909), and *The All-Round and Four Square Man* (Chicago: Sheldon School, 1909). See also "Rotary's Global History Fellowship (An Internet Project) Celebrate History," http://www.rotaryfirst100.org/leaders/sheldon/, section on Arthur "Fred" Sheldon, which has a copy of the advertisement appearing in *The Rotarian* in 1917.

12. Lee Wilson to K. P. Cullom, September 30, 1918, Box Bank of Wilson, 1918, Folder Lee Wilson & Company.

13. This information comes from an evaluation written by Wilson's attorney in 1934. Although undated, the document was almost certainly in response to a question raised by Jim Crain, the successor trustee to Lee Wilson, who asked Coleman for an evaluation. Coleman mentions the request in a letter he wrote to Jim Crain on September 5, 1936. The evaluation is misfiled in a 1941 box. See Charles T. Coleman to James H. Crain, September 5, 1936, Box Lee Wilson & Company, 1936, C–Cor, Folder Charles T. Coleman.

14. Quit Claim Deed, Book 41, Deed Record Books, Osceola District, Mississippi County Court Records, February 5, 1917, p. 543.

15. Manuscript Census of Population, Shelby County, Tennessee, 1910.

16. May 25, 1915, to Bank of Wilson from Charles G. Roth, Resident Manager, Saint Paul Hotel, Box Bank of Wilson Correspondence, 1913–16, Folder Marie Wilson.

17. Mortimer Robinson Proctor, comp., *History of the Class of 1912, Yale College*, vol. 1 (New Haven, Conn.: Yale University, 1912), p. 376; and Richard E. Bishop, Alfred C. Bowman, John H. Eden, James A. Reilly, Greville Rickard, Coulter D. Young, and Theophilus R. Hyde, *History of the Class of Nineteen Hundred and Twelve, Sheffield Scientific School, Yale University* (New Haven, Conn.: Class Secretaries Bureau, 1926), pp. 458–59.

18. The Charter of Incorporation and the Bylaws of the Wilson-Ward Company are in Lee Wilson & Co., 1935, To–Z, Folder: Wilson-Ward Co.

19. Their first correspondent banks in these cities were Security Bank and Trust Company of Memphis, Mechanics American National Bank in St. Louis, and Seaboard National Bank in New York. By 1912, the St. Louis correspondent was Boatman's Bank. Cashier, Bank of Wilson, to J. A. Wardle, Box Bank of Wilson, 1908–1912, Folder J. A. Wardle. The Bank of Wilson cashier was in this correspondence, asking Wardle to make "stamps" of the Security Bank and Trust, Memphis, Mechanics American National Bank in St. Louis, and Seaboard National Bank in New York. By 1912, Boatman's Bank had superseded Mechanics American National Bank in St. Louis: see Bank of Wilson, Daily Transactions, 1912.

20. Jeannie M. Whayne, "Robert E. Lee Wilson and the Making of a Post–Civil War Plantation," in *The Southern Elite and Social Change: Essays in Honor of Willard B. Gatewood, Jr.* (Fayetteville: University of Arkansas Press, 2002), p. 105.

21. See esp. Stuart Schulman, "Origins of the Federal Farm Loan Act: Issue Emergence and Agenda-Setting in the Progressive Era Print Press," in *Fighting for the Farm: Rural America Transformed*, edited by Jane Adams (Philadelphia: University of Pennsylvania Press, 2003), pp. 118–19; George E.

Putnam, "The Land Credit Problem," *Bulletin of the University of Kansas Humanistic Studies* 2, no. 2 (December 1, 1916): 39–53; and Dick T. Morgan, *Land Credits: A Plea for the American Farmer* (New York: Crowell, 1915), pp. 20–34, 53–69. See also American Institute of Banking, *Farm Credit Administration* (New York: American Institute of Banking, 1934); Claude L. Benner, *The Federal Intermediate Credit System* (New York: Macmillan, 1926); David Danbom, *The Resisted Revolution: Urban America and the Industrialization of Agriculture, 1900–1930* (Ames: University of Iowa Press, 1979); Clara Eliot, *The Farmers' Campaign for Credit* (New York and London: Appleton, 1927); Gilbert C. Fite, *American Agriculture and Farm Policy since 1900* (New York: Macmillan, 1964); Virgil P. Lee, *Principles of Agricultural Credit* (New York: McGraw-Hill, 1930); Earl Sylvester Sparks, *History and Theory of Agricultural Credit in the United States* (New York: Crowell, 1932); Alvin S. Tostlebe, *Capital in Agriculture: Its Formation and Financing since 1870* (Princeton, N.J.: Princeton University Press, 1957); and Eli Samuel Troelston, *The Principles of Farm Finance* (St. Louis: Educational Publishers, 1951).

22. For the situation in Arkansas, see Carl H. Moneyhon, *The Impact of Civil War and Reconstruction on Arkansas: Persistence in the Midst of Ruin* (Fayetteville: University of Arkansas Press, 2002), and *Arkansas and the New South, 1874–1929* (Fayetteville: University of Arkansas Press, 1997).

23. Altogether they owned 160 shares worth $1,287.50. Some of his employees, like Blackwell, held executive positions in Lee Wilson & Company and lived in Wilson, Arkansas. William E. Shallow worked as the mill superintendent there, while H. S. Portis ran the Idaho Grocery Company in Bassett, a small Wilson-controlled hamlet in south Mississippi County. W. L. Evans was one of his farm managers, and L. C. Gaty ran the Wilson and Northern, a short-line railroad in south Mississippi County. Others, like Wright H. Smith, were connected to the Memphis office. Wright doubled as a lumber agent and as the general manager for the Jonesboro, Lake City, and Eastern Railroad, another Wilson enterprise that operated in northeastern Arkansas. W. L. Harrison and John Mulroy worked as bookkeepers there, and William Coulson was a lumber agent.

24. Wilson and Rhodes became political allies when Rhodes ran for circuit clerk against Charles Driver (not related to John B. Driver of the St. Francis Levee Board). Driver won the election but a two-year contest ensued, a contest that was ultimately settled by the Arkansas Supreme Court in Driver's favor. The Rhodes and Wilson families had long-standing ties, having settled in south Mississippi County before the Civil War, and John Rhodes had come of age with Lee Wilson. Rhodes owned twenty shares, and two other merchants owned ten and twenty shares, respectively. Attorney J. T. Coston, who had represented Wilson in his effort to secure the drainage enterprises, subscribed to ten shares. Two engineers and three physicians owned altogether eighty-three shares. Shareholders included two planters, C. M. Bell of Scott Township and John E. Uzzell of Pecan Point, who owned twenty shares each.

25. Of the fifty-four initial subscribers, only nineteen (35.2 percent) could not be located on the Manuscript Census of Population or otherwise identified and may have lived outside the Mississippi County/Memphis orbit. It would not be surprising, however, to find that they had a business connection of some sort to Lee Wilson.

26. Wilson employees included W. L. Evans, farm manager; W. E. Shallow, mill superintendent; Wright H. Smith, lumber agent; M. J. Blackwell, secretary-treasurer; and A. B. Hill, store superintendent. R. E. Lee Wilson pushed the number of company-connected directors to six. The other directors were merchant J. W. Rhodes, planters Stephen Ralph and C. M. Bell, and bank cashier William Jett (J. A. Murphy took Jett's place as cashier and director).

27. For Smith's "advice," see Bank of Wilson, 1908–1912, Folder: Wright Smith, October 12, 1908, to W. S. Jett from Wright Smith; November 23, 1908, from Jett to Redman. For the resignation, see Jett to M. J. Blackwell, November 30, 1908.

28. J. A. Murphy was selected to take Jett's place on January 2, 1909 (see folder Mechanics-American National Bank, Bank of Wilson Correspondence, 1908–1912, to M. J. Blackwell, President, Bank of

Wilson from Asst. Cashier, Mechanics American National Bank) but he left to go into business with Dan F. Portis, son of another Wilson employee, M. A. Portis, who ran the Idaho Grocery Company in Bassett. By 1914 he had been replaced by E. A. Mitchell, who stayed until early 1915, when Frank Young took over. By September 1916, L. P. Nicholson was cashier. He was replaced in the spring of 1918 by Kelly P. Cullom, brother to a Wilson company executive who worked as secretary of the Kansas City Shook and Manufacturing Company, another Wilson enterprise.

29. For example, the first president was Wilson employee M. J. Blackwell, and the first vice president was Wilson friend J. W. Rhodes. The second set of officers included J. C. Cullom as president and H. F. Crawford, F. A. Gillette, and A. B. Hill as vice presidents. Cullom, Gillette, and Hill were also employed by Wilson.

30. A series of transactions in the fall of 1910 illustrates the practice. In September 1910, the Bank of Wilson borrowed $5,000 from Mechanics-American National Bank, its correspondent bank in St. Louis, and used as collateral loans it had made to four Wilson entities: two Lee Wilson & Company loans ($925 and $2,000) plus two loans to two Wilson partners, Hill and Wilson ($2,000) and the Idaho Grocery Company ($2,000). When the Bank of Wilson collected those loans in December—a good time for collections in an agricultural economy—it paid off the Mechanics-American Bank. The heavy reliance of Wilson businesses on the bank for operating expenses only intensified over time. In its "Bills Discount" statement on June 8, 1912, the bank listed loans of $46,756.03 made to twenty-eight borrowers, and twenty-four of them were Wilson-related loans totaling $39,321.03; thus 84 percent of the bank's loans were to Wilson executives or Wilson companies.

31. January 17, 1914, to Bank of Wilson from J. D. Covey, Assist. Commissioner, State of Arkansas Bank Department, in Bank of Wilson, 1913–1916, Folder: State of Arkansas Bank Department, Lee Wilson & Company; January 27, 1914, to Bank of Wilson from John M. Davis, Bank Commissioner, in Bank of Wilson, 1913–1916, Folder: State of Arkansas Bank Department, Lee Wilson & Company.

32. May 29, 1914, to the Board of Directors, Bank of Wilson, Wilson, Arkansas, from the State Banking Department, John M. Davis, Commissioner, in Bank of Wilson 1913–1916, Folder: State of Arkansas Bank Department, Lee Wilson & Company. Among the other observations made by the commissioner: one cautioning the bank about overdrafts, another about how to credit furniture and fixtures on their books, one about the necessity to hold monthly board-of-director meetings, one to appoint an examining committee, another to use a stock ledger, and one to procure a Boston General ledger, etc.

33. "Statement of the Condition of the Bank of Wilson, Wilson, Arkansas, At the Close of Business, September 30, 1914," in Boatman's Bank Folder, Bank of Wilson, 1913–1916.

34. September 5, 1918, John M. Davis to Board of Directors, Bank of Wilson, Box Bank of Wilson, 1918, Folder Arkansas State Banking Department.

35. M. J. Blackwell from Wright Smith, January 11, 1911; Wright Smith to Bank of Wilson, January 23, 1911. Both in Folder Lee Wilson & Company, Memphis, Bank of Wilson 1908–12. See also Bank of Wilson to National Car Advertising Company, March 4, 1911, March 10, 1911, May 1, 1911, August 22, 1911, Folder National Car Advertising Company, Bank of Wilson Records, 1908–12. See also in this folder National Car Advertising Company to Bank of Wilson, March 6, 1911, May 3, 1911, Frank A. Gillette, Lee Wilson & Company, to Bank of Wilson and Bank of Wilson to Lee Wilson & Company, H. H. Rapp, Asst. Treas., National Car Advertising Company, to Bank of Wilson. See also J. H. Crain and W. F. Wilson, Executors of the Estate of R. E. Lee Wilson, Sr., Deceased, vs. Commissioner of Internal Revenue, Brief of Petitioner, Lee Wilson & Co., 1935, To-Z.

36. Actually, evidence is not available in sufficient abundance to ascertain for certain that Wilson's percentage of the amount the bank loaned had actually declined. The letters in the files from the bank commissioners make no mention of it, but it is by no means certain that the files are complete.

37. Minutes of Stockholders, January 9, 1923.

38. S. E. Simonson to R. E. Lee Wilson, September 27, 1920, Box Bank of Wilson, 1919, 1921–1922, N–Z, Folder S. E. Simonson.

39. S.E. Simonson to Kelley Cullom, February 6, 1921, in Simonson folder cited in n. 38.

40. Simonson to Kelley Cullom, October 15, 1921. In Simonson folder cited in n. 38. Although this letter was addressed to Kelley Cullom, Simonson apparently assumed Lee Wilson would be reading it (as evidenced by his "You will recall Mr. Wilson . . .").

41. Cullom to Simonson, October 22, 1921. In Simonson folder cited in n. 38.

42. Simonson to Bank of Wilson, Nov. 16, 1921. In Simonson folder cited in n. 38.

43. Cullom to Simonson, April 20, 1922, Bank of Wilson to Simonson May 26, 1922, and Simonson to Bank of Wilson, July 17, 1922. In Simonson folder cited in n. 38.

44. B. M. Gile, "Organization and Management of Agricultural Credit Associations," *Agricultural Experiment Station Bulletin* 259 (1931): 12.

45. Sam L. Thomas to Avondale Mills, Birmingham, Ala., September 10, 1921, Box Lee Wilson & Company, Blytheville, Folder Avondale Mills; Manuscript Census of Population, Tennessee, Shelby County, 20-WD; Memphis, Roll 1764, p. 170.

46. Lee Wilson & Company, Blytheville, 1921–25.

47. R. R. Clabaugh, vice president, Liberty Central Trust Company to Lee Wilson, September 22, 1921, in Box Lee Wilson & Company, Blytheville, 1921–22, Folder Liberty Central Trust Company. See also in this folder Clabaugh to Farmers Bank and Trust Company, Blytheville, Arkansas, September 26, 1921. A fully executed (signed and dated by all parties) copy of the detailed agreement is dated April 24, 1922. It appears to differ in no way from the September 22, 1921, letter.

48. H. B. Deming and Company to Lee Wilson & Company, Blytheville, May 1, 1922.

49. Box Lee Wilson & Company, Blytheville, 1921–1925, Folders Liberty Central Trust Company. and National Bank of Commerce, St. Louis. The Liberty Central Trust Company loans are mentioned in correspondence dated June 6, 1922, to Lee Wilson & Company, Blytheville, from Liberty Central Trust Company. The $26,250 National Bank of Commerce loan is mentioned on December 27, 1921, in National Bank of Commerce to Lee Wilson & Company, Blytheville. The $200,000 insurance policy is mentioned on October 19, 1921, in Lee Wilson & Company, Blytheville to National Bank of Commerce.

50. September 28, 1921, National Bank of Commerce to Lee Wilson & Company, Blytheville, Box Lee Wilson & Company, Blytheville, 1921–25, Folder National Bank of Commerce.

51. See Lee Wilson & Company, Blytheville to National Bank of Commerce, October 6, 18, 25, and 27, 1921; and National Bank of Commerce to Lee Wilson & Company, Blytheville, October 3, 21, 26, and 27, 1921, Box Lee Wilson & Company, Blytheville, 1921–25, Folder National Bank of Commerce.

52. National Bank of Commerce to Lee Wilson & Company, Blytheville, October 28, 1921, Box Lee Wilson & Company, Blytheville, 1921–25, Folder National Bank of Commerce.

53. Note correspondence in the National Bank of Commerce file over a dispute about the "grade" of the cotton being held as collateral. The bank demanded a higher grade and eventually got its way.

54. Donald Crichton Alexander, *The Arkansas Plantation, 1920–1942* (New Haven, Conn.: Yale University Press, 1943), p. 16.

55. Deed of Trust, April 1, 1922, filed for record May 16, 1922, Record Book DD, Page 363, Mississippi County Records, Osceola, Arkansas. The bond issue is described in a letter from Lee Wilson to G. B. Rose, D. H. Cantrell, J. F. Loughborough, A. W. Dobyns, A. F. House, J. W. Barron, William Nash (the Rose law firm in Little Rock), March 24, 1933, in Lee Wilson & Co., 1933, O–R, Folder: Rose, Hemingway, Cantrell & Loughborough. Apparently the Rose law firm handled the matter for Lee Wilson in 1922.

56. Lee A. Dew, *JLC&E: The History of an Arkansas Railroad* (Jonesboro: Arkansas State University, 1968). For recent studies on railroads in the United States, see Thomas J. Misa, *A Nation of Steel: The*

Making of Modern America, 1865–1925 (Baltimore: Johns Hopkins University Press, 1995); and Robert G. Angevine, *The Railroad and the State: War, Politics, and Technology in Nineteenth-Century America* (Stanford, Calif.: Stanford University Press, 2004);

57. Dew, *JLC&E*, p. 3.

58. Ibid., pp. 2–4.

59. Ibid., pp. 13–19.

60. Janet Taylor and Mylinda Nelson, "The Cotton Belt reached Blytheville in 1900." See *Blytheville Plus; 100 Year History of Blytheville* (n.d., n.p.), p. 5.

61. Dew, *JLC&E*, p. 61.

62. Ibid., p. 62.

63. See Wright K. Smith to president and board of directors, Jonesboro, Lake City and Eastern RR Co., June 9, 1913, Special Meeting of the Board of Directors, Jonesboro, Arkansas, June 9, 1913, JLC&E Minute Book Three. The resolution lauding Smith appears in the Minute Book Three on that date.

64. Dew, *JLC&E*, pp. 64–65, for background. See Special Meeting of the Board of Directors, Jonesboro, Arkansas, June 27, 1914, JLC&E Minute Book Three, Frisco Archives, University of Missouri, Rollo.

65. Dew, *JLC&E*, pp. 83, 89.

66. Ibid., pp. 91–93.

67. Ibid., p. 94.

68. Ibid., p. 110–11.

69. For correspondence about freight rates, see September 2, 1927, D. C. Wayne to J. R. Cullom; October 24, 1927, D. C. Wayne to R. E. Lee Wilson, Roy Wilson, and F. A. Gillette; October 26, 1927, D. C. Wayne to R. E. Lee Wilson, and F. A. Gillette; March 23, 1928, D. C. Wayne to F. A. Gillette, all in Box Railroad Negotiations, etc., Folder F. A. Gillette. See also Folder St. Louis, San Francisco Railroad, 1924–1929. On the Mississippi River Western Railway Company, see February 11, 1932, to R. E. L. Wilson from R. E. L.Wilson Jr., Box Lee Wilson & Company Correspondence, 1932, We–Z, Folder R. E. Lee Wilson, Sr., R. E. Lee Wilson Jr., and R. E. Lee Wilson III.

70. Wilson to L. W. Baldwin, November 9, 1932, Lee Wilson & Co., Folder: L. W. Baldwin.

71. W. M. Wallace to Lee Wilson, March 21, 1932, Lee Wilson & Company, Folder: Mill and Motor Company offices.

CHAPTER FIVE

1. Mary S. Hoffschwelle, *The Rosenwald Schools of the American South* (Gainesville and Tallahassee: University Press of Florida, 2006), pp. 262–63.

2. James D. Anderson, *The Education of Blacks in the South* (Chapel Hill: University of North Carolina Press, 1988).

3. The rest consisted of $546 from the Rosenwald fund and $1,000 from the black community. Wilson also paid the greater share of the funds necessary to construct a new residence for the principal, which cost $3,000, of which Wilson contributed $1,600. Two years later, he contributed $2,500 toward the construction of a $3,500 vocational shop building. Hoffschwelle, *The Rosenwald Schools of the American South*, pp. 262–63.

4. WPA workers circulated questionnaires to all the churches in Mississippi County, including both white and black denominations. The twenty-six surviving questionnaires on the black churches included the following denominations: African Methodist Episcopal (5), Church of Christ Divine (2), Colored Methodist Episcopal (3), and Church of God in Christ (12). None of the questionnaires of the Baptist churches survived, although they appear on the WPA list along with their names and locations.

Also missing are the forms for four African Methodist Episcopal churches, two of the Colored Methodist Episcopal churches, and four of the Church of God in Christ churches. WPA Church Records, Special Collections Division, Mullins Library, University of Arkansas.

5. The black auxiliary organizations included the following: ACE League, 6; Bible Band, 5; Bible Study Band, 2; Board of Trustees, 1; Brotherhood Club, 1; Choir, 1; Deacon Board, 1; Epworth League, 1; Florists Aid Society, 1; Home and Foreign Mission, 1; Junior Church, 1; Junior League, 2; Ladies' Missionary Society, 2; Ladies' Sewing Circle, 2; Men's Bible Club, 1; Missionary Band, 1; Missionary Society, 4; Missionary Union, 1; Official Board, 1; Pastor's Aid, 2; Pastor's Club, 2; Prayer Meeting, 3; Serving Circle, 1; Sick Committee, 1; Sisters' Bible Band, 1; Sisters' Prayer and Bible Band, 1; Sisters' Prayer Band, 1; Stewardess Board, 2; Sunbeam Band, 1; Sunday School, 26; Sunshine Band, 4; Usher Board, 2; Usher's Band, 1; Who's Who Club, 1; Willing Workers, 1; Women's Home and Missionary Society, 1; Women's Missionary Band, 1; Women's Missionary Society or Council, 2; Young People's Epworth League, 1; Young People's Society, 1; Young People's Willing Workers, 9.

6. Just as the black community congregated in churches, so too did the county's white population. The records of 108 white churches provide the same kind of detail as that listed above and revealed that they too were relatively impoverished. They included a wider variety of denominations: Pentecostals, mainline Protestant churches such as Methodists and Baptists, as well as three Catholic churches and a Jewish temple. They too were likely to be housed in unpainted frame structures, and pianos and organs were in infrequent use. They adorned their buildings more fully: ten had bells; thirty-six, memorial windows; two held stained glass windows; and three buildings had "art glass." However much they differed in terms of the accouterments they provided their churches, the white congregations resembled the black ones in supporting missionary efforts, holding Bible study groups, and sponsoring activities and clubs for their youth. The denominations included: Assembly of God, 8; American Baptist, 2; General Baptist, 2; Independent Baptist, 1; Southern Baptist, 38; Christian Science, 1; Church of Christ, 11; Church of God, 4; Church of the Lord Jesus Christ, 1; Church of the Nazarene, 2; Disciples of Christ, 2; Independent, 3; Jewish, 1; Lutheran, 1; Methodist 22; Metropolitan Spiritualist, 1; Pentecostal Assembly of Jesus Christ, 1; Pentecostal Assembly of the World, 1; Pentecostal Church of America, 1; Pentecostal Holiness Church of God, 1; Pentecostal Holiness Church, 3; Free Pentecostal Church of God, 2; Pilgrim Holiness Church, 1; Presbyterian Church, 3; Episcopal Church, 2; Catholic Church 3.

7. Lois E. Myers and Rebecca Sharpless, "'Of the Least and the Most': The African American Rural Church," in *African American Life in the Rural South, 1900–1950*, edited by R. Douglas Hurt (Columbia: University of Missouri Press, 2003), pp. 73–74.

8. Ibid., p. 60.

9. Church Records, item 13, p. 1, Church of God in Christ, Box 428, Folder 112, Special Collections, University of Arkansas Library, Fayetteville.

10. WPA Church Records; see item no. 22, pp. 2–3, Colored Episcopal Church, Box 435, Folder 4. Wilson was hardly the first or the only planter to have an association with the black churches in the county. Hickman Chapel, located on the river in the north end of Mississippi County, dated back to the antebellum period and reflected the interest of a plantation owner. According to "Aunt" Lou Buckner, planter Julian Hickman provided his slaves with a log cabin "for a place of worship," and it continued in use until 1890, when "a frame building was erected." In 1907 the members of this African Methodist Episcopal denomination constructed a stone building, which was still in use at the time of the WPA interview in 1941. Hickman was a particularly prosperous planter, claiming $50,000 in real property and $18,000 in personal property in 1860. For details on Julian Hickman, see Manuscript Census of Population and Slave Census of Population, 1860, Canadian Township. For quotes from the WPA Church Records, see item number 22, pp. 2–4, African Methodist Episcopal Church, Box 435, Folder 4.

11. This is in keeping with Myers and Sharpless, who find that only about half of black churches had pianos. See Myers and Sharpless, "'Of the Least and the Most,'" p. 69, for subject of pianos (article is pp. 54–80).

12. Thomas C. Kennedy, *A History of Southland College: The Society of Friends and Black Education in Arkansas* (Fayetteville: University of Arkansas Press, 2009). For educational level of the "present clergyman" see WPA Church Records, item 17(a), p. 2. For information on Rev. Goldsberry, see WPA Church Records, item 17(a), p. 2, CME, Box 436, Folder 10. Southland College had been founded by Indiana Quakers to educate the children of freed people near Helena. . The Methodist ministers were more likely to have more education: 1. Common School: AME, 6; Church of God in Christ (COGIC), 8; Church of Christ Divine (COCD). 2. High School: AME/CME, 5; COGIC, 4; COCD, 0. College: AME/ CME, 1; COGDC and COCD, 0.

13. Rev. Joe Curry of Norman Chapel (CME) and Rev. H. F. Henton of the Wilson Church of God in Christ both indicated common or grammar school education. WPA Church Records, item 17(a) p. 2, CME, Box 435, Folder 4, and Church of God in Christ, Box 428, Folder 112.

14. Myers and Sharpless, "'Of the Least and the Most,'" pp. 62–63, 65–66. See also Gunner Myrdal, *American Dilemma: The Negro Problem and Modern Democracy* (New York and London: Harper & Brothers, 1944).

15. The *Commercial Appeal* (Memphis), April 18, 1918, as cited in Calvin White Jr., *They Danced and Shouted into Obscurity: A History of the Black Holiness Movement, 1897–1961* (Fayetteville: University of Arkansas Press, forthcoming 2011).

16. White, *They Danced and Shouted into Obscurity.*

17. *Twelfth Census of the United States,* Taken in the Year 1900, *Agriculture,* Part I, Farms, Livestock, and Animal Products, Table 10, Farms, June 1, 1900, Classified by Acreage, Tenure and Color, p. 62; and Table 19, Number and Acreage of Farms, p. 268. *Fifteenth Census of the United States: 1930,* Agriculture, Volume II, County Table I, "Farms and Farm Acreage and Cropland Harvested, 1929 and 1924, By Color and by Tenure of Farm Operator," p. 1139; and Volume III, Type of Farm, Table I, "Farms and Farm Acreage, 1930 and 1929, by Type of Farm, p. 782.

18. Jim Giesen, *Boll Weevil Blues: Cotton, Myth, and Power in the American South, 1892–1930* (Chicago: University of Chicago Press, forthcoming).

19. Personal Property Tax Records, Mississippi County, School Districts 16 and 25, 1920, Mississippi County Historical Society, Osceola, Arkansas.

20. Ibid.

21. Appraisement, Estate of O. T. Craig, May 3, 1921, Probate Court, Mississippi County Court, Osceola, Arkansas, Docket 223, Osceola, Arkansas.

22. R. H. Craig (for O. T. Craig & Son) to Bank of Wilson, February 13, 1919, June 3, 1919, August 6 and 16, 1919, December 2, 1919, February 7, 1921, and loan document dated 1921 pledging mules and property of O. T. Craig & Son (eighteen mules, all farming implements, crop of cotton and corn) in return for a loan of $3,000; Bank of Wilson to Craig, December 2, 1919, Bank of Wilson, 1919, 1921–22, A.M. Folder: O. T. Craig & Company. Mississippi County Real Property Records, Osceola Courthouse, 1922, 1933. Richard Craig, who took over management of the plantation, retained ownership of the 689 acres, at least until 1933.

23. According to John Gigge, there were 3,782 Masons in 1904 and 11,085 Odd Fellows and 7,000 Knights of Pythias in Arkansas alone in 1905. He cites W. E. B. DuBois, *Economic Co-operation among Negro Americans* (report of a study made by Atlanta University, under the patronage of the Carnegie Institution of Washington, D.C., together with the Proceedings of the Twelfth Conference for the Study of the Negro . . ., held at Atlanta University, on Tuesday, May 28, 1907), pp. 121–24.

24. Myers and Sharpless, "'Of the Least and the Most,'" pp. 74–75.

25. Manuscript Census of Population, 1920, 1930.

26. Ibid., 1920.

27. *Osceola Times*, March 19, 1915, p. 1; March 26, 1915, p. 1; and April 2, 1915, p. 1.

28. Ibid., April 2, 1915, p. 1.

29. Jeannie Whayne, "Low Villains and Wickedness in High Places: Race and Class in the Elaine Riots," Arkansas Historical Quarterly 57 (Autumn 1999): 285–313, and *A New Plantation South: Land, Labor and Federal Favor in Twentieth Century Arkansas* (Charlottesville: University of Virginia Press, 1996).

30. Grif Stockley, *Blood in Their Eyes: The Elaine Race Massacres of 1919* (Fayetteville: University of Arkansas Press, 2001); Robert Whitaker, *On the Laps of Gods: The Red Summer of 1919 and the Struggle for Justice That Remade a Nation* (New York: Crown, 2008). See also Whayne, "Low Villains and Wickedness in High Places." And see WPA Church Records.

31. *Osceola Times*, October 24, 1919, p. 1; November 7, 1919, p. 8.

32. Speakers included A. B. Poston, pastor of Osceola's AME church, and J. W. Addison, principal of Osceola's black school. C. R. Carrington, a Blytheville preacher, gave the invocation. Ibid, November 7, 1919, p. 8

33. Myrdal, *An American Dilemma*; *Osceola Times*, November 7, 1919, p. 8.

34. *Osceola Times*, August 3, 1923, p. 2

35. Lee Wilson to A. J. Bellinger, May 17, 1932, Lee Wilson & Company Papers, 1932 B, Folder: A. J. Bellinger; Lee Wilson to W. H. Warren, May 17, 1932, Lee Wilson & Company Papers, 1932, Sti–Wa, Folder: W. H. Warren; Warren to Lee Wilson, May 21, 1932, and Lee Wilson to Warren, June 1, 1932, Folder: W. H. Warren. Will Warren shows up in the Manuscript Census of Population in Jackson County in 1930 as a fifty-seven-year-old farm owner claiming $1,000 in assets, a substantial accomplishment for a black farmer in those days. See Manuscript Census of Population, Jackson County, Union Township, 1930.

36. *Osceola Times*, January 21, 1921, p. 91.

37. For a list of the farm units and managers, see Lee Wilson & Co., Folder: Fire Prevention Report, 1932, Lee Wilson & Company Papers. This document details the inspections of the houses occupied by the farm managers (and others) and provides their names and the units to which they were assigned. For Frank Gillette, see Frank A. Gillette to L. T. Defur (crop reporter for the U.S. Weather Bureau), Lee Wilson & Co., 1932, D–E, Folder: L. T. DeFur. Frank Gillette, Wilson's chief executive officer, described the arrangement succinctly: "We work with tenants and rent some by the acre and some for one-third and one-fourth of the crops, and some where we furnish teams, tools, etc. on a 50/50 profit sharing basis. The balance we work by day labor." For the charges incurred by the tenants and sharecroppers, see James H. Crain to J. Jacques, June 29, 1932, Lee Wilson & Co., 1932, I–Mc, Folder: J. Jacques.

38. Skip Wilson, interview with Jeannie Whayne, March 6, 2009.

39. D. N. Morris Report, Week Ending 7/8/33, in Lee Wilson & Co., 1933, O–R, Folder: Pee Wee Morris Weekly Reports.

40. W. Wilson to F. M. Gooding (U.S. Labor Bureau), July 25, 1932, Lee Wilson & Co., 1932, G–H, Folder: F. M. Gooding.

41. C. W. Woodman to Lee Wilson, August 16, 20, and 23, 1932, Lee Wilson & Co., 1932, We–Z, Folder: C. W. Woodman. Lee Wilson to C. W. Woodman, August 18 and 22, 1932, Lee Wilson & Co., 1932, We–Z, Folder: C. W. Woodman.

42. *Delta Historical Review* 52 (Spring 1998): 40–41.

43. William Pickens, "The American Congo—The Burning of Henry Lowery," *The Nation* 112, no. 2907 (March 23, 1921): 426–28. See also Todd E. Lewis, "Mob Justice in the 'American Congo': 'Judge Lynch' in Arkansas during the Decade after World War I," *Arkansas Historical Quarterly*, 52 no. 2 (Summer 1993), 156–84.

44. Pickens, *The American Congo*, p. 427.

45. The Manuscript Census of Population, Golden Lake Township, 1880, indicates that Crate was then two years old. *Osceola Times,* October 2, 1897, p. 4.

46. For various newspaper accounts immediately following the incident, see *Memphis Commercial Appeal,* December 26, 1920, p. 1; *Arkansas Gazette,* December 26 and 29, 1920, p. 1; *Arkansas Democrat,* December 27 and 28, 1920, January 8, 1921, p. 1.

47. *Memphis Press,* January 27, 1921, p. 1.

48. *Osceola Times,* January 28, 1921.

49. *El Paso Herald,* January 19, 1921.

50. Thanks to Will Guzman for information about the El Paso connection. See *El Paso Herald,* January 19, 1921, p. 4, for the quote from Lowery. See also Ibid., January 22, 26, 28, 1921; and Calvin R. Ledbetter Jr., "Thomas C. McRae: National Forests, Education, Highways, and *Brickhouse v. Hill,*" *Arkansas Historical Quarterly* 59, no. 1 (Spring 2000): 16–17. For Nixon and the Supreme Court, see *Nixon v. Herndon,* 273 U.S. 536 (1927) and *Nixon v. Condon,* 286 U.S. 73 (1932).

51. According to one report, Lowery's abductors skirted Memphis after the city's commissioner of police "ordered out the reserves to prevent a parade through the city," *Arkansas Democrat,* January 26, 1921; *New York Times,* January 27, 1921. For the governor's remarks, see *El Paso Herald,* January 27, 1921. For various estimates of the size of the crowd, see *Arkansas Democrat,* January 27, 1921, p. 1 (200–300); *Osceola Times,* January 28, 1921, p. 1 (300 people or more); *Commercial Appeal,* January 27, 1921 (600 or more); and Pickens, The *American Congo,* p. 427 (600).

52. *Memphis Press Scimitar,* January 26, 1921.

53. Pickens, "The American Congo," pp. 427–28.

54. In most studies of lynching and books on the Arkansas delta and race relations, the Lowery lynching is given some treatment. See particularly Nan Elizabeth Woodruff, *American Congo;* Whayne, *A New Plantation South;* Sheldon Avery, *Up from Washington: William Pickens and the Negro Struggle for Equality, 1900–1954* (Newark: University of Delaware Press, 1989), pp. 114–15; Lewis, "Mob Justice in the 'American Congo,'" pp. 156–84; and Richard Buckelew, "Racial Violence in Arkansas: Lynchings and Mob Rule, 1860–1930" (PhD diss., University of Arkansas, 1999). For general studies on lynching, see James Elbert Cutler, *Lynch-Law: An Investigation in the History of Lynching in the United States* (New York: Longmans, Green, 1905); NAACP, *Thirty Years of Lynching in the United States, 1899–1918* (New York: Negro Universities Press, 1969); Robert L. Zangrando, *The NAACP Crusade against Lynching, 1909–1950* (Philadelphia: Temple University Press, 1980); Arthur M. Raper, *The Tragedy of Lynching* (Chapel Hill: University of North Carolina Press, 1933); Jacquelyn Dowd Hall, *Revolt against Chivalry: Jessie Daniel Ames and the Women's Campaign against Lynching* (New York: Columbia University Press, 1979); W. Fitzhugh Brundage, *Lynching in the New South: Georgia and Virginia, 1880–1930* (Urbana: University of Illinois Press, 1993); Walter White, *Rope and Faggott: A Biography of Judge Lynch* (New York: Arno Press, 1969); Herbert Shapiro, *White Violence and Black Response: From Reconstruction to Montgomery* (Amherst: University of Massachusetts Press, 1988); and Stewart E. Tolnay and E. M. Beck, *A Festival of Violence: An Analysis of Southern Lynchings, 1882–1930* (Urbana: University of Illinois Press, 1995).

55. *Memphis Press,* January 27, 1921, p. 2. Other newspapers, however, reported that he admitted he was drunk at the time: *New York Times,*January 27, 1921; *Osceola Times,* January 28, 1921 (quoting a story published in the *Commercial Appeal,* January 27, 1921).

56. *Chicago Defender,* February 5, 1921, , p. 1. Note that William Pickens mentions only a "wife and little daughter" as visiting Lowery at the scene of the burning (Pickens, "The American Congo," *The Nation,* March 23, 1921, p. 427).

57. *Chicago Defender,* February 5, 1921, p. 1

58. *Memphis Press,* January 27, 1921, pp. 1, 2.

59. *Commercial Appeal,* January 28, 1921, pp. 1, 12. According to one report, Williams and Corbin were first moved to Missouri and then later taken to Little Rock (*Arkansas Gazette,* January 28, 1921).

60. C. H. Dennis to Cashier, Bank of Wilson, January 20, 1921, in Bank of Wilson, 1919, 1921–22, A–M.

61. *New York Times*, January 27, 1921.

62. *Memphis Press,* January 27, 1921, p. 1.

63. Ibid.

64. Note that Steve Hahn says there were forty-two in Arkansas; Mary Rolinson says there were forty-five. According to my count based on information at the Schomberg, there were thirty-nine. In addition, Mary Rolinson is mistaken in identifying Calexico as in south Mississippi County. In fact, I cannot locate a Calexico anywhere in Arkansas.

65. Hahn, *A Nation under Our* Feet. For a fine study on the late nineteenth-century movement from Arkansas, see Kenneth Barnes, *Journey of Hope: The Back to Africa Movement in Arkansas in the Late 1800s* (Chapel Hill: University of North Carolina Press, 2004).

66. *Osceola Times,* February 11, 1921, p. 2.

67. Ibid., February 18, 1921, p. 5.

68. Ibid., April 27, 1923, p. 1; and May 4, 1923, p. 1.

69. Ibid., May 4, 1923, p. 1.

70. Scipio Jones to R. E. Lee Wilson, May 25, 1932; and R. E. Lee Wilson to Scipio Jones, May 27, 1932, both in Lee Wilson & Company Correspondence, 1932, I–Mc, Folder Scipio Jones.

71. In May of 1933, Shaver owed Wilson-Ward Company the staggering amount of $7,586.91, and by July Wilson was threatening to foreclose. See R. E. L. Wilson to W. W. Shaver, May 15 and 17, 1933; July 29, 1933, in Wilson Correspondence, Box 1933, S–To, Folder W. W. Shaver.

72. Thomas N. Dysart to Lee Wilson, November 28, 1932; Lee Wilson to Thomas Dysart, November 29, 1932; Lee Wilson to W. W. Shaver, November 29, 1932; and Roy Wilson to Thomas Dysart, November 29, 1932, all in Wilson Correspondence, Box 1932, D–E, Folder Thomas N. Dysart.

73. F. A. Gillette, Wilson's executive officer, wrote to Carpenter asking him to vacate the house he occupied in Wilson (Wilson to H. B. Carpenter, December 14, 1932, Wilson Correspondence, Box 1932 C, Folder H. B. Carpenter), having just read of the hijacking charges being preferred against Mr. Jake Thrailkill and Harry Carpenter. C. H. Welch to F. A. Gillette, December 5, 1932, Box We–Z, Folder C. H.Welch. Records show, moreover, that both men had been having difficulty paying their obligations to Lee Wilson & Company—their rent and their accounts in the company stores—a factor that almost certainly contributed to their harassment of Virgil Branch in the first place. A number of dunning notices had been delivered to both Carpenter and Thrailkill in the months prior to the incident. On February 18, 1932, Lee Wilson & Company executive W. F. "Hy" Wilson (an employee, not a family member), wrote to Carpenter: referring to "quite an old balance and Mr. Wilson is very insistent that you now take care of same." Subsequent correspondence that year indicates that Carpenter was unable to comply. On February 22, 1932, Hy Wilson wrote Thrailkill about a balance owing. Subsequent correspondence indicates that he paid some of it off but remained in debt. See W. F. Wilson to Harry Carpenter: February 18, June 17, July 13 and 29, August 19 and 26, September 1 and 23, 1932, and January 12, 1933, all in Wilson Correspondence, Box 1932 C, Folder Harry Carpenter. See also W. F. Wilson to Jake Thrailkill, February 9 and 22, and August 19, 1932, and January 18 and 26, 1933.Wilson Correspondence, Box 1932, Sti–WA, Folder Jake Thrailkill.

CHAPTER SIX

Chapter title is a line from Randy Newman, "Louisiana, 1927," *Good Ole Boys,* 1974.

1. Pete Daniel, *Deep'n as It Come: The 1927 Mississippi River Flood* (reprint, Fayetteville: University of Arkansas Press, 1994), p. 5.

2. John M. Barry, *Rising Tide: The Great Mississippi River Flood of 1927 and How It Changed America* (New York: Simon & Schuster, 1997), pp. 90–91.

3. St. Francis Levee District, *A History of the Organization and Operations of the Board of Directors, St. Francis Levee District of Arkansas, 1893–1945* (West Memphis: St. Francis Levee District, 1945), pp. 162–67, 179–80.

4. Barry, *Rising Tide*, p. 179.

5. In the end, ten such crests, an unprecedented number, accompanied the 1927 flood. As Pete Daniel puts it, "Statistics tell part of the story: 16,570,627 acres flooded in 170 counties in seven states, $102,562,395 in crop losses, 162,017 homes flooded, 41,487 buildings destroyed, 5,934 boats used in rescue work, 325,554 people cared for in 154 Red Cross camps and 311,922 others fed by the Red Cross in private homes, and between 250 and 500 people killed." *Deep'n as It Come*, pp. 7–8, 177–78.

6. Ibid., p. 7.

7. St. Francis Levee District, *History of the Organization*, pp. 167–68.

8. Elliott B. Sartain, *It Didn't Just Happen* (Osceola, Ark.: Grassy Lake and Tyronza Drainage District No. 9, 197?), pp. 40, 48, 50. See also *Buffalo Island Drainage District No. 16 of Mississippi County . . . then and now* (n.d. n.p.), pp. 7–14.

9. C. E. Collins, "Arkansas, 'Wonder State,' Acclaims 170,000-Acre Project," *National Reclamation Magazine* (March 1922), p. 64.

10. Sartain, *It Didn't Just Happen*, p. 50.

11. *Osceola Times*, April 29, 1927, p. 1. For background information on Otto Kochtitzky, see Otto Kochtitzky, *Otto Kochtitzky: The Story of a Busy Life* (Cape Girardeau, Mo.: Remfre Press, 1957).

12. Ethel C. Simpson, "Letters from the Flood," *Arkansas Historical Quarterly* 55 (Autumn, 1996): 251–85.

13. Ibid., p. 48.

14. *Osceola Times*, April 1, 1927, p. 5.

15. Ibid.; Agricultural Extension Service, County Agent Report, North Mississippi County, J. E. Critz, December 1, 1927 (for period February 1 to December 1, 1927), p. 3; Agricultural Extension Service, County Agent Report, South Mississippi County, Stanley D. Carpenter, December 1, 1927 (for period December 1, 1926, to December 1, 1927), p. 57. Although Carpenter was responsible for south Mississippi County, the need was so great after the breach to the Big Lake levees, he assisted his compatriot J. E. Critz in north Mississippi County for the duration of the emergency.

16. Donna Brewer Jackson, *Manila: 1901–2001, A Centennial Celebration* (n.p., 2001), p. 17.

17. *Osceola Times*, April 29, 1927, p. 1. The Kochtitzky levee was named for the engineer who designed it, Otto Kochtitzky. *Southeast Missourian*, July 4, 1976.

18. George M. Moreland, "Flood Area Pictured by Moreland," *Osceola Times*, May 13, 1927, p. 2.

19. *Buffalo Island, Drainage District No. 16 of Mississippi County Arkansas*, p. 5.

20. Red Cross Files, Box 735, File 224.08, NA, p. 1.

21. Red Cross, Box 736, File 224.08, NA, "Weekly Statistical Report, Arkansas Reconstruction Office, Little Rock," Week ending August 6, 1927. According to this report, 2,276 whites and no blacks sought shelter in camps; 10,351 and 935 African Americans received aid outside of camps.

22. Barry, *Rising Tide*, pp. 234–37. Randy Newman, "1927 Flood."

23. Oscar Alexander appears on the 1920 Manuscript Census of Population as a thirty-six-year-old farmer in Bowen Township. For the two reports by Simonson, see *Osceola Times*, June 3, 1927, p. 1, and July 8, 1927, p. 4. A separate editorial on July 8, 1927, accompanied Simonson's report and lamented the lawlessness. Part of Clear Lake Township was within District 9 boundaries, but the problem, according to Simonson, was caused by those whose lands were in District 17.

24. Dawes's experience is chronicled in an essay written by Ethel C. Simpson, who presented long excerpts from his letters. See Ethel C. Simpson, "Letters from the Flood," *Arkansas Historical Quarterly* 55 (Autumn, 1996): 251–85.

25. Ibid., pp. 252–53.

26. Ibid., pp. 277–78.

27. Ibid., pp. 281–82.

28. Red Cross, Box 735, File DR 224.08, NA, "General Statistics Concerning Extent of Flood and Losses."

29. Red Cross, Box 741, File 224.6201, NA, "List of Persons Drowned and Death by Disease, Mississippi Valley flood—G-224."

30. Red Cross, Box 741, File DR 224.63, NA, "American National Red Cross, Mississippi Valley Flood Relief, Persons cared for in Camp," p. 39.

31. Red Cross, Box 736, File 224.08, NA, "Weekly Statistical Report, Arkansas Reconstruction Office, Little Rock," Week ending August 6, 1927.

32. *Osceola Times,* May 20, 1927, p. 1.

33. He reported, nevertheless, that the five-year averages (1922–26) of cases reported (not deaths) were typhoid, 897; smallpox, 251; and pellagra, 416. The totals for the first eight months of 1927 were typhoid, 450; smallpox, 122; and pellagra, 757. Dr. William DeKleine, Acting Medical Director, American Red Cross, Washington, D.C., from C.W. Garrison, Little Rock, Arkansas, September 3, 1927. In Red Cross, Box 738, Folder 224.11/5, NA.

34. Red Cross, "Report of Mary Emma Smith, Field Nurse, Arkansas State Board of Health Thirty Day Program, Public Health Nursing, Arkansas Flood Relief, 1927." p. 4.

35. Red Cross, Box 735, Folder 224.08, NA, "Narrative Report of Nursing Activities, Week Ending July 30, 1927."

36. Red Cross, Box 735, Folder 224.08, NA, "Weekly Narrative Report for State of Arkansas, State Reconstruction Office, Little Rock," Week ending July 30, 1927.

37. Red Cross document, no title, Box 741, Folder 224.5/04, NA, p. 11.

38. Red Cross, DeWitt Smith, Assistant to Vice Chairman, American Red Cross, to Mr. Fiser, September 2, 1927, Box 740, File 224.5, NA, p. 3. See also Red Cross, Box 735, File 224.08, NA, "Weekly Narrative Report for State of Arkansas, State Reconstruction Office, Little Rock. For distribution of yeast, see "Memorandum of Conference with Doctor W. R. Redden, national medical officer, American Red Cross, Saturday, June 4, 1927," Box 741, Folder 224.5/04, Red Cross Records, NA.

39. Simpson, "Letters from the Flood," p. 255.

40. Red Cross, no title (words scribbled on top say "My file under Disaster Conference"), Box 741, Folder 224.5/04, NA, pp. 6–10.

41. Ibid.

42. Ibid., pp. 7–8.

43. Ibid., pp. 7–9.

44. In northwest Mississippi County alone, 230 horses and mules, 210 cows, 610 hogs, and 12,230 chickens died, either from drowning or from starvation or disease. United States Department of Agriculture, Bureau of Agricultural Economics, Washington, D.C. "Lower Mississippi River Flood, May–July 1927," in Red Cross Records, Box 224, Folder 224.02, NA, p. 10. Note that the Mississippi River Flood Control Association reported vastly different numbers: 25 head of horses and mules, 25 head of cattle, 600 head of hogs and 30,000 poultry. See Mississippi River Flood Control Association, *Losses and Damages Resulting from the Flood of 1927: Mississippi River and Tributaries, in the States of Illinois, Missouri, Kentucky, Tennessee, Arkansas, Mississippi and Louisiana* (Memphis: Mississippi River Commission, n.d.), p. 33. The commission secured its numbers from "men of high standing" within the communities they surveyed. It is likely that the Bureau of Agricultural Economics is the more reliable source.

45. Red Cross, Box 736, Folder DR 224.08, NA, "Cotton Ginnings for the State of Arkansas, 1927–1926."

46. J. E. Critz, "Narrative Report of County Extension Agent," Mississippi County [North], February 1, 1927 through December 1, 1927, pp. 6–7.

47. Of that amount, about "$30,000.00 was spent for seed and hog cholera serum." The seed distributed to the county's farmers by the Red Cross included "3 cars of sudan grass, 3 cars of cowpeas, 2 cars of soy beans, 2 cars of sorghum and sagrain, 1 car seed corn, 2200 packages of garden seed, 1 car rye and rape, thousands and thousands of sweet potato, tomato, and cabbage plants and $2000.00 worth of hog cholera serum." Carpenter argued, in fact, that the farmers who remained on their farms were "in better shape financially . . . than they have been for several years." They were able to make a "fair crop of cotton, a good crop of feed (hay and corn) and from their late fall gardens there is more canned goods on their pantry shelves than they have had for years." Stanley D. Carpenter, Narrative Report of County Extension Agent, Mississippi County [South], December 1926 to December 1, 1927, pp. 67–68.

48. Ibid., pp. 26–27.

49. Cotton prices had been fluctuating wildly since the postwar recession, but after 1927 were to begin a steady decline till they reached rock bottom in 1931: 5.7 cents per pound. *Historical Statistics of the United States: Colonial Times to 1970*, Part I, U.S. Department of Commerce, Bureau of the Census, 1975, p. 517.

50. "Short History of the Arkansas Farm Credit Co.," Box 742, Folder 224.6511, Red Cross Records, NA.

51. W. Wesselius to Robert E. Bondy (Asst. To the Vice Chairman, American National Red Cross, Washington, D.C.), March 10, 1928, Box 742, Folder 224.6511, Red Cross Records, NA.

52. Red Cross, "Arkansas Farm Credit Company, Little Rock, Arkansas, Recapitulation of Losses as Per Detailed Statement of Loans, 1928," p. 1, Box 742, Folder 224.6511, NA.

CHAPTER SEVEN

1. According to county real property tax records, Wilson owned 28,346.01 acres in 1920; he owned 46,458 acres in 1933. Real Property Records, 1920, 1933, Mississippi County Courthouse, Osceola.

2. Cecil Shane to William L. Bourland, March 7, 1935, Oscar Fendler Papers, Box 1, Folder 2, University of Arkansas Library, Fayetteville. Manuscript Census of Population, Stuttgart, Arkansas County, Arkansas, 1930. In 1930 Helen May Vaughan was living with her parents and two young sons in Stuttgart and listed herself as twenty-five years old and divorced.

3. Frank Gillette to J. M. Woods (vice president and trust officer with the Franklin-American Trust Company) on May 25, 1932, Lee Wilson & Co., 1932 F, Folder: Franklin-American Trust Company.

4. R. E. Lee Wilson to Dr. Thomas D. Moore, March 30, 1932 (regarding an "electrical examination" scheduled in early April). See Lee Wilson & Co., Folder: Thomas. D. Moore, M.D. (Memphis); also see R. E. Lee Wilson to Dr. King Wade, April 15, 1932, Lee Wilson & Co., Folder: Dr. King Wade (Hot Springs).

5. Lee Wilson to Marie Wilson Harmon, October 6, 1932, Lee Wilson & Co., Folder: Marie Wilson Harmon.

6. Wilson to Benjamin Lang, May 10, 1932, Lee Wilson & Co., Folder: First National Bank in St. Louis, Mo.

7. Wilson to R. B. Barton, Memphis Bank of Commerce and Trust, December 14, 1932. See Lee Wilson & Co., 1932, B, Folder: Bank of Commerce & Trust.

8. Paul V. Galloway to Lee Wilson, January 17, 1933, Lee Wilson & Co., 1932, G–H, Folder: Rev. Paul V. Galloway. See also Galloway's obituary, *New York Times*, August 7, 1990.

9. Mrs. Walter Wood to R. E. Lee Wilson, August 12, 1932, Lee Wilson & Co., 1932, Folder: Mrs. Walter Wood.

10. Wilson to John B. McFerren, March 18, 1932, Lee Wilson & Co., Folder: J. B. McFerren.

11. Wilson to John H. Brinkerhoff, December 5, 1932, Lee Wilson & Co., Folder: John H. Brinkerhoff; R. E. Lee Wilson to Ben Lang, May 10, 1932, Lee Wilson & Co., Folder: First National Bank; and R. E. Lee Wilson to Clark E. Butler, July 6, 1932, Lee Wilson & Co., Folder: Clark E. Butler.

12. *Wilson Mirror,* April 1, 1930, p. 3.

13. Frank A. Gillette to Cape Girardeau Bridge Company, April 22, 1932, Lee Wilson & Co., Folder: Cape Girardeau Bridge Company.

14. W. E. Miller (Central Shoe Company) to Frank A. Gillette, March 24, 1932; Frank A. Gillette to W. E. Miller, March 11 and March 25, 1932. All three letters are located in Lee Wilson & Co., Folder: Central Shoe Company.

15. There are eleven thick files dealing with Zander's purchases. See Lee Wilson & Co., 1932, Folders: F. L. Zander. One file includes several letters concerning Peter's Island. See Lee Wilson & Co., 1932, Folder: Peter's Island Correspondence.

16. Frank A. Gillette to Jesse Clinton, July 29, 1932, Lee Wilson & Co., Folder: Idaho Grocery Co. For a list of managers, see *Wilson Mirror,* April 1, 1930.

17. Roy Wilson to Oscar Hill, October 26, 1932, Lee Wilson & Co., Folder: Hill & Wilson.

18. Wilson to Charles Crigger, October 15, 1932, Lee Wilson & Co., Folder: Lee Wilson & Co., Armorel; and Wilson to Jesse Clinton, October 15, 1932, Lee Wilson & Co., Folder: Idaho Grocery Co. See also Wilson to Charles Crigger, January 11, 1933, Lee Wilson & Co., Folder: Lee Wilson & Co., Armorel.

19. Wilson to Charles Crigger, January 10, 1933, Lee Wilson & Co., 1932, We–Z, Folder: Lee Wilson & Co., Armorel.

20. *Wilson Mirror,* May 1930, p. 4.

21. Wilson to C. E. Crigger, August 9, 1932, Lee Wilson & Co., Folder: Lee Wilson & Co., Armorel.

22. Ibid., March 17, 1932.

23. Wilson to W. C. Wall, July 15, 1932, Lee Wilson & Co., Folder: Lee Wilson & Co., Armorel.

24. Wilson to E. N. Ahlfeldt, July 14, 1933, July 24, 1933, Lee Wilson & Co., 1933, A, Folder: Ernest Ahlfeldt.

25. Wilson wrote to Futrell on August 24, 1932; see R. E. Lee Wilson to J. M. Futrell, August 24, 1932, Lee Wilson & Co., Folder: J. M. Futrell. He also wrote executives at Chapman and Dewey Lumber Co. (September 5, 1932), Dillard and Coffin Co. (September 1, 1932), Fargason & Co. (September 1), and Hughes and Davis (September 3), asking them to contact Futrell about the appointment of Hidinger.

26. John H. Brinkerhoff to Frank Gillette, March 15, 1932, Lee Wilson & Co., 1932, B, Folder: Brinkerhoff. Brinkerhoff had received a copy of the financial statement and wrote to compliment Gillette on the statement.

27. See F. A. Gillette folder, which contains typewritten notes listing the amount owed to each entity and the collateral pledged to secure the loans. Some of the documents are undated but some of them are dated September 1, 1932. Lee Wilson & Co., 1932, G–H, Folder: F. A. Gillette. For correspondence with the Bank of Commerce concerning the extension of the past due notes and the need for additional collateral, see R. B. Barton to Lee Wilson, November 25, 1932; Lee Wilson to R. B. Barton, November 26, 1932; and R. B. Barton to Lee Wilson, November 28, 1932, in Lee Wilson & Co., 1932, B, Folder: Bank of Commerce & Trust Co.

28. The bank loan was secured by the following items:

744 shares Bank of Wilson stock

498 shares Kansas City Shook & Mfg. Co. stock

344 shares Wilson-Ward Co. stock

30 shares Farmers Bank & Trust Co. stock

30 shares Marked Tree Compress & Warehouse Co. stock

50 shares Caraway Light & Ice Co. stock

33 shares Bank of Keiser stock

100 shares Wilson Motor Co. stock

$7,600 Lee Wilson & Co. bonds

$58,000 Wilson school bonds

$24,500 Keiser school bonds

Deed of Trust on 4,586.97 acres in the Chickasawba District

29. Wilson to H. C. Couch, RFC, Washington, D.C., February 8, 1932; Harvey Couch to Lee Wilson, February 16, 1932, both in Lee Wilson & Co., 1932, C, Folder: Harvey C. Couch. Roy Wilson to R. B. Barton, February 20, 1932, Lee Wilson & Co., 1932 B, Folder: Bank of Commerce.

30. Harry N. Scheiber, Harold G. Vatter, Harold Underwood Faulkner, *American Economic History,* 9th ed. (New York: Harper & Row, 1976), p. 362 ("It was now empowered to lend to states and to public and private agencies to promote self-liquidating projects of public benefit").

31. Wilson to H. C. Couch, December 7, 1932, in Lee Wilson & Co. 1932 C, Folder: H. C. Couch.

32. Wilson to Hattie Caraway, November 5, 1932; Garrett Whiteside from Lee Wilson, November 10, 1932; Elbert Smith to Garrett Whiteside, November 8, 1932; Lee Wilson to Garrett Whiteside, November 12, 1932; Garrett Whiteside to Lee Wilson, November 15, 1932. All in Lee Wilson & Co., 1932, C, Folder: Hattie Caraway.

33. For reference to the suggestion of the RFC, see Lee Wilson telegram to Robenock (or Ribenock) of the RFC, January 24, 1933, referring to "your Jan 19th" communication. In Lee Wilson & Co., 1932, Folder: Western Union Telegrams. For Bachelder references, see Lee Wilson to James L. Ford, January 7, 1933; William C. Bachelder to James L. Ford, January 10, 1933; James L. Ford to Lee Wilson, January 11, 1933; Lee Wilson to William C. Bachelder, January 24, 1933; William C. Bachelder to Lee Wilson, January 26, 1933; William C. Bachelder to James L. Ford, February 6, 1933; James L. Ford to Lee Wilson, February 7, 1933; William C. Bachelder to Lee Wilson, February 19, 1933. All in Lee Wilson & Co., 1932 or 1933, Folders: William C. Bachelder or First National Bank of St. Louis (James Ford is a vice president at that bank). For reference to the Colorado venture, see W. C. Bachelder to Lee Wilson, February 10, 1933, B, Folder: William C. Bachelder.

34. For reference to Sager's attorney friend, see Lee Wilson to C. Leroy Sager, November 12, 1932, and C. Leroy Sager to Lee Wilson, November 19, 1932, in Lee Wilson & Co., 1932, Folder; First National Bank.

35. See, for example, letters to Charles T. Coleman, J. B. & C. E. Daggett, and to his daughter, Victoria, all dated November 9, 1932, in Lee Wilson & Co., 1932, C, Folder: C. T. Coleman; Lee Wilson & Co., 1932, Ni–Ro, Folder: Peter's Island Correspondence; and Lee Wilson & Co., 1932, WE–Z, Folder: Frank H. and Victoria Wilson Wesson.

36. Wilson to Mercantile Commerce Bank, St. Louis, December 30, 1932, in Lee Wilson & Company, 1932, Folder: Mercantile Commerce Bank.

37. Wilson to Mercantile Commerce Bank, St. Louis, December 30, 1932, in Lee Wilson & Company, 1932, Folder: Mercantile Commerce Bank.

38. M. W. H. Collins to Wilson, December 13, 1932, Lee Wilson & Co., 1933, C, Folder: M. W. H. Collins.

39. W. L. Oates to Wilson, February 13, 1933, in Lee Wilson & Co., 1933, Folder: McFadden & Oates. Mention of this loan is also found in Lee Wilson to James Ford, April 1, 1933, in Lee Wilson & Co., 1933, Folder: First National Bank, St. Louis.

40. James L. Ford Jr. to Wilson, April 3, 1933, Lee Wilson & Co., 1933, Folder: First National Bank, St. Louis.

41. Charles T. Coleman to Wilson, February 10, 1933, Lee Wilson & Co., 1933, Folder: Charles T. Coleman.

42. James L. Ford to Wilson, February 20, 1933, in Lee Wilson & Co., 1933, Folder: First National Bank, St. Louis; Wilson to Ben Lang, March 1, 1933, Lee Wilson & Co., 1933, Folder: First National Bank, St. Louis; Wilson to James Ford, April 1, 1933, in Lee Wilson & Co., 1933, Folder: First National Bank, St. Louis.

43. Wilson to James Ford, April 1, 1933, in Lee Wilson & Co., 1933, Folder: First National Bank, St. Louis.

44. Amon Carter to Jesse M. Jones, April 5, 1933, in Lee Wilson & Co., 1933, Folder: George Thompson Jr.

45. W. L. Clayton to Jesse H. Jones, April 6, 1933, in Lee Wilson & Co., 1933, Folder: Anderson Clayton & Co.

46. Wilson to Harvey Couch, April 18 and 19, 1933, in Lee Wilson & Co., 1933, C, Folder: H. C. Couch.

47. Ben Lang to Wilson, May 2, 1933; Frank Gillette to Ben Lang, May 3, 1933; Ben Lang to Frank Gillette, May 4, 1933; Frank Gillette to Ben Lang, May 5, 1933. All in Lee Wilson & Co., 1933, Folder: First National Bank, St. Louis.

48. For a reference to the consummation of the loan, see Ben Lang to Lee Wilson, May 6, 1933, in Lee Wilson & Co., 1933, Folder: First National Bank, St. Louis. See also Wilson to Charles T. Coleman, May 13, 1933, in Lee Wilson & Co., 1933, Folder: Charles T. Coleman; and Wilson to James Ford, May 26, 1933, in Lee Wilson & Co., 1933, Folder: First National Bank, St. Louis.

49. Federal Reserve Bank of St. Louis to W. F. Wilson, June 19, 1933, Lee Wilson & Co., F–Go, Folder: Federal Reserve Bank of St. Louis, Little Rock Branch.

50. For reference to the taxes and improvements bonds, see Lee Wilson to Lee Wilson & Company, April 11, 1933, Lee Wilson & Co., 1933, TR–Z, Folder: R. E. Lee Wilson Sr.

51. James Ford to Wilson, June 19, 1933; Lee Wilson to James Ford, June 20, 1933, in Lee Wilson & Co., 1933, Folder: First National Bank of St. Louis. Lee Wilson to A. O. Merriam (Franklin American Trust Company), June 20, 1933, Lee Wilson & Co., 1933, Folder: Franklin American Trust Company; Lee Wilson to George W. Wilson and Arthur F. Barnes (Mercantile Commerce Bank, Memphis), June 20, 1933; and Lee Wilson to George W. Wilson, June 21, 1933, Lee Wilson & Co., Folder: Mercantile Commerce Bank, Memphis. Ford's bank received the $194,825 owed them just as the Mercantile Commerce Bank and Trust Company took delivery of the $143,663.33 Lee Wilson & Company owed them. The Franklin American Trust Company of St. Louis was the recipient of $120,000, and two banks in Memphis, the Bank of Commerce & Trust and Union Planters Bank, received $120,000 and $125,000 respectively.

52. See especially James Cobb and Michael Namaroto, eds., *The New Deal and the South* (Jackson: University of Mississippi Press, 1984); Anthony J. Badger, *The New Deal: The Depression Years, 1933–1940* (Chicago: Ivan R. Dee, 2001); David E. Hamilton, *From New Day to New South: American Farm Policy from Hoover to Roosevelt, 1928–1933* (Chapel Hill: University of North Carolina Press, 1991); William E. Leuchtenburg, *Franklin D. Roosevelt and the New Deal, 1932–1940* (New York: Harper & Row, 1966); and Theodore Saloutos, *The American Farmer and the New Deal* (Ames: Iowa State University Press, 1982).

53. Wilson telegram to Henry A. Wallace, June 20, 1933; Cully A. Cobb telegram to Wilson, June 22, 1933, in Lee Wilson & Co., 1933, Tr–Z, Folder: Henry A. Wallace. Lee Wilson to Oscar Johnston, September 6, 1933, in Lee Wilson & Co., 1933, Gr–J, Folder: Oscar Johnston. For references made in correspondence concerning newspaper articles in the Memphis *Commercial Appeal* and the *Arkansas Gazette*, see Mrs. Charles B. Box to Lee Wilson, June 28, 1933, in Lee Wilson & Company, 1933, D–E, Folder: Chas. B. Box; Estate of T. Roy Reid to Lee Wilson, June 23, 1933, in Lee Wilson & Co., 1933,

O–R, Folder: T. Roy Reid. For secondary literature on two of the key players in the AAA, see Roy V. Scott and J. G. Shoalmire, *The Public Career of Cully A. Cobb: A Study in Agricultural Leadership* (Jackson: University and College Press of Mississippi, 1973); and Lawrence J. Nelson, *King Cotton's Advocate: Oscar G. Johnston and the New Deal* (Knoxville: University of Tennessee Press, 1999).

54. Bob Robinson to Wilson, July 11, 1933, in Lee Wilson & Co., 1933, K–Mc, Folder: Keiser Supply Co.

55. Stanley D. Carpenter, Narrative Report of County Extension Workers, South Mississippi County, November 30, 1933, p. 6, and J. E. Critz, Narrative Report of County Extension Workers, North Mississippi County, November 30, 1933, pp. 6–8, 10–11.

56. *Historical Statistics of the United States: Earliest Times to the Present, Millennial Edition, Volume III, Part C., Economic Structure and Performance* (Cambridge: Cambridge University Press, 2006). These figures represent the "spot" price. The price rose to twelve cents in 1934–36, dropped back to eleven cents and nine cents in 1937–38, and was at ten cents per pound in 1939–40 before jumping to fifteen cents in 1941.

57. Wilson to J. E. Counts, June 29, 1933, Lee Wilson & Co., 1933, C, Folder: J. E. Counts.

58. Oscar Hill to Wilson, July 11, 1933, Lee Wilson & Co., 1933, Gr–J, Folder: Hill and Wilson; Evadale Mercantile Company, July 10, 1933, signed by George Charlton, in Lee Wilson & Co., 1933, D–E, Folder: Evadale Mercantile Company; Pee Wee Morris to R. E. Lee Wilson, August 12, 1933, in Lee Wilson & Co., 1933, O-R, Folder: Pee Wee Morris Weekly Reports.

59. Dr. J. A. Crisler, Memphis, to W. L. Bourland, Chicago, October 12, 1934; Cecil Shane to William L. Bourland, October 23, 1934; William L. Bourland to Cecil Shane, October 30, 1934; William L. Bourland to Cecil Shane, March 4, 1935; Cecil Shane to William L. Bourland, March 7, 1935. Oscar Fendler Papers, Box 1, Folder 2, Special Collections, University of Arkansas Library, Fayetteville.

60. Crisler to Bourland, October 12, 1934.

CHAPTER EIGHT

1. In 1950 the McFerren plantation, which Lee Wilson had renamed Victoria, was acquired by its namesake, Victoria Wilson Wesson, in an agreement to partially liquidate the trust in 1950. It then consisted of 7,766.52 acres.

2. Lee Wilson's "Last Will," Fendler Papers, Box 1, Folder 4; *Commercial Appeal*, March 7, 1934, p. 3.

3. *Osceola Times*, October 13, 1933.

4. George Thompson Jr. to Roy Wilson, December 26, 1933, Lee Wilson & Co., 1933, Tr–Z, Folder: George Thompson Jr.

5. Roy Wilson to Thompson, December 29, 1933, Lee Wilson & Co., 1933, Tr–Z, Folder: George Thompson Jr.

6. Thompson to Roy Wilson, January 2, 1934, in Lee Wilson & Co., 1933, Tr–Z, Folder: George Thompson Jr.

7. Ibid.

8. "Crain and Enochs Families of Rankin County, Mississippi, Combined and published by Glenn A. Railsback III, 512 South Pine St., Pine Bluff, Ark., Dec. 1996" (copy in possession of author). See also the Manuscript Census of Population, Rankin County, Mississippi, 1900. Jim Crain is listed on that census, recorded on June 4, 1900, as the eleven-year-old son of John and Hattie Crain. Jim Crain is listed as a "farm laborer" who attended six months of school since the previous census.

9. Roy Wilson to Gus Thompson, October 5, 1936, Lee Wilson & Co., 1936, T–V, Folder: Gus Thompson. George Thompson to Roy Wilson, November 13, 1936, Lee Wilson & Co., 1936, T–V, Folder: George Thompson. Roy Wilson to George Thompson Jr., December 29, 1933, Lee Wilson & Co., 1933,

Tr–Z, Folder: George Thompson Jr. In 1939 Roy reserved three stalls again (and may have in the years between); see Roy Wilson to Paducah Fox Hunters, August 11, 1939, Lee Wilson & Co., 1939, D–G, Folder: Paducah Fox Hunters, Inc. See also Roy Wilson to E. S. Mills, October 30, 1939, Lee Wilson & Co., 1939, H–Mo, Folder: E. S. Mills.

10. The bond issue is mentioned in a letter from Lee Wilson to G. B. Rose, D. H. Cantrell, J. F. Loughborough, A. W. Dobyns, A. F. House, J. W. Barron, William Nash (the Rose law firm in Little Rock), March 24, 1933, in Lee Wilson & Co., 1933, O–R, Folder: Rose, Hemingway, Cantrell & Loughborough. Apparently the Rose law firm handled the matter for Lee Wilson in 1922.

11. Lee Wilson had correspondence with several firms during 1933 asking for their help in securing Lee Wilson & Company bonds at fifty to sixty cents on the dollar. See William R. Humphrey (president, Humphrey, Jacques and Company, securities, St. Louis) to Lee Wilson, January 28, 1933; Lee Wilson to Humphrey, February 17, 1933; Lee Wilson to Humphrey, June 21, 1933; Humphrey to Lee Wilson, September 1933, all in Lee Wilson & Co., 193, Gr–J, Folder: Humphrey, Jacques and Company. See also Lee Wilson to Peltason, Tenenbaum and Harris (securities dealers in St. Louis), May 9, 1933; Harry Tenenbaum to Lee Wilson, May 10, 1933; Paul Peltason to Lee Wilson, June 13, 1933; Lee Wilson to Peltason, June 20, 1933; Tanenbaum to Lee Wilson, July 7, 1933, all in Lee Wilson & Co., 1933, O–R, Folder; Peltason, Tenenbaum & Harris. See also Jean Stahl to Lee Wilson (Stahl was secretary to Thomas Dysart of Knight, Dysart and Gamble, Investment Securities, St. Louis), April 4, 1933, Lee Wilson & Co., 1933, K–Mc, Folder: Knight, Dysart & Gamble. See also Lee Wilson to Harris, Upham and Company, St. Louis, June 21, 1933, Lee Wilson & Co., 1933, Gr–J, Folder: Harris, Upham & Co. See also B. A. Lynch (president, Farmers Bank and Trust, Blytheville) to E. C. Coffman (vice president, First National Bank, St. Louis), January 28, 1933, and February 13, 1933, Lee Wilson & Co., 1933, F–Go, Folder; Farmers Bank & Trust Co. See also Lee Wilson to Thomas Dysart, May 7, 1933, May 15, 1933, in Lee Wilson & Co., 1933, D–E, Folder: Thomas N. Dysart. See also Lee Wilson to E. R. Bruce Company, February 22, 1933, Lee Wilson & Co., 1933, B, Folder: E. R. Bruce Co. See also Lee Wilson to John P. Sweeney, May 26, 1933, Lee Wilson & Co., 1933, B., Folder: Bonds, Misc.

12. National Candy Company to Roy Wilson, December 8, 1933, Lee Wilson & Co., 1933, Me–N, Folder: National Candy Company. (The signature of the executive cannot be clearly read.) K. E. Penzler, an investment analyst who represented another bondholder, addressed his concerns to Jim Crain on December 7, 1933, and, like the candy company executive, indicated that Lee Wilson had assured him that the RFC loan "provided for the retirement of these obligations." K. E. Penzler to Jim Crain, December 7, 1933, Lee Wilson & Co., 1933, Me–N, Folder: Mississippi Valley Trust Company.

13. Roy Wilson to H. C. Couch, January 19, 1934, Lee Wilson & Co., 1933, O–R, Folder: Reconstruction Finance Corporation. See also reference to this in Roy Wilson to George Thompson Jr., January 10, 1934, Lee Wilson & Co., 1933, Tr–Z, Folder: George Thompson Jr.

14. James H. Crain to Farquar (Chief Disbursement Office, Cotton Section, Washington, D.C.), September 19, 1935, Lee Wilson & Co. 1935, E–Gin, Folder: Farquar. For the quote, see Hy Wilson to M. K. Noell, February 6, 1939, Lee Wilson & Co., 1939, D–G, Folder: M. R. Noell.

15. James H. Crain to Guthrie King, July 29, 1935, Lee Wilson & Co. 1936, He–Ki, Folder: Guthrie King. Although this letter is in the 1936 folder, it is dated 1935.

16. Decree, Estate of R. E. Lee Wilson, Deceased, March 5, 1934, Probate Court of the Osceola District of Mississippi County, Arkansas. Arkansas Historic Preservation Program, History and Architecture (see http://www.arkansaspreservation.com/historic-properties, the King-Neimeyer-Mathis House). Jack Vilas of Hot Springs and his friend D. D. King, of Chicago, purchased the property in 1917 and built the home in craftsman style in 1917–18. Wilson purchased the home in 1926.

17. "Useful Facts about Historic Bathhouse on Bathhouse Row," National Park, U.S. Department of the Interior, Hot Springs National Park, Arkansas. See Manuscript Census of Population, 1930, Hot

Springs, Garland County, Arkansas, which lists Ernest Latta as a real estate developer and married to a Lenora Latta. According to the manuscript census of population for the town in 1920, he was an express operator and married to Lenora.

18. W. F. Wilson to Jim Crain, October 28, 1935, in Lee Wilson & Co., 1935, Com–D, Folder: J. H. Crain; W. F. Wilson to Credit Bureau of St. Louis, Lee Wilson & Co., 1935, Com–D, Folder: Credit Bureau of St. Louis; Oscar Fendler to W. F. Wilson, August 5, 1936, in Lee Wilson & Co., 1936, R–Sha, Folder: Cecil Shane. F. Wilson to Thomas W. Vinton (Union Planters National Bank), July 29, 1937; Vinton to Crain and Wilson, August 4, 1937; Crain and Wilson to Union Planters, August 5, 1937; Vinton to Wilson, October 1, 1937. All in Lee Wilson & Co., 1937, T–V, Folder: Helen May Vaughan (Latta).

19. In a November letter Helen mentions having suffered the illness earlier that year; see Helen May Latta to Thomas W. Vinton, November 8, 1937, Lee Wilson & Co., 1937, T–V, Folder: Helen May Vaughan (Latta).

20. Latta to W. F. Wilson, telegram, May 9, 1937, in Lee Wilson & Co., 1937, T-V, Folder: Helen May Vaughan (Latta).

21. Latta to Vinton, November 8, 1937, Lee Wilson & Co., 1937, T–V, Folder: Helen May Vaughan (Latta).

22. Thomas W. Vinton to Jim Crain and W. F. Wilson, April 10, 1939, Lee Wilson & Co., 1939, H–Mo, Folder: Latta Trust; Latta to Jim Crain and W. F. Wilson, August 30, 1939, Lee Wilson & Co., 1939, H–Mo, Folder: Latta Trust.

23. Latta to Crain, W. F. Wilson, and Vinton, April 10, 1940, Lee Wilson & Co., 1940, Gi–L, Folder: Helena May Latta Trust.

24. Phil M. Canale (attorney) to Jim Crain, September 25, 1940, Lee Wilson & Co., 1940, Gi–L, Folder: Helen May Latta Trust.

25. For agricultural credit, see J. H. Crain to Pine Bluff Production Credit Association, March 9, 1936, Lee Wilson & Co., 1936, Ne–Q, Folder: Pine Bluff Production Credit Association. See also J. H. Crain to Pine Bluff Production Credit Association, December 9, 1938, Lee Wilson & Co., 1938, Pi–S, 1938, Folder; Pine Bluff Production Credit Association.

26. Jim Crain to Hattie Caraway, January 27, 1938, Lee Wilson & Co., 1938, C–De, Folder: Sen. Hattie W. Caraway. Garrett Whiteside (assistant to Sen. John Miller) to James H. Crain, September 28, 1938; Thomas M. Tatum (for John E. Miller) to James H. Crain, November 5, 1938; telegram from John E. Miller to James H. Crain, November 5, 1938; telegram from Thomas M. Tatum to C. H. Moses, November 7, 1938; telegram from Thomas M. Tatum to Jim Crain, November 7, 1938. All in Lee Wilson & Co., 1938, M–Ph, Folder: John E. Miller. He also corresponded over a variety of other matters; see Hattie Caraway to Jim Crain, March 30, 1939, April 28, 1939, November 1 and 10, 1939, and Crain to Caraway, October 16 and 18, 1939, and November 8, 1939, all in Lee Wilson & Co., 1939, C, Folder: Hattie Caraway.

27. Roy Wilson to F. L. Dunne Company (New York), February 25, 1936, Lee Wilson & Co., 1936, Dix–Fiv, Folder: F. L. Dunne Co.; Roy Wilson to Brooks Brothers, January 16, 1937, Lee Wilson & Co., 1937, W–Z, Folder: R. E. Lee Wilson Jr.; Roy Wilson to F. L. Dunne Company, March 19, 1938, May 4, 1938, Lee Wilson & Co., 1938, Di–G, Folder: F. L. Dunne Co. Regarding the bird dogs, see Roy Wilson to W. H. Stoval, February 5, 1935, Lee Wilson & Co., 1935, S–Th, Folder: W. H. Stoval. For vacations at Woods Lake, Colorado (near Glenwood Springs), see Roy Wilson to E. P. Englebrecht, March 19, 1938, Lee Wilson & Co., 1938, Di–G, Folder: E. P. Englebrecht, and Roy Wilson to Sidney Farnsworth, April 9, 1938, Lee Wilson & Co., 1938, Di–G, Folder: Sidney Farnsworth. For his Hunt and Polo Club membership, see Dunbar Abston to Roy Wilson, October 21, 1937, Lee Wilson & Co., 1938, M–Ph, Folder: Memphis Hunt & Polo Club (note the letter is misfiled in 1938 folder). For fox hunts in Kentucky, see Roy Wilson to Cobb Hotel, November 2, 1939, Lee Wilson & Co., 1939, C, Folder: Irwin Cobb Hotel.

28. Radcliffe "Nad" Romeyn to Roy Wilson, April 19, 1935, Lee Wilson & Co., 1935, O–R, File: Radcliffe Romeyn; Roy Wilson to Joe Cary, August 27, 1935, Lee Wilson & Co., 1935, To–Z, Folder: R. E. Lee Wilson Jr.; Roy Wilson to Dick Bishop, July 9, 1936, Lee Wilson & Co., 1936, B, Folder: Dick Bishop; Joe B. Cary to Roy Wilson, January 19, 1937, Lee Wilson & Co., 1937, W–Z, Folder: R. E. Lee Wilson Jr.

29. Roy Wilson to Mrs. Thelma Engelbrecht, May 23, 1936, Lee Wilson & Co., 1936, Dix–Fiv, Folder: Mrs. Thelma Engelbrecht.

30. Regarding Arkansas State College Board of Trustees meeting, see Roy Wilson to R. Whitaker, June 23, 1938, Lee Wilson & Co., 1938, A–B, Folder: Arkansas State College. For highway dedication, see Roy Wilson to Jack DeLysle, April 29, 1935, Lee Wilson & Co., 1935, Com–D, Folder: Jack DeLysle.

31. Power of Attorney, July 13, 1935, Lee Wilson & Co., 1935, M–N, Folder: Miscellaneous.

32. See, for example, Frank "Lee" Wesson to E. B. Chiles, October 30, 1936 (about a machine shop bill), Wesson to W. F. Collins, October 30, 1936 (about a balance due of $6), both in Lee Wilson & Co., 1936, C–Cor, Folder: E. B. Chiles. See also Wesson to R. C. Branch, July 13, 1938 (about a balance due, amount not specified), Lee Wilson & Co., 1938, A–B, Folder: R. C. Branch; Wesson to Hugh Dillahunty (concerning a $91.85 vet bill), November 23, 1938, Lee Wilson & Co., 1938, Di–G, Folder: Hugh Dillahunty; Wesson to Richard Ferguson, September 9, 1938 (concerning a $16.77 past-due bill), Lee Wilson & Co., 1938, Di–G, Folder: Richard Ferguson; Lee Wesson to Alice Flemmings, September 8, 1938 (concerning an old vet bill of $6.24), Lee Wilson & Co., 1938, Di–G, Folder: Miss Alice Flemmings; Wesson to Charles McDaniels, September 8, 1938 (concerning a $36.26 past-due account), Lee Wilson & Co., 1938, M–Ph, Folder: Charles McDaniels; Wesson to E. M. Murray, September 8, 1938 (concerning an old balance of $28.51), Lee Wilson & Co., 1938, M–Ph, Folder: E. M. Murray; Wesson to Jim Rounsavall, April 9, 1938 (concerning a past-due account), Lee Wilson & Co., 1938, Pi–S; Wesson to Rev. D. D. Segar, April 9, 1938 (concerning a past-due account—mentions seeing him riding around in a new Ford), Lee Wilson & Co., 1938, Pi–S, Folder: Rev. D. D. Segar. For 1939 correspondence, see Wesson to Howard Spicer, July 17, 1939 (concerning a past-due account), Lee Wilson & Co., 1939, So–V, Folder: Howard Spicer.

33. Wesson to James Warren (Stratton-Warren of Memphis), June 25, 1936 (over an air-conditioning unit returned to Stratton-Warren because it was insufficient to "do the job"), Lee Wilson & Co., 1936, W–Z, Folder: James Warren.

34. Isador Lubin (Commissioner of Labor Statistics) to Lee Wilson & Co., December 12, 1936, in Lee Wilson & Co., 1936, Cot–Dis, Folder: Department of Labor. Lee Wesson's name is appended to this form as the contact person for all matters related to these reports.

35. See Lee Wilson & Co., 1939, W–Z, Folder: Lee Wesson.

36. For correspondence Lee Wilson had in 1932 concerning the possibility of constructing an oil mill, see L. W. Baldwin (Missouri Pacific Lines) to Lee Wilson, October 13, 1932; Lee Wilson to L. W. Baldwin, October 28, 1932; Lee Wilson to Baldwin, November 9, 1932; Baldwin to Lee Wilson, November 15, 1932, in Lee Wilson & Co., 1932, B, Folder: L. W. Baldwin. See also Lee Wilson to French Oil Mill Machinery Company (no date but references "your letter of April 5, 1932"), Lee Wilson & Co., 1932, F, Folder: French Oil Mill Machinery Co. For a description of the work of constructing the oil mill in the mid-1930s, see Whayne interview with William "Snake" Toney, July 6, 2002, p. 21. See also A. B. Carr to James H. Crain, May 2, 1936; Crain to A. B. Carr, May 5, 1936, Lee Wilson & Co., 1936, C–Cor, Folder: A. B. Carr; James H. Crain to Charles T. Coleman, Sept. 5, 1936, Lee Wilson & Co., 1936, C–Cor, Folder: Charles T. Coleman.

37. Crain to Coleman, September 5, 1936, Lee Wilson & Co., 1936, C–Cor, Folder: Charles T. Coleman.

38. Hy Wilson to Roy Wilson (in care of American Express Co., London), August 3, 1937, in Lee Wilson & Co., 1937, W–Z, Folder: R. E. Lee Wilson Jr.

39. Victoria Wilson Wesson to Sumner Reginold, June 29, 1936, Lee Wilson & Co., 1936, W–Z, Folder: Victoria Wilson Wesson.

40. Lee Wilson & Co., interoffice memo, December 17, 1935, Lee Wilson & Co., 1936, Ne–Q, Folder: Prudential Insurance Co. of America.

41. Elizabeth Wilson left her shares of her estate to her three children and the grandson she raised, Joe Wilson Nelson (Roy, 666.66 shares; Victoria, 666.66 shares; Marie, 333.33 shares, and Joe, 333.33 shares). Mississippi County Probate Records, Osceola, Arkansas.

42. Victoria Wilson Wesson to James H. Crain, June 20, 1939, Lee Wilson & Co., 1939, W–Z, Folder: Victoria (Mrs. Frank H.) Wesson.

43. "Application for Guardianship," In the Chancery Court for the Osceola District of Mississippi County, Arkansas, Re: R. E. L. Wilson Jr., May 31, 1939; "Guardianship Bond," In the Chancery Court for the Osceola District of Mississippi County, Arkansas, Re: R. E. L. Wilson Jr.; and "Order Appointing Guardian," in the Chancery Court for the Osceola District of Mississippi County, Arkansas, Re: R. E. L. Wilson Jr., , signed by Chancellor and Probate Judge J. F. Gautney. All in Fendler Files, Box 10.

44. It is addressed simply to Mr. Potter but T. W. Potter was the county clerk at time. The document can be found in Fendler Files, Box 10.

45. Ibid., Victoria Wilson Wesson to James H. Crain, June 20, 1939, Lee Wilson & Co., 1939, W–Z, Folder: Victoria (Mrs. Frank H.) Wesson.

46. Dr. Nicholas Gotten to Roy Wilson, February 5, 1941; Gotten to Wilson, February 15, 1941; Wilson to Gotten, February 17, 1941; Wilson to Gotten, April 2, 1941; all in Lee Wilson & Co., 1941, U–Z, Folder: R. E. Lee Wilson Jr.

47. Whayne interview with Eldon Fairley, October 25, 1999, p. 10.

48. Document located in Lee Wilson & Co., 1933, C, Folder: C. F. Coleman et al. Vs. R. E. Lee Wilson Jr. et al.

49. Lee Wilson to Neil Thomason, July 5, 1933; Thomason to Wilson, July 6, 1933 (9:34/9:46 a.m.); Thomason to Wilson, July 6, 1933 (3:46/4:01 p.m.), all in Lee Wilson & Company, 1933, Tr–Z, Folder: Neal Thomason. Wilson's scribbled note appeared on the bottom of the first telegraph from Wilson received on July 6, 1933.

50. Armorel Weekly Reports, Lee Wilson & Co., 1933, A, Folder: Armorel Weekly Report.

51. Hy Wilson to J. B. Daggett (attorney), December 21, 1933, Lee Wilson & Co., 1933, D–E, Folder: Daggett & Daggett (Marianna, Arkansas). See also J. B. Daggett to J. H. Crain and W. F. Wilson, November 27, 1933, in same folder. As of November 27, they were ready to proceed to file an answer and try the lawsuit.

52. J. H. Herron (another Lee Wilson & Company executive) to Ely and Walker Dry Goods Company, November 27, 1933, Lee Wilson & Co., 1933, D–E, Folder: Ely & Walker Dry Goods Company; Herron to Continental Gin Company, December 9, 1933, Lee Wilson & Co., 1933, C, Folder: Continental Gin Co.; Herron to Graybar Electric Company, November 27, 1933, Lee Wilson & Co., 1933, Gr–J, Folder: Graybar Electric Co., Inc.

53. Everett Dewey Henson, "Memories of Dyess Colony," *Delta Historical Review* 2 (Summer 1990): 3–21; Donald Holley, "Trouble in Paradise: Dyess Colony and Arkansas Politics," *Arkansas Historical Quarterly* 32 (Autumn 1973): 203–16; Holley, *Uncle Sam's Farmers: The New Deal Communities in the Lower Mississippi Valley* (Urbana: University of Illinois Press, 1975); Dan W. Pittman, "The Founding of Dyess Colony," *Arkansas Historical Quarterly* 29 (Winter 1970): 313.

54. *Washington Post*, April 12, 1936. A typescript copy of this article is also located in Lee Wilson & Co., 1936, W–Z, Folder: Washington Post.

55. R. N. Elliott to Secretary of Agriculture, July 6, 1937, Lee Wilson & Co., 1941, R–T, Folder: Lloyd Spencer; "A Review ," Lee Wilson & Co., 1939, W–Z, Folder: Farm Program Review.

56. Cully Cobb to W. J. Driver, November 10, 1936, Lee Wilson & Co., 1936, Cot–Dis, Folder: W. J. Driver.

57. C. E. Palmer [head of the commission] to James Crain, August 27, 1936, Lee Wilson & Co., 1936, Dix–Fiv, Folder: Farm Tenancy Commission.

58. "Outline of the Program Submitted," C. T. Carpenter, Lee Wilson & Co., 1936, Dix–Fer, Folder: Farm Tenancy Commission. C. Calvin Smith, "Junius Marion Futrell, 1933–1937," in *Governors of Arkansas*, edited by Timothy P. Donovan, Willard B. Gatewood, and Jeannie M. Whayne (Fayetteville: University of Arkansas Press, 1995), p. 188.

59. Joe T. Robinson to Cully Cobb, December 21, 1936, Lee Wilson & Co., 1936, C–Cor, Folder: Cully Cobb. J. Brewer (sec. to Joe T. Robinson) to Hamilton Moses (Little representative of the AAA), January 6, 1937, Lee Wilson & Co., 1937, R–S, Folder: Joe T. Robinson. After Robinson died, John Miller apparently met with officials in the Department of Agriculture on the company's behalf; see Garrett Whiteside to James H. Crain, September 28, 1938; Thomas M. Tatum to Jim H. Crain, November 5, 1938; John E. Miller to Thomas M. Tatum, November 5, 1938; Thomas M. Tatum to C. H. Moses, November 7, 1938; Thomas M. Tatum to Mim Crain, November 7, 1938, all in Lee Wilson & Co., 1938, M–Ph, Folder: John E. Miller. Hattie Caraway also attempted to intercede: Hattie Caraway to J. H. Crain, November 1, 1939; I. W. Duggan (Director of the Southern Division, U.S. Dept. of Agriculture, AAA) to Hattie Caraway, November 8, 1939; J. H. Crain to Hattie Caraway, November 10, 1939, all in Lee Wilson & Co., 1939, C, Folder: Hattie Caraway. See also Homer C. Adkins to Hattie Caraway, n.d., and P. M. Evans (Administrator, AAA) to Hattie Caraway, March 26, 1940, both in Lee Wilson & Co.,1940, A–Co, Folder: Hattie Caraway.

60. P. M. Evans to Hattie W. Caraway, March 26, 1940, Lee Wilson & Co., 1940, A–Co, Folder; Hattie Caraway. See also C. H. Scott to J. H. Crain, May 15, 1940, Lee Wilson & Co., 1940, M–W, Folder: C. H. Scott.

61. P. M. Evans to Hon. Hattie W. Caraway, March 26, 1940, Lee Wilson & Co., 1940, A–Co, Folder: Hattie Caraway. See also John E. Miller to James H. Crain, February 7, 1941; James H. Crain to Miller, March 9, 1941; Miller to Crain, March 10, 1941; Miller to Crain, September 12, 1941; Crain to Miller, September 15, 1941, Lee Wilson & Co., 1941, L–P, Folder: John E. Miller.

62. S. A. Regenold to George E. H. Goodner, February 20, 1941, Lee Wilson & Co. 1941, E–K, Folder: George E. H. Goodner.

63. Hy Wilson to Neely Bowen, May 2, 1935, Lee Wilson & Co., 1935, B–Col, Folder: Neely Bowen. For reference to the 1933 and 1934 tax relief, see J. T. Coston to J. F. Topkins, J. H. Crain, and C. W. Ramey (directors of Drainage District 9), April 17, 1935, Lee Wilson & Co., 1935, Com–D, Folder: J. T. Coston.

64. Hy Wilson to Neely Bowen, May 1, 1935, Lee Wilson & Co., 1935, B–Col, Folder: Neely Bowen; for reference to the estate owning all the bonds, see Hy Wilson to Sid B. Redding, Lee Wilson & Co., 1936, R–Sha, Folder: Sid B. Redding.

65. Hy Wilson to Neely Bowen, May 1, 1935, Lee Wilson & Co., 1935, B–Col, Folder: Neely Bowen.

66. J. C. Crain to James J. McEntee, August 29, 1941, Lee Wilson & Co., 1941, L–P, Folder: James J. McEntee.

67. "Report on the Flood and Drainage Problem in the Lower Alluvial Valley of the St. Francis River, Arkansas; Report to the East Arkansas Drainage and Flood Control Association," by Arthur E. Heagler, Consulting Engineer, Little Rock, Arkansas, February 1947, p. 51. (Note: C. H. Bond, chairman; J. W. Meyer, secretary; Burke Mann, counselor. Committeemen: J. H. Crain, Mississippi County, Dan Portis, Poinsett County, Arthur Adams, Craighead County, C. N. Houck, St. Francis and Lee counties, C. H. Bond, Crittenden County.)

CHAPTER NINE

1. Alumni Records, Yale University, Information for the Alumni Records, 1941, 1946, Yale University, Records of the Classes of 1701–1978 (RU 830), Manuscripts and Archives, Yale University Library. Alumni Records, Yale University, Richard E. Bishop, the *Alumni Journal* (June 1978), "12 class notes, focusing on the 'class boy.'"

2. According to Yale Alumni Records, Bob Wilson was in Colorado in the ranching business with his Yale roommate, F. Layton McCartney, in January 1947; see alumni form date stamped January 27, 1947. For Lee Wesson's wartime activities: Whayne interview with Lee Wilson Wesson, July 25, 2008.

3. Seymour Freedgood, "The Man Who Has Everything—in Wilson, Ark.," *Fortune*, August 1964, pp. 143–47, 192. Quote on p. 144.

4. Daniel, *Breaking the Land: The Transformation of Cotton, Tobacco, and Rice Cultures since 1880* (Urbana: University of Illinois Press, 1985), and *Standing at the Crossroads: Southern Life since 1900* (New York: Hill & Wang, 1986); ; Jack Temple Kirby, *Rural Worlds Lost: The American South, 1920–1960* (Baton Rouge: Louisiana State University Press, 1987); ; Gilbert C. Fite, *Cotton Fields No More: Southern Agriculture, 1865–1980* (Lexington: University Press of Kentucky, 1984); Jeannie M. Whayne, *A New Plantation South: Land, Labor and Federal Favor in Twentieth-Century Arkansas* (Charlottesville: University of Virginia Press, 1996).

5. The state figures declined from 6,609,833 to 5,994,816.

6. E. H. Burns, "Narrative Report of County Extension Agent," South Mississippi County, December 1, 1940, to November 30, 1941, Agricultural Extension Service, p. 12; E. H. Burns, "Narrative Report of County Extension Agent," South Mississippi County, December 1, 1941 to November 30, 1942, Agricultural Extension Service, p. 10. Soybean prices rose from .89 cents to $2.08 per bushel; cotton prices rose from 9.9 cents to 22.5 cents. U.S. Bureau of the Census of Agriculture, Vol. I, Part 23, Arkansas, Statistics for Counties, County Table II (Part 2 of 3)–Specific Crops Harvested, Census of 1945 and 1940 (Washington, D.C.: U.S. Government Printing Office, 1946), pp. 63, 73. Cropland devoted to soybeans rose in the county from 24,962 to 54,950 acres The acreage also increased statewide but not at the same rate: 223,079 to 275,513 acres. Cotton acreage in the county increased from 183,888 to 194,076 and decreased in the state from 2,056,775 to 1,769,987.

7. *Osceola Times*, October 16, 1942. At first the plants focused on dehydrating Irish potatoes, sweet potatoes, and cabbage for use by the military, but this later expanded to include other vegetables and some fruits.

8. "Application to County Farm Rational Committee," from D. N. Morris, January 7, 1943, Lee Wilson & Co., War Production Board, 1942–45, Box 1.

9. A. R. Shearon to J. M. Rogers, Defense Contract Service, October 3, 1941, Lee Wilson & Co., 1941, C–D, Folder: Defense Contract Service.

10. D. N. Morris to W. H. Howze, April 19, 1943, Lee Wilson & Co., War Production Board, 1942–45, Box 1, Folder: Applications for Allotment of Controlled Materials, 1942–43, 1 of 3.

11. See Lee Wilson & Co., War Production Board, 1942–45, Boxes 1 and 2.

12. "Application for Project Amendment," June 15, 1943, Lee Wilson & Co., War Production Board, 1942–45, Box 1.

13. Homer Adkins to Jim Crain, March 6, 1941, and April 8, 1941, Lee Wilson & Co., 1941, C–D, Folder: Homer Adkins. C. Calvin Smith, *War and Wartime Changes: The Transformation of Arkansas, 1940–1945* (Fayetteville: University of Arkansas Press, 1986).

14. *Sixteenth Census of the United States: 1940*, Population, Vol. II, Table 28 (Washington: Government Printing Office, 1943), pp. 486–87.

15. *Osceola Times*, June 11, 1943, p. 1.

16. E. H. Burns, "Narrative Report of County Extension Agent," South Mississippi County, December 1, 1942, to November 15, 1943, p. 12.

17. *Osceola Times,* July 16, 1943, p. 1; August 6, 1943, p. 1.

18. Ibid., August 20, 1943.

19. Ibid., February 18, 1944.

20. Ibid., April 21, 1944.

21. He placed 1,640 POWs, 135 Bahamians, 90 Mexicans, and 107 hill and local workers. He also filed 600 draft deferments. D. V. Maloch, "Narrative Report of County Extension Agent," South Mississippi County, December 1, 1943, to November 30, 1944, p. 28.

22. D. V. Maloch, "Narrative Report of County Extension Agent," South Mississippi County, December 1, 1944, to November 30, 1945, p. 33. In addition to the POWS, Maloch secured only 61 other farm laborers—51 men, 7 women, and 3 children—to work on south Mississippi County farms. He also applied for 281 draft deferments.

23. Keith J. Bilbrey, "Narrative Report of County Extension Agent," north Mississippi County, December 1, 1944, to November 30, 1945, pp. 46–47.

24. U.S. Bureau of the Census of Agriculture, Vol. I, Part 23, Arkansas, Statistics for Counties, County Table I,–Farms and Farm Characteristics, *Census of 1945 and 1946* (Washington, D.C.: U.S. Government Printing Office, 1946), pp. 34, 42.

25. U.S. Bureau of the Census of Agriculture: 1935, Reports for States With Statistics for Counties and a Summary for the United States, Vol. II, Part 2, The Southern States, County Table I.–Farms, Farm Acreage and Value, by Color and By Tenure (Washington, D.C.: U.S. Government Printing Office, 1936), pp. 670, 674; U.S. Bureau of the Census of Agriculture, Vol. I, Part 23, Arkansas, Statistics for Counties, County Table V (Part 1 of 2)–Farms by Tenure of Operator, *Census of 1945 and 1946,* With Farms by Color and Tenure of Operator, *Census of 1945* (Washington, D.C.: U.S. Government Printing Office, 1946), pp. 119, 122.

26. Donald Holley, *The Second Great Emancipation: The Mechanical Cotton Harvester, Black Migration, and How They Shaped the Modern South* (Fayetteville: University of Arkansas Press, 2000). For an insightful analysis of how tenants and sharecroppers in southeast Missouri viewed the abandonment of this system, see Jarod Roll, *Spirit of Rebellion: Labor and Religion in the New Cotton South* (Chicago: University of Illinois Press, 2010). For another view of resistance to displacement, see Whayne, *A New Plantation South.*

27. J. C. Crain to James J. McEntee, August 29, 1941, Lee Wilson & Co., 1941, L–P, Folder: James J. McEntee.

28. *Osceola Times*, August 2, 1946, p. 1.

29. Rachel Carson, *Silent Spring* (Boston: Houghton Mifflin, 1962).

30. Wayne Rasmussen, *A History of the Emergency Farm Labor Supply Program, 1943–1947,* Agricultural Monograph No. 13, U.S. Department of Agricultural Economics, Washington, D.C., September 1951. See also Whayne, *A New Plantation South,* pp. 224–25, for the use of Mexican laborers in neighboring Poinsett County in the period after World War II.

31. Censuses of Population, 1940, 1960.

32. Censuses of Agriculture, 1934, 1939, 1944, 1949, 1959.

33. Censuses of Agriculture, 1934, 1939, 1944, 1949, 1959.

34. William Snake Toney, interview with Jeannie Whayne, St. Louis, Missouri, July 2004.

35. *Osceola Times*, March 29, 1946, p. 1.

36. Mississippi County Probate Records, Osceola, Arkansas.

37. *Osceola Times*, May 17, 1946, p. 1.

38. *Osceola Times*, October 10, 1946, p. 1.

39. Warranty Deed, R. E. L. Wilson Jr., and J. H. Crain, Trustees of Lee Wilson & Company, to F. L. Wesson, Trustee, June 1, 1948, Record Book 93, pp. 451, 457, 459; September 16, 1948, Record Book 93, pp. 495, 573; December 31, 1948, Record Book 93, p. 578; January 3, 1949, Record Book 93, p. 579; February 1, 1949, Record Book 93, p. 624; December 12, 1949, Record Book 95, p. 207; and May 15, 1950, Record Book 95, pp. 337 and 340. Some of these transactions involved exchanges of property between the parties, probably to equalize the partial liquidation. Lee Wilson Wesson, interview with Jeannie Whayne, July 25, 2008. The Armorel Plantation consisted of 6,520.88 acres as indicated in an "Agreement Not to Disturb," Lee Wilson & Co., 1933, F–Go, Folder: First National Bank. For Victoria's obituary, see *Springfield News*, August 28, 1952.

40. The disposition of property to Crain can be found in Warranty Deed, January 21, 1950, Book 95, p. 247, May 15, 1950, Record Book 95, p. 375, and May 29, 1950, Record Book 95, p. 362.

41. *Commercial Appeal*, February 17, 1950; Warranty Deed, J. H. Crain and RELW Jr., Trustees of Lee Wilson & Co., January 21, 1950, Record Book 95, p. 247. See also Quitclaim Deed, June 12, 1950, Record Book 95, p. 362, and Warranty Deed, May 15, 1950, Record Book 95, p. 375.

42. "Arkansas Business Rankings: Individual/Family Land Owners," http://www.thefreelibrary .com.

43. Mississippi County Censuses of Agriculture, 1949, 1969, 2007.

44. Bishop, *Alumni Journal*.

45. Ibid.

46. Jeannie Whayne telephone interview with Steve Wilson, November 26, 2010.

47. *Arkansas Democrat Gazette*, October 14, 2010, pp. 1,2B.

48. Leslie Meyer, Stephen MacDonald, and Linda Foreman, "Cotton Backgrounder," Outlook Report from the Economic Research Service, U.S. Department of Agriculture (USDA), March 2007, www .ers.usda.gov.; and Mark Ash, Janet Livezey, and Erik Dohlman, "Soybean Backgrounder," Electronic Outlook Report from the Economic Research Service, USDA, April 2006, www.ers.usda.gov.

49. Nathan Childs and Janet Livezey, "Rice Backgrounder," Outlook Report from the Economic Research Service, USDA, December 2006, www.ers.usda.gov.

50. Census of Agriculture, 1982, 2007. For information on the controversy over rice burning, a worldwide phenomenon, see Rebecca Thyer, "India's Burning Rice Bowl," *Partners (November 2007–February 2008*, http://aciar.gov.au/system/files/node/5549/PMg%20Nov07-Feb08%20pg%2019. pdf; Alex Barnum, "Dispute over Burning Rice Fields; Sacramento-area Farmers Want Phase-out Postponed," *San Francisco Chronicle*, November 7, 1995, http://articles.sfgate.com/1995-11-07/ news/17821137_1_rice-straw-rice-farmers-air-pollution; John Dickey, "Burning of Rice Stubble Has Some Hacking and Wheezing," *Appeal-Democrat* (Marysville, Calif.), November 4, 2006, http://www .highbeam.com/doc/1G1-153837398.html.

51. Population Census, 1940, 1960, 2000. For projected figures, see U.S. Census Bureau, American FactFinder, Mississippi County, Arkansas.

52. Alison Sider, "Investors Snapping up Farms, *Arkansas Democrat Gazette*, October 17, 2010, pp. 1, 8G, and "Land Sale at Wilson Is the End of a Line," ibid., October 14, 2010, p. 1, 6B; Jeannie Whayne interview with Tri Watkins, November 26, 2010, Lepanto, Arkansas; and personal communication to author from Tri Watkins, November 28, 2010.

53. J. M. McKimmey, B. Dixon, H. D. Scott, and C. M. Scarlat, *Soils of Mississippi County, Arkansas*, Arkansas Agricultural Experiment Station, Research Report 970, Division of Agriculture, University of Arkansas, p. 36.

53. Ibid., pp. 35–36.

55. Ibid., pp. 12–13.

Index

Acklen, Joseph H., 84–85

African Americans: Agricultural Wheel, 53–85; Black Hawk war, 79; burning of black school, 3, 113, 114; challenge to white authority, 51, 140; demographics, 30, 31, 33, 50, 120, 136, 219–222; disease, 61–62; disfranchisement and segregation, 53, 54–57, 178; education, 3, 113–116, 139; Elaine Race Riot, 114, 121–123, 136; Flood of 1927, 150, 153–154; fraternal organizations, 115, 219, 120, 129, 132, 134; fusion, 54–55; Garvey Clubs, 134–135; Great Migration, 114, 121; homesteading, 79; indebtedness, 3, 4, 127, 139; Scipio Jones representation of, 137–139; Ku Klux Klan, 29–30, 123–124; labor, 1–5, 32–33, 48, 50–53, 113–114, 123, 135–137, 139, 215; lynching, 55–56, 113–114; Lowery, 113–114, 130–137; Memphis, 36–38; nightriders, 4, 33, 57–58, 119–21; political activity, 29, 30, 54–55; and post–World War II transformation of agriculture, 219–220, 227–228; preachers and accommodation, 62–63; property holding on Wilson plantation, 118; racism, 3, 4, 129; religion, 114–117, 119; sharecropping, 51–52, 57–58, 118–119, 121–122, 125, 146, 150, 202, 219; slavery and slaves, 8, 15, 18, 24–29, 124; as stockholder in Bank of Wilson, 96; William "Snake" Toney, 220–223; unification of whites at expense of, 1; violence against, 29–32, 57–58, 117, 134, 137–139; in war industry during World War II, 5; white animosity toward, 27; and Lee Wilson, 3, 34, 58, 112–114, 125, 131, 133, 137–139, 220–223; Wilson legacy to black manservant, 185; Napoleon Wilson, 113; World War II military service, 5, 119–220, 223; Yellow Fever epidemic, 41, 61–62

Agricultural Adjustment Administration, 2, 5, 178–79, 194, 199–200; role of county farm agent, 179–180; sanctions on Lee Wilson & Company, 203–208; Wilson support for, 176, 179–181

Agricultural Extension Service, 136, 158–159, 179–80, 216, 215–216, 224

Agricultural Wheel, 53–54

Ahlfeldt, Ernest, 172–173

Alabama, 8

Alexander, Oscar, 151

Alfalfa production, 214

American Aviation and Lee Wesson, 211

American Trust Company, 104–105

Anderson Clayton and Company, 103, 176

Arkansas Banking Act of 1913, 97–102

Arkansas County, 25

Arkansas Farm Credit Corporation, 160–161

Arkansas Farm Tenancy Commission, 207

Arkansas Highway Department, 170

Arkansas Night Riding Law, 1909, 58

Arkansas Post, Arkansas, 12,

Arkansas Power and Light, 160, 164, 169, 170–171

Arkansas Railroad Commission, 106–107

Arkansas State College, 194

Armorel Plantation, 69, 71, 90–91, 126, 134, 146, 154, 168, 170, 175–177, 179, 180, 188, 202

Armstrong, W. C. (judge), 74

Audubon, John, 14–15

Ayers, C. D., 180

Babson, Roger, 225

Bachelder, William, 174

Baldwin, L. W., 111

Bank of Blytheville, 166